Wireless Power Transmission for Sustainable Electronics

Wireless Power Transmission for Sustainable Electronics

COST WiPE - IC1301

Edited by

Nuno Borges Carvalho
Instituto de Telecomunicações
Universidade de Aveiro
Aveiro
PO

Apostolos Georgiadis
Heriot-Watt University
Edinburgh
UK

Registered Office
John Wiley & Sons, Inc., 111 River Street, Hoboken, NJ 07030, USA

Editorial Office
111 River Street, Hoboken, NJ 07030, USA

For details of our global editorial offices, customer services, and more information about Wiley products visit us at www.wiley.com.

Wiley also publishes its books in a variety of electronic formats and by print-on-demand. Some content that appears in standard print versions of this book may not be available in other formats.

Library of Congress Cataloging-in-Publication Data

Names: Carvalho, Nuno Borges, editor. | Georgiadis, Apostolos, editor. |
 COST WiPE–IC1301 (Project)
Title: Wireless power transmission for sustainable electronics : COST
 WiPE–IC1301 / Nuno Borges Carvalho, Instituto de Telecomunicações,
 Universidade de Aveiro, Apostolos Georgiadis, Heriot-Watt University,
 Edinburgh, UK.
Description: First edition. | Hoboken, NJ : John Wiley & Sons, Inc., 2020.
 | Includes bibliographical references and index.
Identifiers: LCCN 2019041400 (print) | LCCN 2019041401 (ebook) | ISBN
 9781119578543 (hardback) | ISBN 9781119578499 (adobe pdf) | ISBN
 9781119578574 (epub)
Subjects: LCSH: Wireless power transmission. | Electronic apparatus and
 appliances–Power supply. | Green electronics.
Classification: LCC TK3088 .W635 2020 (print) | LCC TK3088 (ebook) | DDC
 621.319–dc23
LC record available at https://lccn.loc.gov/2019041400
LC ebook record available at https://lccn.loc.gov/2019041401

Cover design by Wiley
Cover image: © Jobalou/Getty Images

Set in 9.5/12.5pt STIXTwoText by SPi Global, Chennai, India

Printed in the United States of America

V10017018_011720

Contents

List of Figures

List of Contributors

Daniel Belo
Departamento de Electrónica
Telecomunicações e Informática
Instituto de Telecomunicacoes
Universidade de Aveiro
Aveiro
Portugal

Aggelos Bletsas
School of Electrical and Computer
Engineering
Technical University of Crete
Chania
Crete
Greece

Nuno B. Carvalho
Departamento de Electrónica
Telecomunicações e Informática
Instituto de Telecomunicacoes
Universidade de Aveiro
Aveiro
Portugal

Ricardo Correia
Departamento de Electrónica
Telecomunicações e Informática
Instituto de Telecomunicacoes
Universidade de Aveiro
Aveiro
Portugal

Alessandra Costanzo
Department of Electrical, Electronic
and Information Engineering
"G. Marconi"
University of Bologna
Bologna
Italy

Miroslav Cupal
Department of Radio Electronics
Faculty of Electrical Engineering and
Communication
Brno University of Technology
Brno
Czech Republic

Spyridon N. Daskalakis
School of Engineering and Physical
Sciences
Heriot-Watt University
Edinburgh
Scotland

Antonis G. Dimitriou
Faculty of Technology
School of Electrical and Computer
Engineering
Aristotle University of Thessaloniki
Thessaloniki
Greece

Karol Dobrzyniewicz
SpaceForest
Pomeranian Science and Technology
Park
Gdynia
Poland

Fortunato Carlos Dualibe
Analogue and Mixed-Signal Design
Group
Electronics and Microelectronics Unit
University of Mons (UMONS)
Mons
Belgium

Yvan Duroc
University of Lyon
University Claude Bernard Lyon 1
Ampere Laboratory
Villeurbanne
France

Hugo García-Vázquez
Electronics Department
Instituto de Astrofísica de Canarias
(IAC)
Canary Islands
Spain

Apostolos Georgiadis
School of Engineering and Physical
Sciences
Heriot-Watt University
Edinburgh
Scotland

Ricardo Gonçalves
Instituto de Telecomunicações
Aveiro
Portugal

George Goussetis
School of Engineering and Physical
Sciences
Heriot-Watt University
Edinburgh
Scotland

Jasmin Grosinger
Institute of Microwave and Photonic
Engineering
Graz University of Technology
Graz
Austria

Simon Hemour
IMS
University of Bordeaux
Bordeaux
France

Marina Jordao
Departamento de Electrónica
Telecomunicações e Informática
Instituto de Telecomunicacoes
Universidade de Aveiro
Aveiro
Portugal

Przemyslaw Kant
SpaceForest
Pomeranian Science and Technology
Park
Gdynia
Poland

John Kimionis
School of Electrical and Computer
Engineering
Georgia Institute of Technology
Atlanta
GA
USA

Jaroslav Láčík
Department of Radio Electronics
Faculty of Electrical Engineering and
Communication
Brno University of Technology
Brno
Czech Republic

Caroline Loss
FibEnTech Research Unit
University of Beira Interior
Covilhã
Portugal

and

LabCom IFP
University of Beira Interior
Covilhã
Portugal

Diego Masotti
Department of Electrical
Electronic and Information
Engineering "G. Marconi"
University of Bologna
Bologna
Italy

Hatem El Matbouly
University of Grenoble Alpes
Grenoble INP
LCIS
Valence
France

Jerzy Julian Michalski
SpaceForest Pomeranian Science and
Technology Park
Gdynia
Poland

Loukas Petrou
Faculty of Technology
School of Electrical and Computer
Engineering
Aristotle University of Thessaloniki
Thessaloniki
Greece

Pedro Pinho
Instituto Superior de Engenharia de
Lisboa – ISEL
Lisboa
Portugal

and

Departamento de Electrónica
Telecomunicações e Informática
Instituto de Telecomunicacoes
Universidade de Aveiro
Aveiro
Portugal

Grigory Popov
Analogue and Mixed-Signal Design
Group
Electronics and Microelectronics Unit
University of Mons (UMONS)
Mons
Belgium

Alexandre Quenon
Analogue and Mixed-Signal Design
Group
Electronics and Microelectronics Unit
University of Mons (UMONS)
Mons
Belgium

Zbyněk Raida
Department of Radio Electronics
Faculty of Electrical Engineering and
Communication
Brno University of Technology
Brno
Czech Republic

Rita Salvado
LabCom IFP
University of Beira Interior
Covilhã
Portugal

Mazen Shanawani
Department of Electrical
Electronic and Information
Engineering
"G. Marconi"
University of Bologna
Bologna
Italy

Stavroula Siachalou
Faculty of Technology
School of Electrical and Computer
Engineering
Aristotle University of Thessaloniki
Thessaloniki
Greece

Jan Špůrek
Department of Radio Electronics
Faculty of Electrical Engineering and
Communication
Brno University of Technology
Brno
Czech Republic

Smail Tedjini
University of Grenoble Alpes
Grenoble INP
LCIS
Valence
France

Manos M. Tentzeris
School of Electrical and Computer
Engineering
Georgia Institute of Technology
Atlanta
USA

Emmanouil Tsardoulias
Faculty of Technology
School of Electrical and Computer
Engineering
Aristotle University of Thessaloniki
Thessaloniki
Greece

Erika Vandelle
IMEP-LaHC
Institut Polytechnique de Grenoble
(Grenoble INP)
Université Grenoble Alpes
Grenoble
France

Jan Vélim
Department of Radio Electronics
Faculty of Electrical Engineering and
Communication
Brno University of Technology
Brno
Czech Republic

Georgios Vougioukas
School of Electrical and Computer
Engineering
Technical University of Crete
Chania
Crete
Greece

Konstantinos Zannas
University of Grenoble Alpes
Grenoble INP
LCIS
Valence
France

Preface

Wireless power transmission (WPT) is an emerging technological area that is changing our lives and will change the paradigm how we use electrical equipment. In Europe, this area was considered interesting enough to discuss and to explore further, thus the European Union brought together a group of scientists and professionals to explore WPT. This European research consortium was called COST IC1301 "WiPE (**W**ireless **P**ower **T**ransmission for **S**ustainable **E**lectronics)."

This COST action IC1301 aims to address efficient WPT circuits, systems, and strategies specially tailored for battery-less systems: namely energy-autonomous sensors, passive radio frequency identification (RFID), and near-field communications (NFC). All these systems implement closely related concepts that make use of WPT and energy harvesting systems to remotely power up mobile devices or to remotely charge batteries, contributing to development and foster the Internet of Things (IoT) evolution.

Devoted to the main COST mission, that is to strengthen Europe's scientific and technical research capacity by supporting cooperation and interaction between European researchers, IC1301 WiPE regroups active researchers in the domain of WPT from Universities, Research Institutes, and Companies located in 25 European countries. The expected benefits of WiPE include the creation of a wide network of experts both from academia and industry that can address the existing and upcoming challenges in WPT scenarios in an interdisciplinary manner paving the way for the future generations of wireless power transmission solutions and the associated regulation.

In this framework, we have brought together the main publications related to the WiPE activities. The chapters in this book present interesting novel solutions provided by single and by joint research groups in the exploitation of the near- and far-field techniques of wireless power transfer. Joint research results were possible thank to the financial support of the COST organization in realizing short term scientific missions (STSM) among the members of the WiPE COST action. Furthermore, contributions from outside of Europe have been considered, and they were possible due to the relationships established during the COST action meetings, by inviting outstanding researchers in this field. A total of 12 chapters are here presented.

We would like to extend special thanks to all the author contributors for their time and effort. We also would like to express our sincere gratitude to Prof. Raquel Castro Madureira that helped us in the editing of the overall book.

Nuno Borges Carvalho is a Full Professor and a Senior Research Scientist with the Instituto de Telecomunicacoes, Universidade de Aveiro and an IEEE Fellow. He coauthored Intermodulation in Microwave and Wireless Circuits (Artech House, 2003), Microwave and Wireless Measurement Techniques (Cambridge University Press, 2013), and White Space Communication Technologies (Cambridge University Press, 2014). He has been a reviewer and author of over 200 papers in magazines and conferences. He is associate editor of the IEEE Transactions on Microwave Theory and Techniques, IEEE Microwave Magazine and Cambridge Wireless Power Transfer Journal.

He is the co-inventor of four patents. His main research interests include software-defined radio front-ends, wireless power transmission, nonlinear distortion analysis in microwave/wireless circuits and systems, and measurement of nonlinear phenomena. He has recently been involved in the design of dedicated radios and systems for newly emerging wireless technologies. He is the chair of the COST Action IC1301.

Apostolos Georgiadis is an Associate Professor at the School of Engineering and Physical Sciences, Heriot-Watt University, UK, an IEEE Senior Member and an URSI Fellow. He co-authored Coupled Oscillator Based Active Array Antennas (Wiley, 2012) and co-edited Microwave and Millimeter Wave Circuits and Systems: Emerging Design, Technologies and Applications (Wiley, 2012) and Wireless Power Transfer Algorithms, Technologies and Applications in Ad Hoc Communication Networks (Springer, 2016). He has authored over 200 papers in magazines and conferences. He has been Associate Editor of the IEEE Journal on RFID, IEEE Microwave and Wireless Components Letter and Editor-in-Chief of Cambridge Wireless Power Transfer Journal.

He is the co-inventor of two patents. His main research interests include energy harvesting and wireless power transmission, active antennas and arrays and inkjet and 3D printed microwave electronics. He is the vice-chair of the EU COST Action IC1301.

Acknowledgments

We would like to extend special thanks to all the author contributors for their time and effort. We also would like to express our sincere gratitude to Prof. Raquel Castro Madureira that helped us in the editing of the overall book.

Finally, we should extend this acknowledgement to all COST IC1301 – WIPE colleagues for their support and hard work to bring Wireless Power Transmission to Europe scientific agenda.

1

Textile-Supported Wireless Energy Transfer

Miroslav Cupal, Jaroslav Láčík, Zbyněk Raida, Jan Špůrek, and Jan Vélim

Department of Radio Electronics, Faculty of Electrical Engineering and Communication, Brno University of Technology, Brno, Czech Republic

1.1 Introduction

In daily use, wired technologies are extensively replaced by wireless ones. Whereas wireless communication has been reaching its maturity, wireless energy transfer has been still developing.

Using a conventional coaxial cable for transmitting electromagnetic energy, attenuation about 6 dB can be reached at $f = 2$ GHz for the distance $R = 100$ m [1]. If the energy is transmitted at the same frequency for the same distance in free space using two half-wavelength dipoles, attenuation can be evaluated by Friis equation [2]

$$\frac{P_2}{P_1} = G_1 G_2 \frac{c^2}{4\pi f^2} \frac{1}{4\pi R^2} \tag{1.1}$$

where c is velocity of light and $G_1 = G_2 = 1.64$ is gain of a half-wavelength dipole. Substituting numerical values to (1.1), we obtain the corresponding attenuation 75 dB if no losses and perfect matching are assumed.

The presented simplified calculation shows that low efficiency is the main disadvantage of a far-field wireless energy transfer:

In wired transmission, a plane wave is propagating with constant amplitude in a single direction (along a transmission line). Attenuation is therefore mainly caused by losses in metallic conductors and dielectrics.

In wireless transmission, antennas radiate a spherical wave, which intensity is reciprocally proportional to the distance, to all directions in space. Attenuation caused by losses can be usually neglected.

Whereas reciprocal proportion between the field intensity and the distance cannot be influenced, energy propagation can be limited to a selected subspace, and

Wireless Power Transmission for Sustainable Electronics: COST WiPE - IC1301,
First Edition. Edited by Nuno Borges Carvalho and Apostolos Georgiadis.

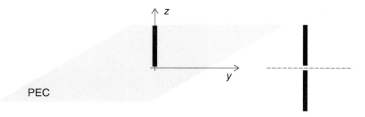

Figure 1.1 On-surface wireless energy transfer: replacing dipole antenna by monopole one.

efficiency of the wireless energy transfer can be improved that way. If a half-wavelength dipole is replaced by a quarter-wavelength monopole (Figure 1.1), only the half-space above the reflector is filled-in by energy (radiation resistance of the antenna is half-sized), and transmission can be improved for +3 dB.

Let us assume a perfectly electrically conductive (PEC) reflector being coated by a layer of dielectrics. Then, this coated structure can play the role of a single-wire transmission line [3]:

- The conductive shield of a coaxial line is projected into the plane;
- The inner dielectric insulator is projected into the coating;
- The conductive core is going to vanish.

Since the single-wire transmission line is an open structure, wave propagation stays attenuated due to radiation losses. On the other hand, resistive and dielectric losses are much smaller than in closed structures because the field is not confined to a small area [3].

So, the single-layer transmission line seems to be a good candidate for wireless energy transfer along coated conductive surfaces. Since bodies of cars, buses, and airplanes are conductive, such an energy transfer can be used for energy distribution along the surface of vehicles.

In Section 1.2, we investigate exploitation of a three-dimensional (3D) knitted fabric for coating of conductive surfaces. Next to creating the single-wire transmission line, the knitted coating can provide additional functions. The knitted coating can:

- Play the role of a microwave substrate for manufacturing transmission lines, antennas, and electronic circuits.
- Be used for thermal insulation of vehicle interior.
- Attenuate vibrations and other mechanical phenomena.
- Provide functional properties of textile components (covers of seats, textile upholstery, etc.).

Integration of several functions into a knitted coating is the main idea of so called intelligent fabrics. Such fabrics can obviously reduce fabrication costs, weight of vehicle, and consequently, fuel consumption, and CO_2 emissions. Examples of integrated components and subsystems are described in Section 1.3.

A conductive surface of a vehicle covered by intelligent fabrics can play the role of the distribution network for an in-vehicle wireless communication (or wireless energy transfer). Antennas integrated into the fabrics can be understood as an interface in between this network and interior of the vehicle. Wireless communication between the on-surface network and interior of a vehicle is discussed in Section 1.4.

The chapter is concluded by Section 1.5.

1.2 Textile-Coated Single-Wire Transmission Line

Conventional upholstery used by vehicle manufacturers is usually composed as a sandwich structure, which is designed to meet mechanical and thermal requirements dominantly (Figure 1.2a). Nevertheless, the same objectives can be met by a three-dimensional fabric (Figure 1.2b).

The fabric is knitted from polyester yarns creating a structure similar to a conventional microwave substrate:

Firm top and bottom surfaces can be metallized to create a ground plane and a metallic layout of a planar circuit, or an antenna.

In between, polyester yarns ensure the constant distance between surfaces. Since the majority of volume between the surfaces consists of an air, losses of the fabric are low.

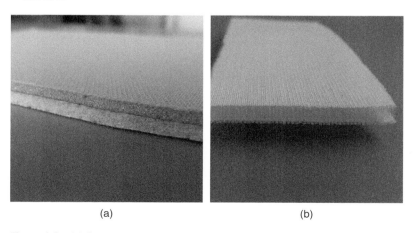

(a)   (b)

Figure 1.2   (a) Conventional upholstery from a car. (b) Knitted fabric SINTEX T2–3D 041.

When knitting, mechanical properties of the fabric can be optimized. High density of yarns makes the fabric more mechanically stable. Height of the fabric can vary from 0.5 to 3.5 mm, and lower fabrics are again more stable.

In order to use the fabric for vehicular applications, textile has to be flame-retardant, and a fireproof finishing has to be applied.

From the electrical viewpoint, all the versions of the textile substrate are of very similar parameters; the dielectric constant varies from 1.10 to 1.22.

Initial experiments, which were aimed to verify suitability of the 3D fabric for coating the single-wire transmission line, were performed in 60 GHz ISM band. Using two open-ended WR-15 waveguides in a distance of 180 mm, frequency response of transmission coefficient was measured in free space, above an uncovered conductive plate, above a conductive plate covered by the conventional upholstery, and the 3D knitted fabric [4]. Results of measurements are shown in Figure 1.3.

Obviously:

The uncovered plate increases transmission for +6 dB; thanks to the in-phase interference of the direct wave and the reflected one.

Whereas the conventional car upholstery increases transmission for additional +4 dB, the 3D fabric improves transmission for additional +8 dB.

We can therefore conclude that the conductive plate covered by the 3D fabric behaves like the single-wire transmission line and can be used for wireless energy distribution along a surface of a vehicle.

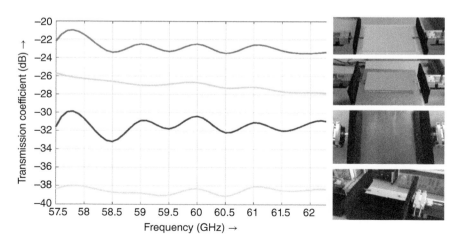

Figure 1.3 Wireless energy transfer. From bottom: wave in free space, wave above conductive plate, wave in conventional upholstery, and wave in 3D knitted fabric.

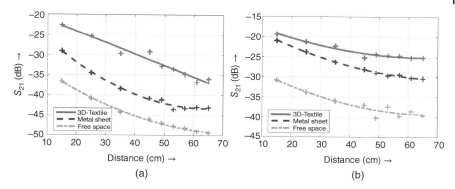

Figure 1.4 Dependence of transmission on the distance between antennas. (a) 60 GHz, (b) 8 GHz.

Unfortunately, electronics operating at 60 GHz is currently too expansive for commercial exploitation in vehicular applications. In the second step, the initial experiment was therefore downscaled to the band-group VI of ultra-wideband (UWB) frequencies ranging from 7392 to 8976 MHz (the center frequency is 8184 MHz).

Measured frequency responses of the transmission are depicted in Figure 1.4: Dependence of the transmission on the distance between antennas is shown left for operation frequency 60 GHz and right for frequency 8 GHz. Conclusions are obvious:

Transmission is higher at 8 GHz due to lower attenuation.

If frequency is decreased, influence of 3D fabric is weakened. Comparing to the uncovered conductive plate, the 3D fabric improves transmission for +8 dB at 60 GHz and for +2 dB to +5 dB at 8 GHz, approximately.

For measurements at 8 GHz, substrate-integrated horn antennas operating with the fundamental TE_{10} mode were used (see Figure 1.5). Simulations of the horn antenna above a finite conductive plate show that maximum of radiation is elevated, and the textile coating is over-radiated. The elevation angle is increasing with decreasing electrical dimensions of the plate (i.e. with the decreasing frequency). And this is the reason why the transmission with frequency drops.

Rising elevation angle is a result of scattering caused by edges of the reflector. If the described problem is going to be solved, the single-wire transmission line has to be replaced by a waveguide integrated into the textile substrate, which is fed by a line-to-waveguide transition. The described step converts a semi-wireless energy transfer to a wired one: energy is delivered to given directions only, and efficiency of energy transfer is increased. A lower flexibility is not an issue if

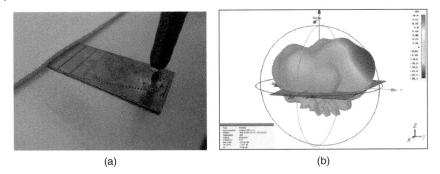

(a)   (b)

Figure 1.5   Wireless energy distribution by single-wire transmission line at 8 GHz.

electronic components (sensors, transmitters, receivers) are integrated into the textile substrate as we will describe in Section 1.3.

In Section 1.4, textile integrated waveguides (TIWs) will be used to feed antennas which transmit electromagnetic wave from roof upholstery to passengers. So, electromagnetic energy is distributed along the conductive surface by a guided wave, and a fully wireless transmission is exploited in between the conductive surface and seats of passengers as we will describe in Section 1.4.

Exploitation of surface waves for electromagnetic energy transfer was intensively investigated in fifties of the twentieth century [3, 5]. Nevertheless, the concept of the surface-wave energy transfer has appeared in the open literature in last years as well [6–8] and seems to become popular again.

1.3   Textile-Integrated Components

When integrating electromagnetic components into a textile substrate, the concept of the substrate integrated waveguide (SIW) is applied. The concept of SIW is related to electromagnetic band-gap (EBG) structures [9, 10].

EGB is a periodic structure, which can be created by metallic vias in between metallic surfaces on the top side and the bottom side of a dielectric substrate. EBG can support wave propagation in frequency bands of constructive interferences of waves scattered by vias and can suppress wave propagation in bands of destructive interferences.

Behavior of an EBG is illustrated by outputs of a computer simulation shown in Figure 1.6. Removing a row of vias, a wave-guiding channel is created, and electromagnetic energy propagates in the longitudinal direction. In the transversal direction, wave propagation is suppressed thanks to the band gap of the periodic structure. If the number of vias at sides is decreased, penetration of electromagnetic energy to the transversal direction is rising, but is still kept well-suppressed.

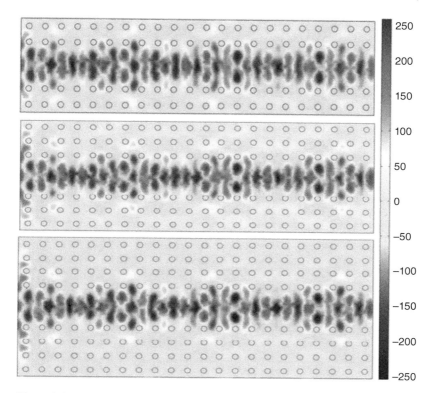

Figure 1.6 Wave-guiding channel inside a periodic structure: penetration of energy to transversal directions depending on the number of vias.

If sides of the wave-guiding channels are created by a single row of vias, the SIW is created [11]. SIW is an equivalent of a rectangular waveguide filled in by dielectrics with the dielectric constant ε_r and the loss factor $\tan \delta$. The height of the waveguide is identical with the height of the substrate h, and the effective width of the waveguide w_{eff} has to be determined to have the cutoff frequency of the waveguide f_m below the frequency of operation (Figure 1.7).

The effective width of the SIW can be evaluated according to [11]

$$w_{\text{eff}} = \frac{c}{2f_m \sqrt{\varepsilon_r}} \tag{1.2}$$

where c is the velocity of light in free space (vacuum).

The width of SIW depends on the effective width w_{eff}, the distance between vias b, and the diameter of vias D [11]

$$w = w_{\text{eff}} + \frac{D^2}{0.95b} \tag{1.3}$$

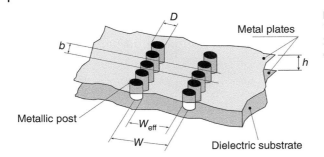

Figure 1.7 Substrate integrated waveguide. Source: Adapted from Wu et al. 2003 [11].

The diameter of vias D and the distance between vias b depend on the technology of manufacturing on one hand and have to meet conditions to provide the band gap on the other hand [11]

$$\frac{D}{w} < \frac{1}{8} \tag{1.4a}$$

$$\frac{b}{w} < 2 \tag{1.4b}$$

Obviously, the design formulas (1.2)–(1.4) neglect losses both in the metal and in the dielectrics.

Now, let us use the 3D knitted fabrics for manufacturing SIW. Such a structure is called a TIW (textile integrated waveguide). Implementing TIW, following technological problems have to be solved [12].

1.3.1 Fabrication of the Top Conductive Layer and the Bottom One

Exploitation of a self-adhesive copper foil is the simplest solution. Very good electrical conductivity of such surfaces is an advantage. Laborious manufacturing, problematic accuracy, low mechanical stability, and worse utility properties are the disadvantages.

Exploitation of screen printing for creating conductive surfaces seems to be the solution. First, surfaces are covered by the Digiflex-Master foil to smoothen the surface and eliminate penetration of a conductive printing paste. Second, layout of the waveguide is screen-printed on the Aurel C880 screen printer using the ESL 1901-S polymer silver paste. Third, the structure is cured at 80 °C for 30 minutes.

1.3.2 Fabrication of Conductive Vias of Side Walls

Side walls of TIW can be sewed by using a proper thread with a high electrical conductivity and appropriate mechanical properties (e.g. properties allowing

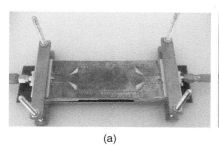 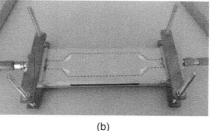

(a)   (b)

Figure 1.8 (a) Textile integrated waveguide fabricated from a self-adhesive copper foil, and (b) screen-printed one. Source: Adapted form Cupal et al. 2018 [12].

machine sewing). After the testing of more than 20 commercially available conductive threads, Bekinox VN 14.1.9.200Z by Bekaert was selected as the optimum one.

In Figure 1.8, the testing of TIW is shown [12]. Two identical waveguides were fabricated from the 3D knitted fabrics SINTEX T2–3D 041 ($h = 3.41$ mm, $\varepsilon_r = 1.22$) using the conductive thread Bekinox VN 14.1.9.200Z ($\sigma = 70\,\Omega/m$). At the input and the output of the transmission line, coaxial SMA connectors were soldered. The transition between the coaxial line and the TIW waveguide exploits a coplanar waveguide with widened slots to obtain a wider bandwidth of the transition [13].

The described TIW was designed for the cutoff frequency $f_m = 5$ GHz. Considering the diameter of the thread $D = 0.8$ mm and the distance between neighboring threads $b = 1.5$ mm, the width of TIW can be evaluated using (1.2) and (1.3); i.e. $w = 28$ mm.

Next, an experiment for characterization of electrical parameters of conductive threads was prepared. In order to eliminate parasitic effects accompanying fabrication of textile waveguides, TIW was manufactured from a conventional FR4 substrate ($h = 1.52$ mm, $\varepsilon_r = 4.2$, and $\tan \delta = 0.02$) for the cutoff frequency $f_m = 5$ GHz, diameter of conductive threads $D = 1$ mm, and the distance between threads $b = 1$ mm.

In Figure 1.9, CST simulation of a reference structure is shown. Side walls of the reference structure were constructed from conventional metallic vias, and all conductive parts were assumed being PEC. The structure was not optimized for the maximum transmission because the verification of electrical properties of conductive threads was aim of the experiment. For this reason, a simple coaxial-microstrip-SIW transition [14] was used instead of the previous coaxial-CPW-SIW one.

In the following step, metallic vias were replaced by tested conductive threads, and frequency response of transmission coefficient was measured

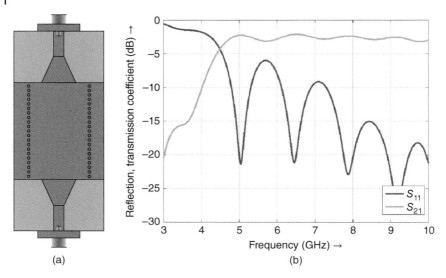

(a)   (b)

Figure 1.9 (a) Numerical simulation of the reference transmission line. (b) Frequency response of the transmission coefficient S_{21} and the reflection coefficient S_{11}.

(see Figure 1.10). The previously used Bekinox VN 14.1.9.200Z corresponds to the Thread 2.

In previous paragraphs, a coplanar waveguide (Figure 1.8) and a tapered microstrip line (Figure 1.9) were used to implement a transition between a coaxial connector and a TIW. Both the transitions were developed for conventional microwave substrates initially.

Since the mechanical stability of the 3D textile substrate is significantly lower compared to conventional substrates and transitions are sensitive to the mechanical stability, we have developed special multilayered transitions combining both types of substrates [15]; see Figure 1.11. The coaxial connector is soldered to the microstrip on a conventional Teflon-based substrate of limited dimensions. This microstrip is coupled then to TIW or microstrip on a textile substrate.

In case of the microstrip, a capacitive coupling is used i.e. the capacitance between the narrow microstrip on the Teflon substrate and the wide microstrip on the textile substrate plays the role of the coupling element.

In case of the TIW, an inductive coupling is used i.e. the inductance of via connecting the open end of the narrow microstrip on the top substrate and the bottom ground surface of TIW plays the role of the coupling element.

Both the transitions were optimized for the Teflon-based substrate ARLON 25N ($\varepsilon_r = 3.38$, $h = 0.762$ mm) and the 3D textile substrate with relative permittivity

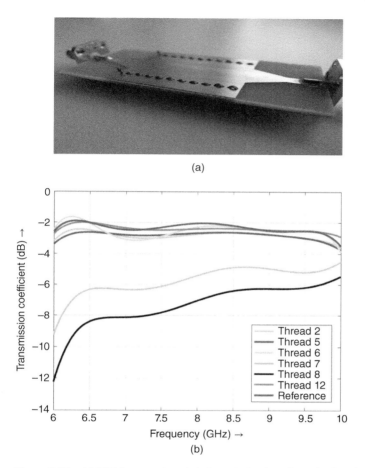

Figure 1.10 (a) SIW for experimental characterization of conductive threads. (b) Frequency responses of transmission coefficient for different threads.

$\varepsilon_r = 1.2$ and height $h = 2.6$ mm. The coax-to-TIW transition was designed for the center frequency of UWB band-group VI, and the coax-to-microstrip was designed for 5.8 GHz ISM band.

Transitions have been fabricated. Then, frequency responses of reflection and transition coefficient have been measured. Conductive layers on the 3D textile were fabricated from a self-adhesive copper foil. The shapes were manufactured by printed circuit board (PCB) technology. Conductive walls of TIW and conductive walls close to the end of microstrip were fabricated from a conductive thread. The conductive yarns of the thread are from silver, and conductivity of this thread

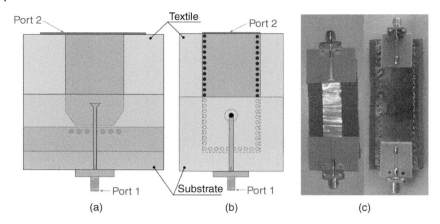

Figure 1.11 Multi-layer coaxial-to-TIW and coaxial-to-microstrip transitions.

is 12.5 Ω/m. Since all conductive walls are hand-made, accuracy of manufacturing is 0.5 mm.

Frequency responses of scattering parameters of simulated structures (lossless, losses neglected) and measured ones (lossy) are shown in Figure 1.12. Whereas reflection coefficient is below −10 dB at operating frequencies both simulated and measured, the difference between the simulated and measured transmission coefficient is more than 5 dB due to the losses.

As an example of a component integrated into the textile substrate, a conventional microwave filter can be shown (Figure 1.13).

We designed a band-pass filter for the operation band 7.5–8.0 GHz and fabricated the structure from the 3D textile substrate 3D045 ($\varepsilon_r = 1.2$, $h = 3.4$ mm), the conductive thread Imbut ELITEX 440 dtex (resistivity 12.5 Ω/m) and the self-adhesive copper foil. Simulations (Figure 1.13a) show a good correspondence with measurements (Figure 1.13b). Obviously, significant losses in the structure are a problem.

Another electromagnetic component, which can be integrated into a 3D textile substrate, is an antenna. Nevertheless, antennas are discussed in Section 1.4, which is devoted to wireless communication and wireless power transfer between the distribution network integrated into the upholstery of the vehicle and seats of passengers.

Dealing with the integrated distribution network discussed in this Section, high losses seem to be the most serious problem:

Contribution of the 3D knitted fabrics itself is not significant because most of the volume is an air, and a minor part consists of polyester only. Moreover, the loss tangent of polyester itself should be lower than tan $\delta < 0.003$ as demonstrated in [16].

Figure 1.12 (a) Frequency responses of reflection and transmission coefficients of coaxial-to-TIW transition and (b) coaxial-to-microstrip transition.

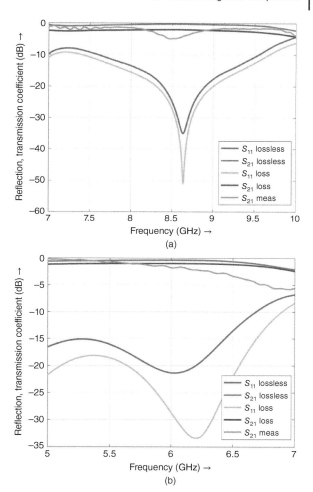

(a)

(b)

Contribution of sewed side walls is stronger. Threads with extremely high conductivity (e.g. silver wires) cannot be used due to the difficult use for sewing. Moreover, a part of energy penetrates walls because a single wall of vias from the EBG structure is exploited only to make the fabrication of the structure as simple as possible.

Contribution of metallic surfaces is important. Surfaces are large, and interface layers (glue, films smoothing the surface and eliminating penetration, etc.) might increase losses.

So, attention has to be turned to improving technologies of manufacturing metallic parts of textile-integrated components.

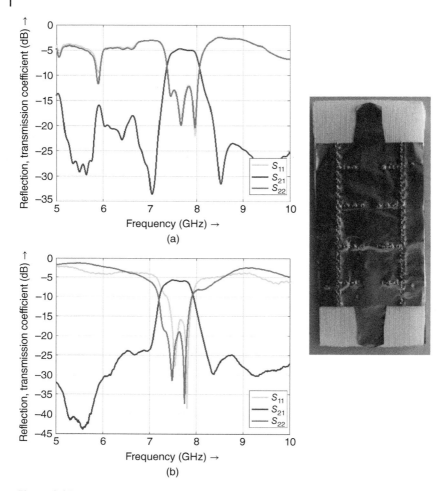

Figure 1.13 (a) Frequency responses of textile-integrated microwave filter: simulation and (b) measurement.

In order to use the textile-based distribution network in vehicular application:

The textile has to be flame-retardant. Hence, the fireproof finishing was applied on the fabric 3D 106 FR, and the test CS 23.853 (enclosure F-(d)(3)(ii) the vertical test) was performed [12].

In case of an airplane, the textile-based distribution network has to be installed under glass-laminate panels covering the inner surface of the airplane (see Figure 1.14). When designing the network, influence of these panels has to be considered.

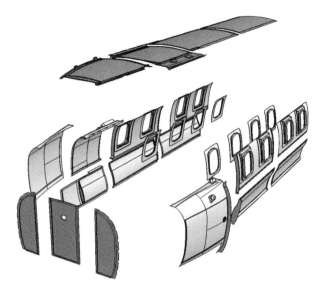

Figure 1.14 Glass-laminate panels used for fixing the textile substrate. By the courtesy of EVEKTOR. Source: Adapted from Cupal et al. 2018 [12].

1.4 In-Vehicle Wireless Energy Transfer

In Section 1.3, a textile-based distribution network was presented. The network was intended to feed and connect electronic subsystems (sensors, harvesters, communication units) integrated into a textile layer covering the inner surface of a vehicle.

If the network is expected to integrate also components and subsystems, which are not located on the surface of the vehicle (e.g. phones of passengers), textile integrated antennas have to be implemented. Such antennas play the role of an interface in between the textile-integrated distribution network and an inner space of the vehicle.

The antennas should meet following requirements:

The antenna should be conceived as a planar structure so that the textile layer can be installed in between the metallic (or composite) surface of a vehicle and a glass-laminate panel covering the inner surface.

Since the distribution network is based on TIWs, antenna elements should be integrated into the top metallic surface to make the feeding as simple as possible.

Structure of the antenna should be as simple as possible so that the fabrication costs can be minimized.

Radiation pattern of the antenna should be shaped properly. If seats of passengers are going to be covered by the signal then lobes of the pattern should be oriented toward the seats.

Antennas should radiate a circularly polarized wave preferably. That way, a potential problem with improperly oriented antennas can be eliminated.

Gain of the antenna should be as high as possible to ensure a good efficiency of the wireless link. On the contrary, a wider beam-width might be requested to cover several seats by a single antenna.

Dealing with operation frequency of antennas, we consider:

The ISM band 5.8 GHz: This frequency band is a good compromise between dimensions of antennas on one hand, and losses and costs on the other hand.

UWB frequencies, the band-group VI: This frequency band is preferred by vehicle manufacturers due to the availability of commercial chipsets.

The ISM band 24 GHz: An antenna at this frequency was designed to study the frequency limit of implementation of textile-integrated antennas given by increasing losses and higher requested manufacturing precision.

Initial design of a textile-integrated antenna was related to wearable body-centric communication systems operating at 5.8 GHz [17]. The antenna consisted of a circular slot etched to the top surface of TIW. The metallic via in the center of the antenna forms the field distribution inside TIW properly to optimize antenna parameters. At the input, a microstrip-to-TIW transition is used.

Conductive layers of the first antenna were fabricated from a self-adhesive copper foil. Walls of SIW were created from the conductive thread Shieldex® 235/34 dtex 4ply HC+B. In the second antenna, conductive layers were manufactured from the conductive fabric material NI/CI NYLON RIPSTOP FABRIC produced by Laird Technology (see Figure 1.15).

In Figure 1.16, measured frequency responses of reflection coefficient at the input of antennas are compared with simulation. Higher losses in textile

Figure 1.15 Textile-integrated slot loop antennas. Source: Based on data from Hubálek et al. 2016 [17].

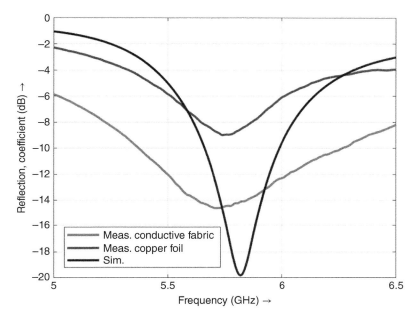

Figure 1.16 Frequency response of reflection coefficient at the input of textile-integrated slot antenna.

substrates, conductive layers, and SIW walls made the impedance bandwidth wider compared to simulation. SIW walls were made by hand, and therefore, the textile antennas are slightly detuned.

The described antenna operated with linear (vertical) polarization. Since the antenna was designed for on-body communication initially, the radiation pattern was optimized to follow the surface of the antenna. For vehicular applications, the antenna was redesigned:

- The operation frequency was shifted to the band-group VI of UWB frequencies.
- Radiation patters were reshaped to cover seats of passengers.
- Polarization was changed from linear to circular.

In order to implement the antenna, three different technologies were used [18]:

In order to create a reference structure, metallic parts of the antenna were created from a self-adhesive copper foil.

Metallic parts were screen-printed on the 3D textile substrate directly. For screen-printing, we used the Aurel C880 screen printer and the ESL 1901-S polymer silver paste. The printed structure was cured at 80 °C for 30 minutes.

The textile substrate SINTEX 3D041 ($h = 3.41$ mm, $\varepsilon_r = 1.22$) was covered by a plastic film Digiflex-Master (Aplhaset) to make the fabric material smooth and impenetrable. Screen-printing was applied in the second step.

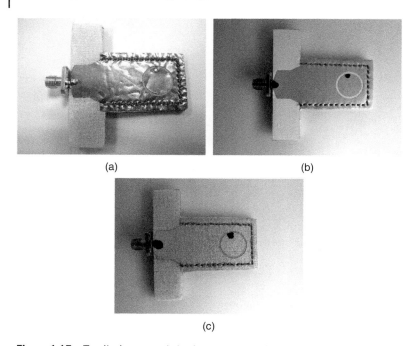

(a)   (b)

(c)

Figure 1.17 Textile-integrated slot loop antennas for vehicular applications [18].
(a) Self-adhesive copper foil, (b) printing on plastic film, (c) printing on textile.

Technology of sewing side walls of TIW stayed unchanged.

Fabricated antennas are shown in Figure 1.17. In Figure 1.18, frequency responses of reflection coefficient at the input of implemented antennas are compared. The simulated antenna shows the worst impedance matching; all losses were neglected in this case to obtain the worst case. The simulated results are approached by the copper-foil antenna, and the antenna printed on a film. Antenna printed on the textile directly shows the highest losses due to the penetration of the paste into the textile substrate.

Via inside the antenna was shifted from the center to excite circular polarization; via was connected to the screen-printed surface by a conductive glue. Moreover, the radiation pattern was shaped so that the maximum radiation is oriented to neighboring seats in an airplane. Radiation patterns are shown in Figure 1.19.

In Figure 1.20, an experiment in a small airplane Evektor VUT 100 is shown. The transmitting screen-printed antenna is fixed to the glass-laminate panel covering the roof, and the receiving antennas are placed to a seat of passengers [19].

Frequency responses of transmission between the transmit antenna and the receive one are shown in Figure 1.21. The transmit antenna was in a fixed position

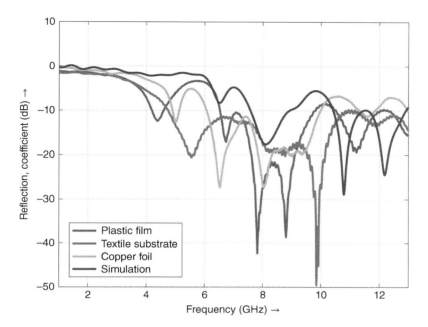

Figure 1.18 Frequency response of reflection coefficient at the input of textile-integrated, slot loop antennas for vehicular applications [18].

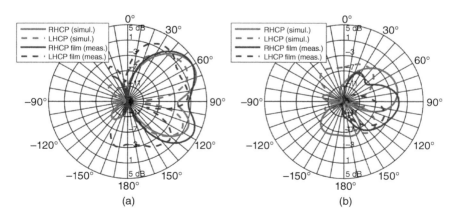

Figure 1.19 Directivity patterns of textile-integrated slot loop antenna for vehicular applications [18] in vertical plane (a) and horizontal plane (b). Source: Based on data from Špůrek et al. 2016 [18].

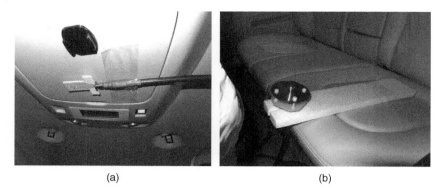

(a)   (b)

Figure 1.20 Measurements in the airplane Evektor VUT 100. (a) Transmit antenna on roof, (b) receive antenna on seat. Source: Based on data from Raida et al. 2016 [19].

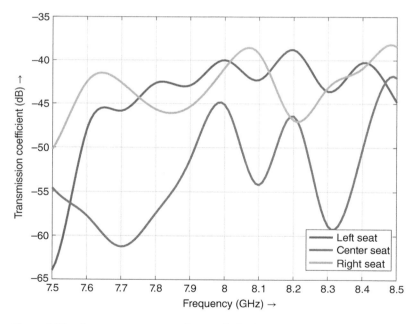

Figure 1.21 Frequency response of transmission between printed slot ring antenna on roof and reference antenna on seat. Source: Based on data from Raida et al. 2016 [19].

Figure 1.22
Textile-integrated
four-element antenna
array [20]. Screen-printing
on plastic film.

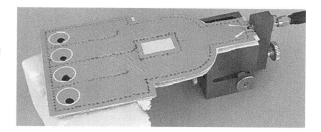

in the center of the roof, and the receive antenna was placed to the position of a left passenger, a right passenger, and in between passengers. In the position between passengers, the transmission is significantly lower due to the minimum of the radiation pattern of the transmit antenna in the corresponding direction. The transmission covers the band-group VI of UWB frequencies with variations caused by the multipath propagation [19].

In order to improve directive properties of the textile-integrated slot loop antennas for vehicular applications, a possibility of grouping slot elements into an array was studied. Structure of the array is obvious from Figure 1.22 [20]:

A coaxial connector feeds the transition between the coplanar waveguide and TIW.

The first power divider divides the wave into two branches with the same amplitude and the same phase. Following two dividers create four waveguides with identical amplitudes and phases, which feed four slot elements.

Directivity patterns of the array are shown in Figure 1.23. The main lobe with maximum of the gain of 10 dBi is oriented perpendicularly to the surface of the

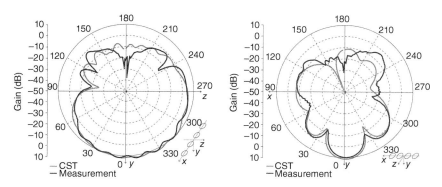

Figure 1.23 Radiation patterns of the antenna array; right-hand circular polarization. Source: Based on data from Špůrek et al. 2017 [20].

substrate. The 3 dB beam width is 25° in the *xy* plane and 93° in the *yz* plane. Thus, the whole inner space of a vehicle can be covered by one properly oriented antenna array.

Simulating the antenna array, operation with $|S_{11}| < -10$ dB was achieved within the band from 7.93 to 8.08 GHz. The measured operation band was from 7.95 to 8.25 GHz. A wider bandwidth of the prototype is caused by losses which were neglected in simulations.

Polarization properties of the designed antenna were characterized by axial ratio (AR). At the operation frequency, the simulated and measured AR is 0.5 and 4.0 dB, respectively. In order to reach a better AR and impedance matching at the operation frequency, the fabrication process of the antenna needs further optimization. Especially, the sewing of the TIW has still an insufficient accuracy because it is done by a hand.

Parametric analysis shows that the design is sensitive to the position and manufacturing of shorting pins of antenna elements also. Inaccuracy of the location of shorting pins therefore contributes to differences between measurements and simulations.

The discussed manufacturing aspects should have even stronger impact when increasing operation frequency of the antenna. In order to study this impact, a textile integrated antenna for the operation in the 24 GHz ISM band was designed, simulated, manufactured, and measured [21].

At higher operation frequencies, antennas cannot be designed for the so-far used textile substrate SINTEX 3D041 ($h = 3.41$ mm, $\varepsilon_r = 1.22$). Due to a relatively high height of 3D041, higher modes can be excited in TIW. To overcome this difficulty, the textile substrate SINTEX 3D097 ($h = 2.59$ mm, $\varepsilon_r = 1.22$) was therefore considered.

At 24 GHz, transitions from microstrip or CPW to TIW showed a significant parasitic radiation. For this reason, the antenna was fed by a coaxial probe (the dashed circle in Figure 1.24).

Finally, via used at lower frequencies for the excitation of circular polarization was not applicable at 24 GHz due to a strong sensitivity of antenna parameters on its position. Via was therefore replaced by crossed slots in the center of the loop (Figure 1.24).

Simulated and measured impedance characteristics are shown in Figure 1.25a. Frequency responses of simulation and measurement agree well:

Lower values of measured $|S_{11}|$ are caused by higher losses of the prototype compared to the simulation model.

Both the simulation and the measurement indicate very similar resonant frequency close to 22 GHz instead of requested 24 GHz. Later-on experiments showed that the described frequency shift was caused by the under-etching with high probability.

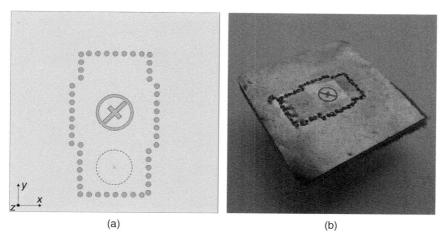

(a)   (b)

Figure 1.24 Textile-integrated slot loop antenna for the 24 GHz ISM band. (a) Schematics, (b) manufactured prototype. Source: Based on data from Cupal and Raida 2017 [21].

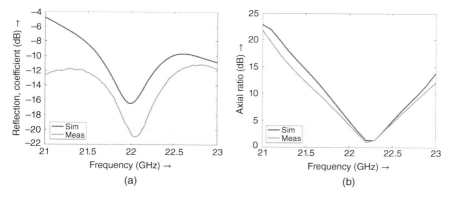

(a)   (b)

Figure 1.25 Textile-integrated slot loop antenna for the 24 GHz ISM band [21]. Frequency response of reflection coefficient (a) and axial ratio (b). Source: Based on data from Cupal and Raida 2017 [21].

The axial ratio (Figure 1.25b) met the condition $|AR| < 3$ dB for the bandwidth 285 MHz.

Radiation patterns of the antenna are shown in Figure 1.26. The antenna has a well-defined single lobe, which is perpendicular to the surface of the antenna. At the frequency 22.3 GHz, the gain of the antenna is 5.12 dBi.

Obtained results showed that operation frequency of textile-integrated antennas can be increased toward millimeter-wave frequencies if technological problems are solved.

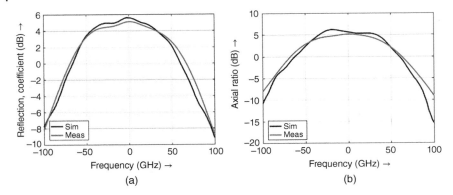

Figure 1.26 Radiation patterns of textile-integrated slot loop antenna for the 24 GHz ISM band [21]. (a) the *yz* plane, (b) the *xz* plane. Source: Based on data from Cupal and Raida 2017 [21].

1.5 Summary

In this chapter, exploitation of wireless energy transfer in vehicles was discussed. The research was motivated to substitute wired connections by wireless ones, potentially, to reduce the amount of cabling or increase reliability by wireless backup of wired services.

Low efficiency of wireless energy transfer was shown to be one of most severe problems of wireless energy transfer. In order to increase efficiency, the concept of a single-wire transmission line was applied, and the energy transfer was supported by the conductive surface of a skin of the vehicle. Further increase of efficiency was achieved by coating the surface by a textile layer.

The textile coating could be understood as a dielectric substrate suitable for the implementation of a network distributing electromagnetic energy along the surface of a vehicle. The substrate can be also used for the integration of sensors, subsystems, and antennas.

Antennas integrated in the textile play the role of an interface between the textile-integrated network and the inner space of the vehicle. Thanks to these antennas, electronic on-board equipment can be connected to the network.

Simulations and experiments showed that the described wireless energy transfer can be implemented up to the frequency 24 GHz. At higher frequencies, fabrication tolerances are too strict to be implemented using a textile substrate. And moreover, losses become too high.

References

1 TIMES Microwave Systems. (2019). LMR-1700: flexible low loss communications coax (11 September). www.timesmicrowave.com.

2 Jordan, E.C. and Balmain, K.G. (1968). *Electromagnetic Waves and Radiating Systems*, 2e. Englewood Cliffs: Prentice Hall.

3 Gobau, G. (1951). Single-conductor surface-wave transmission lines. *Proceedings of the IRE* 39 (6): 619–624.

4 Velim, J., Raida, Z., Lacik, J. et al. (2015). On-roof wireless link operating at 60 GHz. In: *IEEE-APS Topical Conference on Antennas and Propagation in Wireless Communications (APWC 2015)*, 153–156. Torino: Polytecnico di Torino.

5 Brick, D.B. (1955). The excitation of surface waves by a vertical antenna. *Proceedings of the IRE* 43 (6): 721–727.

6 Peterson, G. (2015). The application of electromagnetic surface waves to wireless energy transfer. *IEEE Wireless Power Transfer Conference (WPTC)* https://doi.org/10.1109/WPT.2015.7139133.

7 Podilchak, S.K., Tornero, J.L.G., and Goussetis, G. (2015). Microwave power transmission by electromagnetic surface wave propagation for wireless power distribution, *URSI Atlantic Radio Science Conference (URSI AT-RASC 2015)*, Las Palmas, Spain (16–24 May 2015). doi: https://doi.org/10.1109/URSI-AT-RASC.2015.7303004,

8 Corum, K.L., Corum, J.F., and Miller, M.W. (2016). Surface waves and the 'crucial' propagation experiment – the key to efficient wireless power delivery. *Texas Symposium on Wireless and Microwave Circuits and Systems (WMCS)* https://doi.org/10.1109/WMCaS.2016.7577497.

9 Sievenpiper, D.F., Zhang, L., Broas, R.F.J. et al. (1999). High-impedance electromagnetic surfaces with a forbidden frequency band. *IEEE Transactions on Microwave Theory and Techniques* 47 (11): 2059–2074.

10 Clavijo, S., Diaz, R.E., and Mckinzie, W.E. (2003). Design methodology for Sievenpiper high-impedance surfaces: an artificial magnetic conductor for positive gain electrically small antennas. *IEEE Transactions on Antennas and Propagation* 51 (10): 2678–2690.

11 Wu, K., Descantes, D., and Cassivi, Y. (2003). The substrate integrated circuits – a new concept for high-frequency electronics and optoelectronics. *TELSIKS* 2003, Nis (Serbia and Montenegro), Nis, Yugoslavia (1–3 October 2003), pp. P-III–P-X. doi: https://doi.org/10.1109/TELSKS.2003.1246173.

12 Cupal, M., Dřínovský, J., Götthans, T. et al. (2018). Textile-integrated electronics for small airplanes. In: *European Conference on Antennas and Propagation (EuCAP 2018)*. London: European Association for Antennas and Propagation.

13 Taringou, F. and Bornemann, J. (2011). New interface design from substrate-integrated to regular coplanar waveguide. *Asia-Pacific Microwave Conference*, Melbourne, VIC, Australia (5–8 December 2011), pp. 403–406.

14 Deslandes, D. and Wu, K. (2001). Integrated microstrip and rectangular waveguide in planar form. *IEEE Microwave and Wireless Components Letters* 11 (2): 68–70.

15 Cupal, M., Raida, Z., and Vélim, J. (2017). Transition adapters for 3D textile substrates. In: *Microwave and Radio Electronics Week (MAREW 2017)*. Brno: IEEE.

16 Torres, G.A. (1980). The microwave loss tangent of nylon 6 and polyester as a function of water content. Theses and Dissertation. Bethlehem (USA): Lehigh University.

17 Hubálek, J., Láčík, J., Puskely, J. et al. (2016). Wearable antennas: comparison of different concepts. In: *The 10th European Conference on Antennas and Propagation (EuCAP 2016)*. Davos: European Association on Antennas and Propagation.

18 Špůrek, J., Vélim, J., Cupal, M. et al. (2016). Slot loop antennas printed on 3D textile substrate. In: *The 21st International Conference on Microwave, Radar and Wireless Communications (MIKON 2016)*. Krakow: Microwave and Radiolocation Foundation.

19 Raida, Z., Vélim, J., Cupal, M., and Krutílek, D. (2016). Wireless power transmission in small airplanes. In: *Wireless Power Transfer Conference (WPTC 2016)*. Aveiro: IEEE.

20 Špůrek, J., Raida, Z., Láčík, J. et al. (2017). Circular slot antenna array printed on 3D textile substrate. *Elektrorevue* 19 (5): 141–144. Available: http://www .elektrorevue.cz.

21 Cupal, M. and Raida, Z. (2017). Circularly polarized substrate integrated textile antenna for ISM band 24 GHz. In: *International Conference on Electromagnetics in Advanced Applications (ICEAA 2017)*, 1154–1157. Verona: IEEE.

2

A Review of Methods for the Electromagnetic Characterization of Textile Materials for the Development of Wearable Antennas

Caroline Loss[1,4], Ricardo Gonçalves[2], Pedro Pinho[2,3], and Rita Salvado[4]

[1]*FibEnTech Research Unit, University of Beira Interior, Covilhã, Portugal*
[2]*Departamento de Electrónica, Telecomunicações e Informática, Instituto de Telecomunicacoes, Universidade de Aveiro, Aveiro, Portugal*
[3]*Instituto Superior de Engenharia de Lisboa - ISEL, Lisboa, Portugal*
[4]*LabCom IFP, University of Beira Interior, Covilhã, Portugal*

2.1 Introduction

Nowadays, the socioeconomic development and lifestyle trends indicate an increasing consumption of technological products and processes, powered by emergent concepts such as Internet-of-Things (IoT), where everything is connected in a single network [1]. According to Cisco [2], the IoT describes a system where items in the physical world, and sensors within or attached to these items, are connected to the Internet via wireless and wired Internet connections. These sensors can use various types of local area connections such as radio frequency identification (RFID), near field communication (NFC), Wi-Fi, Bluetooth, and Zigbee. Sensors can also have wide area connectivity such as global system for mobile communication (GSM), global positioning system (GPS), 3G, and long-term evolution (LTE).

Following the European Commission Report [3], in the next years, the IoT will be able to improve the quality of life, especially in the health monitoring field. The development of smart objects for IoT applications, include the capacity of these objects to be identifiable, to communicate and to interact [4]. In this context, wearable technology has been addressed to make the person, mainly through his clothes, able to communicate with and be part of this technological network.

Wireless communication systems are made up of several electronic components, which over the years have been miniaturized and made more flexible, such as batteries, sensors, actuators, data processing units, interconnectors, and antennas [5]. Turning these systems into wearable systems is a demanding research subject. In the systems for on-body applications, the antennas have been challenging

Wireless Power Transmission for Sustainable Electronics: COST WiPE - IC1301,
First Edition. Edited by Nuno Borges Carvalho and Apostolos Georgiadis.
© 2020 John Wiley & Sons, Inc. Published 2020 by John Wiley & Sons, Inc.

because they are conventionally built on rigid substrates, hindering their efficient and comfortable integration into the garment. However, embedding antennas into clothing allows expanding the interaction of the user with some electronic devices, making them less invasive and more discrete. Since 2001, when P. Salonen et al. [6] proposed the first prototype of flexible antenna made of fabrics, the use of textile materials to develop wearable antennas has increased exponentially.

Textile antennas combine the traditional textile materials with new technologies. They emerge as a potential interface of the human-technology-environment relationship. They are becoming an active part in the wireless communication systems, aiming applications such as tracking and navigation [7–10], mobile computing [11–13], health monitoring [14–17], energy harvesting [18–20], and others [21–24].

To achieve a low profile and unobtrusive integration of the antenna into the garment, antennas have to be thin, lightweight, robust, and easy to maintain. Moreover, they must be low cost for manufacturing and commercializing. In this way, planar antennas have been proposed for wearable applications, because this type of antenna topology combines all these characteristics, and is also adaptable to any surface. Such antennas are usually formed by assembling conductive (patch and ground plane) and dielectric (substrate) layers [25], as shown on Figure 2.1. Furthermore, planar antennas, such as the microstrip patch antenna, radiate perpendicularly to a ground plane, which shields the antenna radiation, ensuring that the human body is exposed only to a very small fraction of the radiation.

In order to efficiently design textile antennas, the electromagnetic properties of the materials used as substrates must be known, since the relative permittivity (ε) and the dissipation factor (tan δ) (also known as loss tangent) of the material influence the propagation characteristics and, therefore, must be considered during the design of a circuit for microwave applications. Besides the dielectric constant of the substrate, the choice of the conductive fabric for the patch and the ground planes

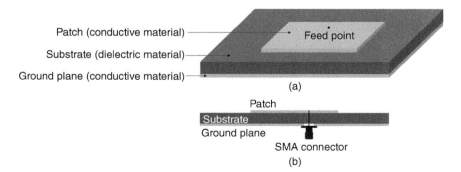

Figure 2.1 Layout of microstrip patch antenna: (a) front view and (b) cross section.

is also very important to assure a good performance of the antenna [26]. For these reasons, the knowledge of the electromagnetic properties of the textiles is crucial.

2.2 Electromagnetic Properties of Materials

The dimensions of the microstrip patch antennas are directly related to the desired resonance frequency, which are defined as portions of the wavelength of propagation [25]. The wavelength is dependent on the permittivity and permeability of the propagation medium (e.g. the dielectric substrate). For this reason, the knowledge of the electromagnetic properties of the textile materials is crucial to design wearable antennas [27].

The response of the material to the electromagnetic fields is described by the constitutive parameters of the material: permittivity (ε), permeability (μ), and conductivity (σ). These parameters also determine the spatial extent to which the electromagnetic field can penetrate the material at a given frequency. The relationship between these constitutive parameters and the electromagnetic fields is described through the Maxwell's equations [28]. In this way, materials can be classified based on their conductivity. Materials with high conductivity ($\sigma \gg 1$) are classified as conductors, also referred as *metals*, and materials with low conductivity ($\sigma \ll 1$) as insulators, also referred as *dielectrics*. What lies in between are referred as *semiconductor*.

2.2.1 Conductive Fabrics

Generally, the fabrics are insulating materials. However, there are materials with high electrical conductivity, which incorporate fibers, filaments, or coatings of metals or conductive polymers [29–34]. There is a large range of commercially available electrotextiles (e.g. Less EMF Inc. at http://www.lessemf.com or Shieldex Trading at http://www.shieldextrading.net) with high conductivity, enabling the development of antennas with acceptable performance.

Fabrics are planar materials and, therefore, their electrical behavior may be quantified by the surface (or sheet) resistance (R_s) and characterized by the surface (or sheet) resistivity (ρ_s). The resistance, which unit is (Ω), is the ratio of a DC voltage (U) to the current flowing between electrodes that are in contact with the same face of the material under test (MUT) in a specific configuration, as shown on Figure 2.2. In a conductivity measuring apparatus, the resistance may be dependent on the geometry of the electrodes used for measurement. Nevertheless, resistance can be related to the resistivity, which is independent of the geometry of the conductor. The sheet resistance is the ratio of the DC voltage drop per unit length (L) to the surface current (I_s) per unit width (D). It is thus

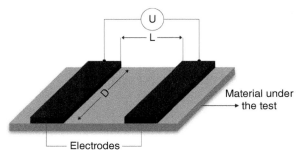

Figure 2.2 Basic setup for surface resistance and surface resistivity measurement. Source: Adapted from ASTM Standards 1999 [35].

a property of the material, not depending on the configuration of the electrodes used for the measurement [36]. The results of sheet resistance (R_s) are given in ($\Omega/\square$) [35]. The sheet resistivity is then determined by multiplying the sheet resistance by the thickness of the material.

The conductivity of the fabric, which unit is Siemens per meter (S/m), is related to the sheet resistance by Eq. (2.1), where h is the thickness of the fabric:

$$\sigma = \frac{1}{(R_s \cdot h)} \tag{2.1}$$

The sheet resistance is usually given by the manufacturer and may be measured by standard methods, such as, ASTM Standard D 257-99 Standard Test Method for DC Resistance or Conductance of Moderately Conductive Materials [35], ASTM Standard F 1896 – Test Method for Determine the Electrical Resistivity of a Printed Conductive Material [37], AATCC Test Method 76-2011: Electrical Surface Resistivity of Fabrics [38], among others [39, 40]. Despite the existence of several standard methods, an accurate characterization of highly conductive fabrics demands specific techniques, such as for instance the ones based on transmission lines and waveguide cavities [41–44].

Besides the dielectric constant of the substrate, the choice of the conductive fabric for the patch and the ground planes is also very important to assure a good performance of the antenna. In general, the conductive fabrics must have a very low electrical sheet resistance, $\leq 1\ \Omega/\square$, in order to minimize the electric losses and thus increase the antenna efficiency.

Despite the fact that the surface resistance value should be constant over the area of the antenna [45], the fabric may present some heterogeneities, such as for instance some discontinuities in the electric current. If these discontinuities are parallel to the surface current, then they will not interfere with the electromagnetic fields [11], but if discontinuities prevent the flow of the electrical current, the fabric resistance will increase [45].

Other factors, as for instance the float, which is the length of the conductive yarns laying on the surface of the woven, also influence the electrical behavior of the material. The side with the longest floats of conductive electric wires exhibit lower sheet resistance than the other side [45]. Likewise, a higher density of conductive wires also causes lower sheet resistance [42]. Furthermore, the humidity content in the material is also an important factor to consider when determining the electrical resistivity of textile materials because the presence of moisture in the fibers significantly decreases the electrical resistivity [43, 46–48].

2.2.2 Dielectric Fabrics

As in the dielectric materials, the conductivity is very small or null, their electromagnetic behavior is mainly determined by the permittivity and the permeability. Permittivity describes the interaction of the material with the electric field applied on it, whereas permeability describes the interaction of the material with the magnetic field applied on it.

Permittivity is a complex value that generally depends on frequency, temperature, and moisture [28, 49, 50]. Furthermore, permittivity is usually expressed as a relative value (ε_r), given by (2.2), where ε_0 is the permittivity of vacuum, equal to 8.854×10^{-12} F/m.

$$\varepsilon = \varepsilon_0 \varepsilon_r = \varepsilon_0 \left(\varepsilon_r' - j\varepsilon_r'' \right) \tag{2.2}$$

The relative permittivity is often called dielectric constant. The real part of the relative permittivity (ε_r') is a measure of how much energy from an external electric field is stored in the material. The imaginary part of the relative permittivity (ε_r''), or loss factor, is a measure of how dissipative a material is to an external electric field.

The ratio between the imaginary and the real part of the relative permittivity is the loss tangent ($\tan \delta$) – often called the material dissipation factor (D_f), expressed by (2.3). In the ideal scenario, the perfect dielectric material has $\tan \delta = 0$.

$$\tan \delta = \frac{\varepsilon_r'}{\varepsilon_r''} \tag{2.3}$$

When designing antennas, the key parameter for the performance of the dielectric substrate is the relative permittivity as well as the loss tangent. In general, textiles present a very low dielectric constant as they are very porous materials, and the presence of air approaches the relative permittivity to one. As reference, Table 2.1 shows the dielectric constant of different material.

For the development of wearable antennas, several conventional textile fabrics have been applied as dielectric substrate, exhibiting very low ε_r and $\tan \delta$, which reduce the surface wave losses and improve the impedance bandwidth of the antenna. Surface waves are connected to the guided wave propagation within the substrate. Hence, by reducing the dielectric constant, the contribution of the spatial waves increases, and consequently, the impedance bandwidth of the

Table 2.1   Relative permittivity of different type of materials.

References	Material	ε_r
[51]	Alumina	10.1
[52]	FR-4	4.50
[53]	Paper	7
	Glass	5–10
[41]	Quartzel$^{®}$ Fabric	1.95
	100% cotton fabric	1.60

antenna increases, allowing the development of antennas with higher gain and acceptable efficiency [8, 54–56]. A sufficiently wideband and efficient planar textile antenna is realized by selecting a substrate with a low dielectric constant (preferably ≤ 4) and a low loss tangent ($<10^{-2}$) [13].

2.3   Dielectric Characterization Methods Applied to Textile Materials and Leather: A Survey

In one side, specific electrically conductive textiles have been successfully used in the radiating components of the microstrip patch antennas. For other side, conventional textile fabrics have been used as substrates; however, little information can be found on the electromagnetic properties of these regular textiles.

During the past century, several textile engineers have studied the dielectric properties of single fibers [46, 57]. The main challenge founded on these works was the heterogeneity of air and fibers. Later, in 2010, Bal and Kothari have reported a comparison of formulas for the air fiber mixture, used in the dielectric characterization [58]. Besides the large reports on the fiber context, only few researches on the characterization of fabrics in the context of textile antennas were founded.

The dielectric behavior of textiles depends on the properties of the component fibers and the structure of the yarns and/or of the fabrics, and on the fiber packing density in the fibrous material [46, 59]. Also, being textiles highly porous materials, the presence of air and moisture influences their dielectric characterization.

Despite the growth of research studies on textile antennas, the accurate characterization of the dielectric properties of the textile materials is still a challenge due to the intrinsic inhomogeneity and deformability of textiles [60], that will be discussed later in Section 2.4. Some factors that affect the measurement of dielectric properties of textiles. Up today, there is no standard method to measure the dielectric properties of textiles. In this section will be reviewed only the methods that have already been proposed and applied to characterize textile materials.

The main objective of the methods for the electromagnetic characterization of dielectric materials is to measure the relative permittivity of the specimen for a specific frequency, or bandwidth frequency, and field orientation [50]. The methods for the characterization of the dielectric properties that have been used are generally subdivided into two main categories: resonant and nonresonant methods [28]. Each of these categories includes several procedures, as will be presented in the following subsections.

2.3.1 Resonant Methods

Resonant methods usually provide higher accuracy and sensitivity than nonresonant methods for low-loss materials [49], even though they only allow material characterization for a single frequency. They include the resonator method and the resonance-perturbation method. The resonator method is based on the fact that the resonant frequency and quality factor of a dielectric resonator with given dimensions are determined by its permittivity and permeability. These methods are usually applied to measure low-loss dielectric materials. The resonance-perturbation method is based on resonant perturbation theory. For a resonator with given electromagnetic boundaries, when part of the electromagnetic boundary condition is changed by introducing a sample, its resonant frequency and quality factor (Q-factor) will also change. Measuring these shifts in the frequency and in the Q-factor is possible to extract the permittivity value, as presented in [28]. The Q-factor [28] is a parameter often used to describe an electromagnetic material, according to Eqs. (2.4–2.6):

$$Q_e = \frac{\varepsilon_r'}{\varepsilon_r''} = \frac{1}{\tan \delta_e} \tag{2.4}$$

$$Q_m = \frac{\mu_r'}{\mu_r''} = \frac{1}{\tan \delta_m} \tag{2.5}$$

where Q_e is the electric quality factor and Q_m is the magnetic quality factor. Based on this, it is possible to calculate the total quality factor (Q) of the material:

$$\frac{1}{Q} = \frac{1}{Q_e} + \frac{1}{Q_m} \tag{2.6}$$

2.3.1.1 Cavity Perturbation Methods
There are two main standard methods based on resonant perturbation techniques, one is the cavity perturbation method and the other is the SDR (split dielectric resonator). The first is defined by the ASTM D2520-13 standard, and used by insulator material manufacturers for permittivity estimation. The second has been defined by the IPC as TM-650 2.5.5.13 and is used by substrate manufacturers.

For textiles, the most common resonance techniques are those based on resonant cavities formed by a rectangular or circular waveguide. The MUT is inserted

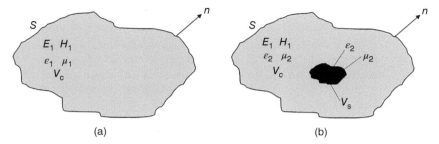

(a)   (b)

Figure 2.3 Cavity perturbation (a) original cavity (b) perturbed cavity after the material insertion. E_1 and H_1 are the electric and magnetic fields, respectively; ε_1 and μ_1 are the permittivity and permeability of the cavity; V_c is the volume of the cavity; ε_2 and μ_2 are the permittivity and permeability of the material; and V_s is the volume of the sample. Source: Adapted from Chen et al. 2004 [28].

into the resonant cavity, and their electromagnetic properties are calculated from the changes in the resonant frequency and in the quality factor of the cavity caused by the introduction of the material [28, 61–63]. This phenomenon is illustrated on Figure 2.3, and the theoretical equations can be found in [28].

These methods are difficult to apply as in every measurement the sample has to be placed in the same position. Also, the cavity needs to be dismantled and reassembled every time a new sample is tested [28], consuming time and maybe introducing errors on the measurements. Moreover, the cavity resonator methods measure the permittivity in the plane of the sample and, therefore, cannot determine any anisotropy in the measured plane [49].

In [63], Kumar and Smith present the electromagnetic characterization of yarns and fabrics using a cylindrical split cavity resonator. In this work, the averaged result obtained for five yarns with different diameters, $\varepsilon_r = 3.1141$, is compared to the averaged result for three fabrics made with these yarns $\varepsilon_r = 2.7851$. As the textile materials are a mixture of nylon fibers and air, the authors concluded that fabrics exhibit higher porosity and, for this reason, a lower permittivity value than the yarns measured stand-alone.

Besides the cavity methods, there are also the printed circuit characterization methods. In these, the material to be tested is used as substrate to a given printed resonant structure. Although presenting less precision in the characterization, have an advantage of characterizing the material under the operation circumstances. That is, the permittivity component being measured is exactly the same as the one that is used during circuit operation. Therefore, even if the material has some kind of anisotropy, this can be safely discarded. The main disadvantage of this method is the fact that the sample used for testing is lost and cannot be reused for other purposes. For textile materials, different techniques of the printed circuit characterization methods were applied and will be discussed in Sections 2.3.1.2 and 2.3.1.3.

2.3.1.2   Microstrip Resonator Patch Method

In [64], Sankaralingam and Bhaskar proposed a novel microstrip patch radiator method that consists in designing a patch antenna using an estimated permittivity value found by literature review. After manufacturing the antenna and measuring its S_{11}, the real value of the dielectric constant is calculated based on the shift of the resonant frequency. Six different textile materials were characterized and the extracted values can be found in Table 2.2. To validate this method, the authors designed some textile antennas for 2.45 GHz, using three tested materials as a dielectric substrate, and copper sheets for the conductive parts. Despite that these antennas exhibit good results, confirming the suitability of this method, the results can be influenced by the manufacturing technique. Indeed, the use of an interface to assembly the layers, such as glue or an adhesive sheet, and the inaccuracies when soldering the SMA connector to the microstrip patch radiator, can introduce errors in the final values.

2.3.1.3   Microstrip Resonator Ring Method

The microstrip resonator methods are based in the resonance-perturbation method. The sample of the MUT is placed over the microstrip resonator affecting its resonant frequency and quality factor. Measuring the shift in the frequency and in the Q-factor enables to extract the permittivity values [28]. The microstrip resonator methods are subdivided into three types: straight ribbon resonator, ring resonator, and circular resonator, as shown on Figure 2.4.

The microstrip ring resonator does not have open ends, decreasing the radiation loss and consequently increasing the quality factor. For this reason, the ring resonator is more accurate and sensitive than the straight and circular resonators. Details about the theoretical equations of microstrip resonator can be found in [28]. For the characterization of the textile materials, the procedure of the microstrip ring resonator method only involves attach the conductive parts, made of copper foil, to the textile dielectric probe under test. In [65], the characterization of a "Bakhram" textile fabric at two different frequencies (831.940 MHz and 1.6890 MHz) is presented. After the measurements, the permittivity and loss tangent extracted values were $\varepsilon_r = 2.031$ and $\tan \delta = 0.0038$, and $\varepsilon_r = 1.965$ and $\tan \delta = 0.0024$, respectively. To validate the method, the authors designed a textile patch antenna to resonate at 831.940 MHz, also made of copper foil in the conductive parts. Despite the deviation in the measured frequency, the authors considered the characterization method well suited for the characterization of textile materials, considering this deviation as part of the design and manufacturing process of the antenna.

2.3.1.4   Microstrip Patch Sensor

The microstrip patch sensor consists of a patch antenna covered with the dielectric MUT, called superstrate. As for the microstrip ring resonator method, this technique relies on the resonance-perturbation method, where the calculation of the

Table 2.2 Summary of resonant methods to characterize textile materials and leather.

References	Frequency application	Material	Thickness (mm)	ϵ_r	tan δ	Method
[63]	9.8 GHz	100% Nylon 6.6 fabric	—	2.82	0.02681	Resonant microwave cavity
			—	2.75	0.02420	
			—	2.78	0.02831	
[8]	GPS (1.5 GHz)	100% PA, Cordura® fabric	0.5	Between 1.1 and 1.7	—	Cavity perturbation technique
[12]	WLAN (2.4 GHz)	Fleece fabric	3	1.04	—	
[65]	831.940 MHz	"Bakhram" fabric	0.37	2.031	0.00038	Microstrip ring resonator method
	1.6890 GHz			1.965	0.0024	
[64]	2.45 GHz	100% washed cotton fabric	3.0	1.51	—	Microstrip patch radiator
		100% cotton, denim	2.84	1.67	—	
		65% PES 35% CO fabric	3.0	1.56	—	
		100% CO, fabric for Curtain	3.0	1.47	—	
		100% polyester (PES)	2.85	1.44	—	
		100% CO, bed sheet/floor spread fabric	3.0	1.46	—	
[23]	ISM (2.45 GHz) and 4.5 GHz	Original cowhide leather	0.7	1.76	0.0009	Agilent 85070E Dielectric Measurement Probe Kit
		Original sheepskin leather	0.7	2.5	0.0035	
		Original oiled sheepskin leather	0.7	2.66	0.085	
		Original scratched cowhide leather	0.7	3.13	0.15	
		Original oiled cowhide leather	0.7	2.3	0.04	
[66]	2.22–2.59 GHz	Denim fabric	1.6	1.7	0.085	

Ref.	Frequency	Material		Microstrip patch sensor	
[67]	2.25 GHz	100% polyester 3D fabric I	2.650	1.10	0.005
		100% polyester 3D fabric II	3.068	1.10	0.006
		100% polyester 3D fabric III	2.821	1.12	0.017
		100% polyester 3D fabric IV	2.410	1.13	0.018
		100% polyester 3D fabric V	4.140	1.11	0.004
		Cordura I – Plain weave 100% PA 6.6, PU coated	0.503	1.58	0.008
		Cordura II – Plain weave 100% PA 6.6, TF coated	0.501	1.56	0.008
		Neoprene I – Neoprene laminated with jersey 100% Polyester on both sides	5.000	1.37	0.0010
		Neoprene II – Neoprene laminated with jersey 100% Polyester (face side) and 100% nylon (reverse side)	3.095	1.30	0.001
		Fake leather I – Carded knit 100% polyester, PU coated	0.831	1.45	0.017
		Fake leather II – Carded knit 100% polyester, PU coated	0.923	1.43	0.012
[68]	1.9 GHz	PES + (LMF-PES), stitched nonwoven	10	1.013	—
		P84® +(LMF-PES), stitched nonwoven	8	1.012	—
		Kermel® + (LMF-PES), stitched nonwoven	8	1.014	—
		(PES + LMF-PES)/T, stitched nonwoven, thermal processed	1.4	1.175	—
		(P84 + LMF-PES)/T, stitched nonwoven, thermal processed	4	1.036	—
		(Kermel + LMF-PES)/T, stitched nonwoven, thermal processed	4	1.050	—

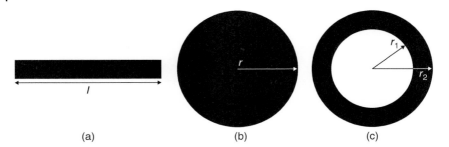

(a)   (b)   (c)

Figure 2.4 Types of microstrip resonators: (a) straight ribbon resonator, (b) circular resonator, and (c) ring resonator, where l is the length of straight ribbon resonator; r is the radius of the circular resonator; and r_1 and r_2 are the inner and outer radius of the resonator ring, respectively. Source: Adapted from Chen et al. 2004 [28].

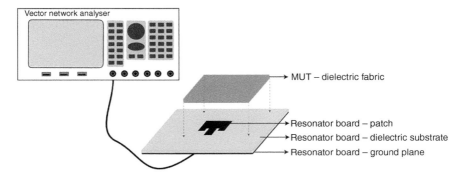

Figure 2.5 Dielectric characterization procedure through the microstrip patch sensor.

permittivity value is based on the shift in frequency caused by the introduction of the superstrate. Figure 2.5 illustrates the characterization procedure, and more information about theoretical equations can be found in [68–70].

In [70], this method is presented to measure solids and liquids. In [68], the method is used to test six nonwoven materials, manufactured by the stitching method, at 1.9 GHz. The results of the electromagnetic characterization are presented in Table 2.2.

In [67], the authors were used this method, also called resonator-based experimental technique, to characterize 11 fabrics, divided in four groups, at 2.25 GHz:

- *Five 3D fabrics, 100% polyester (PES):* (1) $\varepsilon r = 1.10$ and tan $\delta = 0.005$, (2) $\varepsilon r = 1.10$ and tan $\delta = 0.006$, (3) $\varepsilon r = 1.12$ and tan $\delta = 0.017$, (4) $\varepsilon r = 1.13$ and tan $\delta = 0.018$, and (5) $\varepsilon r = 1.11$ and tan $\delta = 0.018$.
- *Two Cordura* ®*fabrics:* (6) a plain weave 100% polyamide (PA) 6.6, polyurethane (PU) coated, showing $\varepsilon_r = 1.58$ and tan $\delta = 0.008$ and (7) a plain weave 100% PA 6.6, Teflon (TF) coated, presenting $\varepsilon_r = 1.56$ and tan $\delta = 0.008$.

- *Two Neoprene fabrics*: (8) Neoprene laminated with jersey 100% PES on both sides, showing $\varepsilon_r = 1.37$ and tan $\delta = 0.001$ and (9) Neoprene laminated with jersey 100% PES (face side) and 100% PA (reverse side), presenting $\varepsilon_r = 1.30$ and tan $\delta = 0.001$.
- Two fake leathers, carded knit 100% PES, PU coated, showing (10) $\varepsilon_r = 1.45$ and tan $\delta = 0.017$ and (11) $\varepsilon_r = 1.43$ and tan $\delta = 0.001$.

To validate the method, the authors designed one textile patch antenna for each characterized substrate material, to resonate at the same frequency used in the characterization process (2.25 GHz). Despite some deviation in the measured frequency, due to the influence of the surface roughness and porosity, the authors considered the characterization method well suited for the characterization of textile materials. In the Section 2.4.4 will be discussed these influences of surfaces features. In summary, nondestructive method can be a solution for a quick, easy, and low-cost characterization procedure.

2.3.1.5 Agilent 85070E Dielectric Measurement Probe Kit
The Agilent 85070E Dielectric Measurement Probe Kit is an equipment by Agilent Technologies, available on the market to measure the dielectric properties of several types of materials [71]. In [23], the characterization of textile and leather was performed using the open-ended coaxial probe method. In this method, the permittivity value is calculated only from the S_{11} parameter from the coaxial probe, measured with the vector network analyzer (VNA) or impedance analyzer, as illustrated in Figure 2.6.

Five types of leather were characterized at 2.45 GHz: (i) original cowhide, $\varepsilon_r = 1.76$ and tan $\delta = 0.0009$, (ii) original sheepskin $\varepsilon_r = 2.5$ and tan $\delta = 0.035$, (iii) oiled sheepskin $\varepsilon_r = 2.66$ and tan $\delta = 0.085$, (iv) scratched cowhide $\varepsilon_r = 3.13$ and tan $\delta = 0.15$, and (v) oiled cowhide $\varepsilon_r = 2.3$ and tan $\delta = 0.04$. Among the

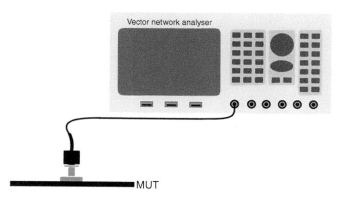

Figure 2.6 Agilent 85070E dielectric measurement probe kit.

characterized leathers, considering the operating frequency, leather (ii) was chosen to develop a dual-band wearable antenna for 2.45 and 4.5 GHz applications. Due to the structure of five-layered substrate, the simulated and measured results deviate. For this reason, the authors included a new simulation using four air gaps of 0.0875 mm thickness to correct the results. The size of the air gaps was obtained through the difference between the simulated substrate thickness (3.5 mm) and the measured thickness of the final substrate (3.85 mm). The gain and the radiation efficiency at the lower frequency (2.49 GHz) are −0.29 dBi and 15.2%, respectively; and at the higher frequency (4.52 GHz) 3.05 dBi and 41.4%, respectively.

In [66], a denim fabric was characterized using the open-ended coaxial probe method, as well. The denim fabric, with 1.6 mm of thickness, has shown $\varepsilon_r = 1.7$ and tan $\delta = 0.085$ and was used as dielectric substrate for a slotted antenna for radio frequency (RF) energy harvesting between 2.22 and 2.59 GHz.

On the one side, the advantages of this method are nondestructive technique, quick and easy to perform. On the other side, this method is very expensive (price of probe kit + software) and a complex calibration of the probe is required, using standard materials such as distilled water or methanol.

2.3.1.6 Summary of the Characterization of Textile Materials by Resonant Methods

Table 2.2 sums up the results obtained in the several studies previously referred.

2.3.2 Nonresonant Methods

The nonresonant methods mainly include procedures based on reflection and transmission/reflection measurements. As the reflection-based techniques, the dielectric properties of the MUT are extracted based on the reflection of the electromagnetic waves in free space by the sample. In transmission/reflection methods, the dielectric properties are calculated based on the reflection from and transmission through the sample [28]. For the characterization of textile materials, several nonresonant methods have been tested, such as transmission lines [27, 72–76], metallic and dielectric waveguides [41], and free space [77, 78].

2.3.2.1 Parallel Plate Method

The parallel plate method is the oldest way to measure the dielectric properties in fiber materials [57]. In this method, the MUT is placed between two parallel plates. In this case, the structure creates a capacitor whose capacity is measured by an LCR meter (LCR meter is an electronic equipment used to measure the inductance (L), capacitance (C), and resistance (R) of an electronic component). Owing to the simplicity of the procedure and the power measured by the LCR, this method is

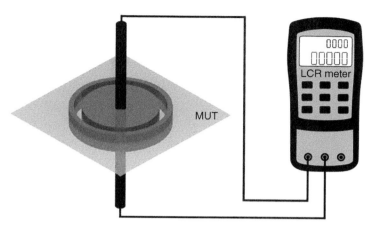

Figure 2.7 Setup of the parallel plate dielectric characterization method. Source: Adapted from Lesnikowski 2012 [76].

limited to the maximum 1 MHz frequency. Figure 2.7 shows the setup procedure, and the details about the theoretical equations can be found in [28, 46]. In [76], nine textile fabrics were characterized at 200 kHz, under controlled ambient conditions to avoid the influence of moisture. The obtained results are summarized in following Table 2.3.

2.3.2.2 Free Space Methods

This method typically consists in placing the MUT between two horn antennas and by measuring the S-parameters – transmission (S_{11}) and reflection (S_{21}) coefficients – of the antennas with a VNA, the dielectric constant is estimated, as shown in Figure 2.8. The main advantage of the free space method is that it is contactless and nondestructive. The main drawback of this method is the required precise calibration of the horn antennas.

In [80], this method was used to measure the dielectric properties of textile composites. The results of these characterizations can be found on Table 2.3. Furthermore, in [77], the authors present a variation of this method, based on an angular-invariant approach. Due to the typical nonregular surface of the textiles and leathers, in this angular-invariant approach based on Rayleigh scattering, the authors have measured the scattering parameters of the antenna by varying the angle of the material under the test. All materials tested in this work have shown variations in the complex permittivity as a function of the different incidence angles, due to irregular thickness, surface roughness, and texture of the materials. Also, the free-space method can be used to characterize the conductive textiles [78].

Table 2.3 Summary of nonresonant methods to characterize textile materials and leather.

References	Frequency application	Material	Thickness (mm)	ϵ_r	tan δ	Method
[76]	200 kHz	100% cotton, twill weave	0.62	2.23	0.0366	Parallel plate
		100% cotton, plain weave	0.48	2.07	0.0314	
		100% wool plain weave	0.42	1.86	0.0079	
		Wool, twill weave fabric	0.64	2.05	0.0076	
		Wool, plain weave fabric	1.26	1.67	0.0073	
		Wool + polyamide, twill weave	1.47	1.53	0.0053	
[76]	200 kHz	100% polyester (PES) plain weave	0.36	1.74	0.0044	
		Viscose + PES twill weave	0.52	1.70	0.0079	
		100% polyester plain weave	0.08	2.12	0.0035	
[72]	2.45 GHz	100% cotton, Denim by Santista	0.49	2.11	0.01	Transmission line
[74]	Nonspecified single frequency	Felt fabric	4.0	1.22	0.016	Stub resonator
		Denim woven	—	1.6	0.05	
	1–6 GHz	Felt fabric	4.0	1.215–1.225	0.016	Stripline
		Denim woven	—	1.6–1.65	0.05	
[75]	UWB (3.1–10.6 GHz)	100% PAN fabric	0.5	2.6	—	Two-line method
[79]	2.45 GHz	65% PES 35% CO	—	3.23	0.06	
[73]	2.45 GHz	98% PAR 2% Carbon Woven 1	0.6	1.57	0.007	Matrix-Pencil
		98% PAR 2% Carbon Woven 2	0.4	1.91	0.015	Two-line method
		100% PP Nonwoven	3.60	1.18	0.025	
		Fleece fabric	2.56	1.25	0.007	

[80]	8–11 GHz	E-Glass G7628	0.210	4.8–5.0	0.003–0.11	Free space
		E-Glass G880	0.152	3.84–4.0	0.003–0.11	
		Kevlar K141	0.254	3.97–4.05	0.003–0.11	
		Kevlar K151	0.254	3.88–3.04	0.003–0.11	
[77]	330 GHz	Denim woven	0.8	2.73	0.073	Free space: angular-invariant approach
		Textile 1	0.45	2.74	0.031	
		Textile 2	0.25	2.72	0.068	
		Textile 3	0.7	3.66	0.042	
		Textile 4	0.25	2.54	0.066	
		Wool fabric from scarf	1.6	3.18	0.15	
		Stockinet fabric	0.65	3.22	0.08	
		Satin fabric	0.2	2.74	0.075	
		Natural leather	1.15	3.4	0.127	
		Artificial leather	0.95	3.104	0.079	
		Artificial leather	1	3.008	0.08	
		Artificial leather	0.6	2.17	0.066	
		Artificial leather	0.8	3.02	0.084	
		Artificial leather	0.6	2.49	0.085	

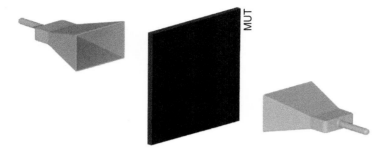

Figure 2.8 Setup of the free space method. Source: Adapted from Bakar et al. 2014 [80].

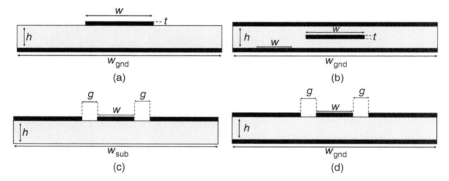

Figure 2.9 Types of transmission lines: (a) microstrip, (b) stripline, (c) coplanar waveguide, and (d) coplanar waveguide with ground plane, where h and t are the thickness of the dielectric and conductive materials, respectively; w is the length of the line, and g is the gap between the line and the ground plane in coplanar lines; W_{gnd} is the length of the ground plane, respectively. Source: Adapted from Chen et al. 2004 [28].

2.3.2.3 Planar Transmission Lines Methods

Planar transmission lines methods are the most common methods applied to characterize textile materials. Based on the scattering (S-) parameters, the advantage of this method is that it can also be applied to determine conductive properties, as presented in [28, 49]. The planar transmission lines are subdivided into three types: microstrip, coplanar, and stripline ones, as illustrated in Figure 2.9.

As shown on Figure 2.9d, the stripline consists of upper and lower grounding planes, and a central conductive line. The dielectric MUT is placed between the grounding planes and the central line. The advantage of this structure of transmission line is that the radiation losses are negligible. In [72], a stripline prototype was built to characterize a denim fabric, 100% cotton, at 2.45 GHz. For the conductive parts, Flectron® fabric was applied, and to design the transmission line, they used an estimated $\varepsilon_r = 1.2$. To achieve the required stripline height ($h = 2.45$ mm), five layers of denim fabric were superposed. The measurements with the VNA yielded $\varepsilon_r = 2.117$ and $\tan \delta = 0.01$. To validate this method, a microstrip patch antenna for 2.45 GHz was designed and manufactured using the same materials. The antenna has shown good performance, and 6.1594 dBi of gain was obtained.

In [74], Mantash et al. propose a comparison between a stripline and an open stub resonator, described in [81], to characterize felt and denim fabric. For the stub resonator, a copper tape was used for the conductive parts. Measuring at a nonspecified single frequency, for the denim, they extract $\varepsilon_r = 1.6$ and $\tan \delta = 0.05$, and for felt, $\varepsilon_r = 1.22$ and $\tan \delta = 0.016$. As to the broadband measurements, the authors have used a stripline fixture resonator for the dielectric material. The results between 1 and 6 GHz have shown a range of values from $\varepsilon_r = 1.215$ to $\varepsilon_r = 1.225$ for the felt material and from $\varepsilon_r = 1.6$ to $\varepsilon_r = 1.65$ for the denim fabric. Comparing the obtained results from the two methods, both are acceptable.

In order to validate the results, two patch antennas for 2.45 GHz were proposed, using the permittivity value extracted at single frequency, and Shieldit® Super fabric ($R_s = <0.5 \, \Omega/\square$) for the conductive parts. These antennas have shown a good agreement between the measured and simulated S_{11} parameters. Also, 6 and 2 dBi of gain was obtained for the antenna made with felt and denim, respectively. Both results from the return loss and gain confirmed the reliability of the retrieved data.

The propagation characteristics of two microstrip transmission lines with different lengths were measured in [75]. The lines were made using Nora® fabric, with $R_s = 0.03 \, \Omega/\square$, for the conductive parts, and an acrylic fabric with 0.5 mm of thickness for the dielectric substrate. Knowing the length difference and measuring the S-parameters between 3 and 10 GHz, the permittivity value for the acrylic fabric was calculated: $\varepsilon_r = 2.6$. In order to validate the method, two ultra wide-band (UWB) (3.1–10.6 GHz) wearable antennas were designed. The measured antenna parameters have shown to be in agreement with simulation estimations. Besides, the antennas have been tested for transmission of UWB pulses into the human body in order to evaluate reflectivity, which can be used in diagnosis applications. Moreover, they have proven to be reliable when in close contact with the human body, proving the usefulness of the development of such antennas in textiles.

In [79], the authors have used the two-line technique as well. They characterize a fabric, whose fibrous composition is 65% PES and 35% CO, to use as a dielectric substrate of patch antenna for 2.45 GHz RF power transfer and harvesting. The fabric showed $\varepsilon_r = 3.23$ and $\tan \delta = 0.06$. The measurement of the S_{11} of screen printed antenna shows an excellent agreement with the simulated results.

In [73], the authors have combined the two-line method for microstrip lines with the matrix-pencil technique, in order to reduce the perturbations in the parameters of the transmission lines. They have characterized four different materials: (i) woven fabric, 98% aramid and 2% carbon, with $h = 0.60$ mm and $\varepsilon_r = 1.57$; (ii) woven fabric, 98% aramid and 2% carbon, with $h = 0.40$ mm and $\varepsilon_r = 1.91$;

(iii) nonwoven polypropylene fabric with $h = 3.60$ mm and $\varepsilon_r = 1.18$; and (iv) fleece with $h = 2.56$ mm and $\varepsilon_r = 1.25$. For the conductive parts, copper sheet and Flectron fabric were used to manufacture the transmission lines. The authors concluded that the lines using copper sheet enabled to estimate the loss tangent of the substrates under test, given that the losses in copper sheet are much smaller than in the Flectron. To validate this method, the authors designed three textile antennas for 2.45 GHz, using the tested materials (i, ii, and iii) as dielectric substrate. For the conductive parts, copper sheets and Flectron fabric were used, and both copper and Flectron based antennas exhibited good results.

2.3.2.4 Summary of the Characterization of Textile Materials by Nonresonant Methods

Table 2.3 sums up the results obtained in the several studies previously referred.

2.4 Some Factors that Affect the Measurement of Dielectric Properties of Textiles

As explained before, textile materials incorporate some intrinsic singularities due to their inhomogeneity and instability, being a challenge in terms of accurate characterization. The dielectric behavior of textile materials are dependent on frequency, temperature, and humidity [46, 82–84]; it also depends on the properties of the component fibers and the structure of the yarns and/or of the fabrics, and on the fiber packing density in the fibrous material [46, 58, 59, 64]. In this subsection, we will discuss some factors that influence the dielectric characterization of textiles.

2.4.1 Influence of the Moisture Content

Textiles always establish a dynamic equilibrium with the temperature and humidity of the air surrounding they are in contact with, as the fibers are constantly exchanging water molecules with the air.

The amount of water that a textile material takes until reaching this equilibrium depends on the type of fiber. The extent to which a material is sensitive to moisture is described by its regain, which is defined, by the ratio of the mass of absorbed water in specimen to the mass of dry specimen, expressed as a percentage [46]. In [46] shows the relation between regain and relative humidity of the air (RH), for various textile fibers, compiling studies made by several authors. Indeed, for the same RH conditions, there are textile fibers with largely different moisture contents. For instance, at 65% RH, wool fiber might present a regain of 14.5%, cotton might present a regain of 7.5%, and polyester fiber might present a regain of 0.2%.

In general, the moisture absorption changes the properties of the fabrics. Because of this, textile metrology is performed always in a conditioned environment (20 °C and 65% RH). Also, for commercial transactions, there are national legislations setting nominal values of moisture content that are values close to the regain obtained at 65% RH.

Water has a dielectric constant of $\varepsilon_r = 78$ at 2.45 GHz and 25 °C [83]. Although this value depends on the salinity, temperature, and frequency, water has a much higher and more stable dielectric constant than textile fabrics, whose dielectric constant is generally in the range $\varepsilon_r = 1$–2, as previously presented. Therefore, when water is absorbed by the textile fibers or is trapped into the fabric structure, it changes the electromagnetic properties of the fabric, increasing its dielectric constant and loss [45, 46, 59, 82, 83]. Charts presenting the relationship between the RH of the air or the moisture content of various fibers and their dielectric properties can be found in [46]. Additionally, several authors have been correlating the electromagnetic properties of fabrics to their regain, showing that the higher permittivity of the water increase the permittivity of the fabrics, driving the performance of the antenna, reducing its resonance frequency and bandwidth [48, 82, 83, 85, 86]. Therefore, to avoid these influences, the measurements should be run in a conditioned atmosphere.

2.4.2 Influence of the Material Anisotropy

As textile materials are anisotropic materials, their characterization also depends on the electric field's orientation. This anisotropy is fully described by a permittivity tensor, although in most practical applications like the ones herein surveyed, one specific component of this tensor is enough to characterize the behavior of the textile material for a specific application.

However, anisotropy is intrinsic in textile materials, where the fabric possibly presents different weaving densities in warp and weft directions. Additionally, different fibers may compose warp and weft yarns. In [44], the authors presented the dielectric characterization of nine textile materials, using a closed cavity resonator. In this work, the authors have measured the textiles in all planes (x, y, and z), and the obtained results have showed the influence of different alignments of the material with the electromagnetic field. The anisotropy caused by the different density of yarns in the warp and weft directions is clearly seen.

2.4.3 Influence of the Bulk Porosity

Recently, in [67], the authors have related the results of the dielectric characterization obtained through the microstrip path sensor (see Section 2.3.1.4), to the bulk porosity (∅) of the characterized fabrics. The studied materials that showed higher bulk porosity also showed lower dielectric constant values, which is due to the higher amount of air inside the fabric structure. Also, a linear regression with $R^2 = 0.971\,69$ was presented, describing very well the relationship between ε_r and $\emptyset$.

2.4.4 Influence of the Surface Features

Further in [67], the authors have also correlated the results of the dielectric characterization to some surface features of the materials and conclude that the surface features of the MUT influence the dielectric results obtained with the resonator-based experimental technique. Analyzing the measured ε_r values, it was possible to verify that the dielectric constant extracted when placing the rougher face turning down contacting the resonator board presented a lower value. Also, it was reported that the ε_r exponentially decreases with the superficial porosity increase. This was observed for all samples. Both of these observed facts occur because of the presence of the air on the surface, trapped in the superficial pores. Besides the values of the dielectric constant measured through the resonator-based experimental technique differ when placing the face or reverse sides in contact to the resonator board and are thus influenced by the structure of the surface of the fabrics, this does not seem to be significant for the development of textile antennas. When designing microstrip patch antennas, the average of ε_r is well suited to ensure a reliable performance of the antenna.

2.5 Conclusions

The use of textiles in wearable antennas requires the characterization of their properties. Specific electrical conductive textiles are available on the market and have been successfully used. Ordinary textile fabrics have been used as substrates.

The conductive fabrics for the patch and the ground planes must have a very low electrical sheet resistance in order to minimize the electric losses and so increase the antenna efficiency.

In general, textiles present a very low dielectric constant, between 1 and 2, as they are very porous materials, and the presence of air approaches the relative permittivity to one. The low dielectric constant reduces the surface wave losses that are tied to guided wave propagation within the substrates.

In addition, the ability of the fibers to absorb moisture must also be considered in the characterization of the dielectric behavior of textiles. Water has a much higher and more stable dielectric constant than textile fabrics. Therefore, when water is absorbed by the textile fibers or is trapped in the fabric structure, it changes the electromagnetic properties of the fabric, increasing its dielectric constant and loss

Table 2.4 Summary of the advantages and drawbacks of resonant and nonresonant methods.

Method	Measured planes	Required equipment	Advantages	Drawbacks
Parallel plate	Z plane	Parallel plates and LCR	Nondestructive method	Maximum frequency of measurements: 1 MHz
Transmission lines	Z plane	VNA	Quick to perform	Requires a complex sample preparation, due to the mechanical instability of the fabrics, such as fraying and deformation in/after the cut process. The measured values are influenced by the conditions of some variables, as for example, the type of e-textile or conductive metal that is used, the glue/adhesive sheet, the connector and the manufacturing technique to make the probe, which can lead to nonrepeatability of the measurements and introduce errors in the final values
Free-space	Z plane	Anechoic chamber	Nondestructive method	Requires a precise calibration of the horn antennas
Resonance cavities	X, Y, and Z plane	Resonance cavities and VNA	Nondestructive method and possibility to measure the dielectric properties on several planes	In every measurement, the sample has to be placed in the same position. Also, the cavity needs to be dismantled and reassembled every time a new sample is tested
Microstrip resonator ring and microstrip patch sensor	Z plane	Resonator antenna and VNA	Nondestructive method, easy and quick to perform. Also no need sample preparation	Requires a precise calibration of the VNA
Microstrip patch radiator	Z plane	VNA	Does not require a sophisticated equipment	Complex sample preparation and estimation of an initial permittivity value. Also, the results are influenced by the inaccuracies of the manufacturing process of the antenna
Agilent 85070E dielectric measurement probe kit	Z plane	Agilent 85070E dielectric measurement probe kit and VNA or impedance analyzer	Easy and quick to perform	Very expensive equipment and a complex calibration of the probe is required

tangent. Therefore, materials with low regain values are preferable for use as substrates and as conductive components of the antenna.

This chapter has also presented a review of the methods to characterize the dielectric properties of textile materials, to be used as dielectric substrates of wearable antennas and systems. The resonant and nonresonant techniques were presented. Despite the resonant techniques only characterize the material in a single frequency, generally, the resonant methods provide higher accuracy and sensitivity than nonresonant methods for low-loss materials.

In this review, all methods have shown good results when characterizing textile materials, which are summarized on Tables 2.3 and 2.2. These materials are suitable for use as dielectric substrate in antennas. To finalize, Table 2.4 summarizes the presented methods and describes their advantages and drawbacks.

Acknowledgments

The authors wish to thank The European COST Action IC1301 WiPE for financing the STSM 18709; and the Project TexBoost (with Nr. 024523 [POCI-01-0247-FEDER-024523] framework FEDER, through the COMPETE 2020 of the Portugal 2020) for the Post-Doctoral Fellowship.

This work is funded by FCT/MEC through national funds and when applicable co-funded by FEDER-PT2020 in partnership agreement under the projects UID/EEA/50008/2013 and UID/Multi/00195/2013.

References

1 Kopetz, H. (2011). Internet of things. In: *Real-Time Systems: Design Principles for Distributed Embedded Applications*, 307–323. Cambridge: Springer.

2 Lopez Research LLC (2013). *An Introduction to the Internet of Things (IoT)*, 1–6. CISCO.

3 European Union (2009). *Internet of Things – An action plan for Europe*. Brussels: European Union.

4 Miorandi, D., Sicari, S., De Pellegrini, F., and Chlamtac, I. (2012). Ad hoc networks Internet of things: vision, applications and research challenges. *Ad Hoc Networks* 10 (7): 1497–1516.

5 Khaleel, H. (ed.) (2015). *Innovation in Wearable and Flexible Antennas*. Southampton: WIT Press.

6 Salonen, P., Keskilammi, M., Rantanen, J., and Sydanheimo, L. (2001). A novel Bluetooth antenna on flexible substrate for smart clothing. *IEEE International Conference on Systems, Man, and Cybernetics* 2: 789–794.

7 Vallozzi, L., Vandendriessche, W., Rogier, H. et al. (2010). Wearable textile GPS antenna for integration in protective garments. *4th European Conference on Antennas and Propagation (EUCAP)*, Barcelona (April 2010), pp. 1–4.

8 Salonen, P., Rahmat-samii, Y., Schafhth, M., and Kivikoski, M. (2004). Effect of textile materials on wearable antenna performance: a case study of GPS antenna. *IEEE Antennas and Propagation Society Symposium (APS)*, Monterey (20–25 June 2004), pp. 459–462.

9 Doi, T., Kinugasa, T., Okugawa, M. et al. (2013). Development of rescue vest using ICT. *IEEE International Symposium on Safery, Security, and Rescue Robotics (SSRR)*, Linkoping (21–26 October 2013).

10 A. A. Serra, P. Nepa, and G. Manara (2011). A wearable multi antenna system on a life jacket for Cospas Sarsat rescue applications. *IEEE Antennas and Propagation Society Symposium (APS)* (3–8 July 2011), pp. 1319–1322.

11 P. Salonen, Y. Rahmat-samii, H. Hurme, and M. Kivikoski (2004). Effect of conductive material on wearable antenna performance: a case study of WLAN antennas. *IEEE Antennas and Propagation Society Symposium (APS)*, Monterey (20–25 June 2004), pp. 455–458.

12 Salonen, P. and Hurme, H. (2003). A novel fabric WLAN antenna for wearable applications. *IEEE Antennas and Propagation Society International Symposium* 2: 100–103.

13 Brebels, S., Ryckaert, J., Boris, C. et al. (2004). SOP integration and codesign of antennas. *IEEE Transactions on Advanced Packaging* 27 (2): 341–351.

14 Pantelopoulos, A. and Bourbakis, N.G. (2010). A survey on wearable sensor-based systems for health monitoring and prognosis. *IEEE Transactions on Systems, Man, and Cybernetics, Part C (Applications and Reviews)* 40 (1): 1–12.

15 Ha, J. O., Jung, S. H., Park, M. C. et al. (2013). A fully integrated 3–5 GHz UWB RF transceiver for WBAN applications. *IEEE International Microwave Workshop Series on RF and Wireless Technologies for Biomedical and Healthcare Applications (MTT-S)*, Singapore (9–11 December 2013) pp. 1–3.

16 Curone, D., Secco, E.L., Tognetti, A. et al. (2010). Smart garments for emergency operators: the ProeTEX project. *IEEE Transactions on Information Technology in Biomedicine* 14 (3): 694–701.

17 Osman, M.A.R., Rahim, M.K.A., Samsuri, N.A. et al. (2011). Embroidered fully textile wearable antenna for medical monitoring applications. *Progress in Electromagnetics Research* 117: 321–337.

18 Ivši, B., Babi, M., Galoi, A., and Bonefa, D. (2017). Feasibility of electromagnetic energy harvesting using wearable textile antennas. *11st European Conference on Antennas and Propagation (EuCAP)*, Paris (19–24 March 2017), pp. 485–488.

19 Gonçalves, R., Carvalho, N., Pinho, P. et al. (2013). Textile antenna for electromagnetic energy harvesting for GSM900 and DCS1800 bands. *Antennas and Propagation Society International Symposium (APSURSI)*, Orlando (7–13 July 2013), pp. 1206–1207.

20 Lemey, S., Rogier, H., Declercq, F. et al. (2014). Textile antennas as hybrid energy-harvesting platforms. *Proceedings of the IEEE* 102 (11): 1833–1857.

21 Paul, D.L., Giddens, H., Paterson, M.G. et al. (2013). Impact of body and clothing on a wearable textile dual band antenna at digital television and wireless communications bands. *IEEE Transactions on Antennas and Propagation* 61 (4): 2188–2194.

22 Massey, P. (2001). Fabric antennas for mobile telephony integrated within clothing. *11th International Conference on Antennas and Propagation*, Manchester (17–20 April 2001).

23 Tak, J., Member, S., Choi, J., and Member, S. (2015). An all-textile Louis Vuitton logo antenna. *IEEE Antennas and Wireless Propagation Letters* 1225: 3–6.

24 Kennedy, T.F., Fink, P.W., Chu, A.W., and Studor, G.F. (2007). Potential space applications for body-centric wireless and e-textile antennas. *IET Seminar on Antennas and Propagation for Body-Centric Wireless Communications*, London (24–24 April 2007).

25 Balanis, C.A. (2005). *Antenna Theory Analysis and Design*, 3e. Hoboken, NJ: Wiley Interscience.

26 Salvado, R., Loss, C., Gonçalves, R., and Pinho, P. (2012). Textile materials for the design of wearable antennas: a survey. *Sensors* 12: 15841–15857.

27 Gonçalves, R., Magueta, R., Pinho, P., and Carvalho, N.B. (2016). Dissipation factor and permittivity estimation of dielectric substrates using a single microstrip line measurement. *Applied Computational Electromagnetics Society Journal* 31 (2): 118–125.

28 Chen, L.F., Ong, C.K., Neo, C.P. et al. (2004). *Microwave Electronics: Measurement and Materials Characterization*. Wiley: Chichester.

29 Bonaldi, R.R., Siores, E., and Shah, T. (2010). Electromagnetic shielding characterization of several conductive fabrics for medical applications. *Journal of Fiber Bioengineering and Informatics* 2 (4): 245–253.

30 Brzeziński, S., Rybicki, T., Karbownik, I. et al. (2012). Textile materials for electromagnetic field shielding made with the use of nano- and micro-technology. *Central European Journal of Physics* 10 (5): 1190–1196.

31 Brzeziński, S., Rybicki, T., Karbownik, I. et al. (2009). Textile multi-layer systems for protection against electromagnetic radiation. *Fibres and Textiles in Eastern Europe* 17 (2): 66–71.

32 Gimpel, S., Mohring, U., Muller, H. et al. (2004). Textile-based electronic substrate technology. *Journal of Industrial Textiles* 33 (3): 179–189.

33 Kaushik, V., Lee, J., Hong, J. et al. (2015). Textile-based electronic components for energy applications: principles, problems, and perspective. *Nanomaterials* 5 (3): 1493–1531.

34 Zeng, W., Shu, L., Li, Q. et al. (2014). Fiber-based wearable electronics: a review of materials, fabrication, devices, and applications. *Advanced Materials* 26: 5310–5336.

35 D 257-99 ASTM Standards (1999). Standard test methods for DC resistance or conductance of insulating materials.

36 Maryniak, W.A., Uehara, T., and Noras, M.A. (2003). Surface resistivity and surface resistance measurements – using a concentric ring probe technique. *Trek Application Note* 1005: 1–4.

37 F 1896, ASTM Standards (2004). Test method for determining the electrical resistivity of a printed conductive material.

38 Test Method 76-2011, AATCC (2011). Electrical surface resistivity of fabrics. pp. 1–3.

39 ISO 10965:2011 (2011). Textile floor coverings – determination of electrical resistance. pp. 1–5.

40 ISO 21178:2013 (2013). Light conveyor belts – determination of electrical resistances. pp. 1–20.

41 Shawl, R., Longj, B., Werner, D., and Gavrin, A. (2007). The characterization of conductive textile materials intended for radio frequency applications. *IEEE Antennas and Propagation Magazine* 49 (3): 28–40.

42 Ouyang, Y. and Chappell, W.J. (2008). High frequency properties of electrotextiles for wearable antenna applications. *IEEE Transactions on Antennas and Propagation* 56 (2): 381–389.

43 Cottet, D., Gryzb, J., Kirstein, T., and Troster, G. (2003). Electrical characterization of textile transmission lines. *IEEE Transactions on Advanced Packaging* 26 (2): 182–190.

44 Lilja, P. Salonen, P., Maagt, P. D., and Zell, N. K. (2009). Characterization of conductive textile materials for softwear antenna. *Antennas and Propagation Society International Symposium (APSURSI)*, Charleston (1–5 June 2009), pp. 1–5.

45 Locher, I., Klemm, M., Kirstein, T., and Tröster, G. (2006). Design and characterization of purely textile patch antennas. *IEEE Transactions on Advanced Packaging* 29 (4): 777–788.

46 Morton, W.E. and Hearle, W.S. (2008). *Physical Properties of Textile Fibres*, 4e. Cambridge: Woodhead Publishing in Textiles.

47 Lilja, J. and Salonen, P. (2009). Textile material characterization for softwear antennas. IEEE Military Communications Conference (MILCOM), Boston (18–21 October 2009), pp. 1–7.

48 Wang, X., Xu, W., and Wenbin, L. (2009). Study on the electrical resistance of textiles under wet conditions. *Textile Research Journal* 79: 753–760.

49 Baker-Jarvis, J., Geyer, R., Grosvenor, J. et al. (1998). Dielectric characterization of low-loss materials: a comparison of techniques. *IEEE Transactions on Dielectrics and Electrical Insulation* 5 (4): 571–577.

50 Baker-Jarvis, J., Janezic, M.D., and DeGroot, D.C. (2010). High-frequency dielectric measurements. *IEEE Instrumentation and Measurement Magazine* 13 (2): 24–31.

51 Thorp, J.S., Akhtaruzzaman, M., and Evans, D. (1990). The dielectric properties of alumina substrate for microelectronic packaging. *Journal of Materials Science* 25 (9): 4143–4149.

52 Coonrod, J. (2011). Understanding when to use FR-4 or high frequency laminates. *Onboard Technology Magazine*: 26–30.

53 Sadiku, M.N.O. (2007). *Elements of Electromagnetics*, 4e. New Delhi: Oxford University Press.

54 Grupta, B., Sankaralingam, S., and Dhar, S. (2010). Development of wearable and implantable antennas in the last decade: a review. *10th Mediterranean Microwave Symposium*, Guzelyurt (25–27 August 2010), pp. 251–267.

55 Tronquo, A., Rogier, H., Hertleer, C., and Van Langenhove, L. (2006). Applying textile materials for the design of antennas for wireless body area networks. *1st European Conference on Antennas and Propagation (EuCAP)*, Nice (6–10 November 2006), pp. 1–5.

56 Hertleer, C., Tronquo, A., Rogier, H., and Van Langenhove, L. (2008). The use of textile materials to design wearable microstrip patch antennas. *Textile Research Journal* 78 (8): 651–658.

57 Balls, W.L. (1946). Dielectric properties of raw cotton. *Nature* 158: 9–11.

58 Bal, K. and Kothari, V.K. (2010). Permittivity of woven fabrics: a comparison of dielectric formulas for air-fiber mixture. *IEEE Transactions on Dielectrics and Electrical Insulation* 17 (3): 881–889.

59 Bal, K. and Kothari, V.K. (2009). Measurement of dielectric properties of textile materials and their applications. *Indian Journal of Fibre and Textile Research* 34: 191–199.

60 Hasar, U.C. (2009). A new microwave method for electrical characterization of low-loss materials. *IEEE Microwave and Wireless Components Letters* 19 (12): 801–803.

61 Gershon, D., Calame, J.P., Carmel, Y. et al. (2007). Adjustable resonant cavity for measuring the complex permittivity of dielectric materials. *Review of Scientific Instruments* 3207 (2000): 12–15.

62 Faz, U., Siart, U., Eibert, T.F. et al. (2015). Electric field homogeneity optimization by dielectric inserts for improved material sensing in a cavity resonator. *IEEE Transactions on Instrumentation and Measurement* 64 (8): 2239–2246.

63 Kumar, A. and Smith, D.G. (1977). Microwave properties of yarns and textiles using a resonant microwave cavity. *IEEE Transactions on Instrumentation and Measurement* 26 (2): 95–98.

64 Sankaralingam, S. and Bhaskar, G. (2010). Determination of dielectric constant of fabric materials and their use as substrates for design and development of antennas for wearable applications. *IEEE Transactions on Instrumentation and Measurement* 59 (12): 3122–3130.

65 Roy, B. and Choudhury, S. K. (2013). Characterization of textile substrate to design a textile antenna. *International Conference on Microwave and Photonics (ICMAP)*, Dhanbad (13–15 December 2013), pp. 1–5.

66 Ismail, M.F., Rahim, M.K.A., Hamid, M.R., and Majid, H.A. (2013). Circularly polarized textile antenna with bending analysis. *IEEE International RF and Microwave Conference (RFM)*, Penang (9–11 December 2013), pp. 460–462.

67 Loss, C., Gonçalves, R., Pinho, P., and Salvado, R. (2018). Influence of some structural parameters on the dielectric behavior of materials for textile antennas. *Textile Research Journal* 0: 1–13.

68 Hausman, S., Januszkiewicz, Ł., Michalak, M. et al. (2006). High frequency dielectric permittivity of nonwovens. *Fibres and Textiles in Eastern Europe* 14 (5): 60–63.

69 Bahl, I.J. and Stuchly, S.S. (1980). Analysis of a microstrip covered with a lossy dielectric. *IEEE Transactions on Microwave Theory and Techniques* 28 (2): 104–109.

70 Bogosanovich, M. (2000). Microstrip patch sensor for measurement of the permittivity of homogeneous dielectric materials. *IEEE Transactions on Instrumentation and Measurement* 49 (5): 1144–1148.

71 Agilent Technologies (2014). Measuring Dielectric Properties Using Agilent's Materials Measurement Solutions, USA, pp. 1–4.

72 Moretti, A., Malheiros-silveira, G. N., Hugo, E., and Gonçalves, M. S. (2011). Characterization and validation of a textile substrate for RF applications. *SBMO/IEEE MTT-S International Microwave & Optoelectronics Conference (IMOC)*, Natal (29 October–1 November 2011), pp. 546–550.

73 Declercq, F., Rogier, H., and Hertleer, C. (2008). Permittivity and loss tangent characterization for garment antennas based on a new matrix-pencil two-line method. *IEEE Transactions on Antennas and Propagation* 56 (8): 2548–2554.

74 Mantash, M., Tarot, A.C., Collardey, S., and Mahdjoubi, K. (2012). Investigation of flexible textile antennas and AMC reflectors. *International Journal of Antennas and Propagation*: 1–10.

75 Klemm, M. and Troster, G. (2006). Textile UWB antennas for wireless body area networks. *IEEE Transactions on Antennas and Propagation* 54 (11): 3192–3197.

76 Lesnikowski, J. (2012). Dielectric permittivity measurement methods of textile substrate of textile transmission lines. *Electrical Review* 88 (3): 148–151.

77 Kapilevich, B., Litvak, B., Anisimov, M. et al. (2012). Complex permittivity measurements of textiles and leather in a free space: an angular-invariant approach. *International Journal of Microwave Science and Technology*: 1–7.

78 Hakansson, E., Amiet, A., Kaynak, A. et al. (2007). Dielectric characterization of conducting textiles using free space transmission measurements: accuracy and methods for improvement. *Synthetic Metals* 157: 1054–1063.

79 Adami, S., Zhu, D., Li, Y., Mellios, E., Stark, B. H., and Beeby, S. (2015). A 2.45 GHz rectenna on screen-printed on polycotton for on-body RF power transfer and harvesting. *IEEE Wireless Power Transfer Conference (WPTC)*, Boulder (13–15 May 2015), pp. 1–4.

80 Bakar, A. S. A., Misnon, M. I., Ghodgaonkd, D. K. et al. (2014). Comparison of electrical physical and mechanical properties of textile composites using microwave nondestructive evaluation. *RF and Microwave Conference*, Selangor (5–6 October 2014), pp. 164–168.

81 Himdi, O. and Lafond, M. (2009). Printed millimeter antennas – multilayer technologies. In: *Advanced Millimeter-Wave Technologies: Antennas, Packaging and Circuits* (eds. D. Lui, U. Pfeiffer, J. Gryzb and B. Gaucher). Chichester: Wiley Interscience.

82 Pourova, M., Zajicek, R., Oppl, L., and Vrba, J. (2008). Measurement of dielectric properties of moisture textile. *14th Conference on Microwave Techniques*, Prague (23–24 April 2008), pp. 1–4.

83 Hertleer, C., Laere, A.V., Rogier, H., and Van Langenhove, L. (2009). Influence of relative humidity on textile antenna performance. *Textile Research Journal* 80 (2): 177–183.

84 Lilja, J., Salonen, P., Kaija, T., and De Maagt, P. (2012). Design and manufacturing of robust textile antennas for harsh environments. *IEEE Transactions on Antennas and Propagation* 60 (9): 4130–4140.

85 Kaija, T., Lilja, J., and Salonen, P. (2010). Exposing textile antennas for harsh environment. *The Military Communications Conference*, San Jose (31 October–3 November 2010), pp. 737–742.

86 Declercq, F., Couckuyt, I., Rogier, H., and Dhaene, T. (2013). Environmental high frequency characterization of fabrics based on a novel surrogate modelling antenna technique. *IEEE Transactions on Antennas and Propagation* 61 (10): 5200–5213.

3

Smart Beamforming Techniques for "On Demand" WPT

Diego Masotti, Mazen Shanawani, and Alessandra Costanzo

Department of Electrical, Electronic and Information Engineering, "G. Marconi", University of Bologna, Bologna, Italy

3.1 Introduction

When dealing with a link devoted to a wireless power transfer (WPT), there are many and different power contributions to be taken into account. With reference to Figure 3.1, these power contributions are both in dc and at radiofrequency (RF): on the transmitter side, the first term is the dc power spent to properly bias the nonlinear device of the oscillator (P_{BIAS}), followed by the RF term at the input port of the amplifier (P_{IN}): these powers are related by the dc-to-RF efficiency of the oscillator. By taking into account the RF-to-RF efficiencies of both the power amplifier and the antenna, one can estimate the power at the input port of the antenna (P_{ANT}) and the power radiated by the antenna (P_{TX}). The link propagation is responsible for the power (P_{RX}) received by the receiving antenna (of the rectifying-antenna or rectenna system) that makes it available at the rectifying section (matching network + rectifier). This is of course the last RF contribution: after this and the rectification process, the dc power at the output of the rectenna port (P_{dc}) is quantified by the RF-to-dc conversion efficiency of the rectifier itself and is managed by the power management unit (PMU), typically designed to guarantee the maximum power point (MPT) condition [1, 2]. The other dc power factors are all within the PMU and are mainly due to the dc-to-dc efficiency of the dc–dc converter: they correspond to the energy provided by the dc–dc converter to the output storage capacitor (P_{ST}) and finally to the final dc power to be exploited by the user/battery (P_{BAT}).

As regards the aforementioned efficiencies, they all have typical values known from the literature, and mostly depending on the frequency range: for the transmitting system, the oscillator, amplifier, and antenna efficiencies can easily reach values significantly greater than 50% (80%, 70%, 90% in the low-GHz range,

Wireless Power Transmission for Sustainable Electronics: COST WiPE - IC1301,
First Edition. Edited by Nuno Borges Carvalho and Apostolos Georgiadis.

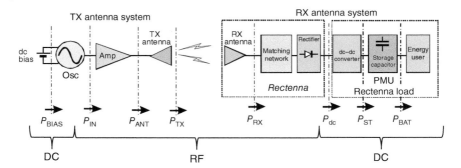

Figure 3.1 Block scheme of a WPT link with the involved power contributions.

respectively, down to 60%, 40%, 80% above 10 GHz). Similarly, for the receiving side, the RF-to-dc efficiency has a known behavior strongly dependent on the P_{RX} amount [3]: in low-power conditions, typical of environmental energy harvesting scenarios (-20 to -10 dBm), the efficiency can be around 30–40%, reaching 60–70% for higher power levels ($0-10$ dBm) typical of colocated and dedicated RF sources [4, 5]. A less critical block is the one performing the dc-to-dc conversion, where around 90% of efficiency can be guaranteed [6].

The most critical path is the one in between the transmitting and the receiving blocks: the link. By simply taking into consideration the free space attenuation, the RF-to-RF efficiency of this block is the lowest, rapidly approaching 0% for increasing values of the frequency. Being the overall link efficiency the product among all these terms, the effect of the weakest contribution is dramatic.

For this reason, the maximization of this part of the link is of strategic importance and a more detailed investigation on it is worthwhile. As reported in Figure 3.2, a very important power contribution is the one indicating the amount of transmitted power that is actually received or collected by the receiving antenna (P_{REC}). In order to quantify it, an additional and intermediate power factor is introduced: P_{TGT} represents the amount of the transmitted power that is effectively directed onto the target. Through the introduction of this additional term, two new efficiencies can be taken into account to maximize the overall link efficiency:

$$\eta_{TX\_beam} = \frac{P_{TGT}}{P_{TX}}; \quad \eta_{RX\_beam} = \frac{P_{REC}}{P_{TGT}} \tag{3.1}$$

The first one, named beam transmission efficiency, represents the capability of the radiating system to focus the power in the region where the receiving rectenna (the target) is. The second factor, named beam reception efficiency, takes into account the aptitude of the receiving antenna/array to efficiently deploy the power made available by the RF source.

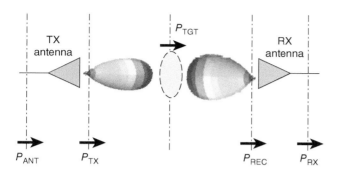

Figure 3.2 Detail of the wireless link with the involved power contributions.

As per the last term, it is mainly a matter of the rectenna arrangement: a low-directive behavior, given by a single antenna or by more antennas, dc-combined after rectification, leads to a low beam reception efficiency; conversely, a multiantenna combination at RF (i.e. an array) provides a higher beam reception efficiency because of the higher directivity, provided that the rectenna is pointing in the direction of the incoming field. The two previous choices are typically a matter of environmental energy harvesting and intentional energy transmission, respectively.

The beam transmission efficiency plays, vice versa, a strategic role in the link performance optimization: the capability of the radiating system to focus the energy in the region where the targets represent the major issue in far-field wireless power transfer scenarios.

The solution to this problem can be faced with different approaches, each of them with advantages and drawbacks.

As regards solutions with direction-only selectivity,

Phased arrays [7]: These arrays represent a widely adopted solution, where each element of the array is fed with a proper phase. In this way, the main beam can be steerable in any desired direction, but the complexity of the phase-shifter design and piloting is not trivial.

Reflect (or retrodirective) arrays: Both in the passive (or Van Atta [8]) and active (or Pon [9]) architectures, the layout design is a delicate task. Moreover, the mechanism of reflection of incident RF signals back in the direction of arrival is better suited for applications with relaxed pointing accuracy and automatic beam forming needs.

Series-fed arrays [10]: Despite of a demanding design task, the steering capability is reduced in a (limited) frequency band.

Leaky-wave antennas [11]: They can be an alternative and effective solution to series-fed ones, offering promising radiation properties, but their design is difficult, and they are bulky systems [10].

More advanced solutions can additionally provide ranging capability, therefore the possibility to focus the energy in specific direction and distance:

Nonuniformly excited arrays [12]: Where complex optimization algorithms can provide the excitation condition able to focus the energy in the broadside direction only, at the expense of a complex feeding network realization

Frequency-diverse arrays [13]: These are promising, but up to now purely theoretical architectures, where each radiator makes use of a different RF signal to be transmitted. This additional degree of freedom makes the array factor dependent not only on the direction (playing with the excitation phase as per the phased-arrays) but also on the distance. However, the synthesis of coherent sinusoidal signals at different frequencies is not an easy task, at all.

Retrodirective frequency diverse arrays [14]: Combining two previously discussed strategies and offering very promising performance of direction-range selectivity, but this remains, up to know a purely theoretical conjecture.

There is also the possibility, on board of the transmitting antenna, to exploit a pilot signal sent from the target to discover the target position form the direction-of-arrival of the pilot signal [15]. A more energy-aware and simpler solution of this approach is represented by the exploitation of the second harmonic generated on board by the rectifier of the target/rectenna as pilot signal [16].

All the proposed solutions share, in different ways, common drawbacks: the complexity of their design, often combined with a low-level of reconfigurability. The purpose of this chapter is to present a promising solution of a smart RF energy source, that has already demonstrated not to be simply a theoretical conjecture: the time-modulated arrays (TMAs). These radiating systems offer an unrivaled level of reconfigurability in almost real time, despite their architectural simplicity: each antenna of the array is driven by a switch, controlled in time through a low modulation law. The exploitation of time (duration of the ON interval of each switch and its ON instant) as additional degree of freedom open the way to an almost infinite set of driving sequences. Up to now, these arrays have been exploited for application where the sole direction (not the range) is involved, because the time-dependent array factor do not depend on the distance: however, the time-modulation strategy could, in near future, be combined with one of the aforementioned range-dependent solutions, in order to have a versatile direction-range selection capability.

However, the main characteristic of this family of radiating systems is the possibility to simultaneously radiate at different close frequencies (the so-called sideband radiation phenomenon) [17]. In recent years, this TMA capability has been considered as an added value for specific applications, instead of an unwanted phenomenon to be taken under control. In this chapter, the exploitation of sideband

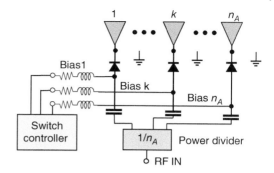

Figure 3.3 Schematic representation of a linear n_A-element TMA. Source: Masotti and Costanzo 2017 [23]. Reproduced with permission of IEEE.

radiation of TMA for direction finding [18, 19] and harmonic-beamforming [20, 21] will be combined to describe a smart WPT procedure [22].

3.2 Basics of Time-modulated Arrays

Let us consider the circuit described in Figure 3.3, where a linear TMA with n_A antennas is reported. This circuit can be described by its steady-state regime under sinusoidal excitation, by means of a set (say n_H) of harmonics of the fundamental (angular) frequency ω_0.

The n_A-dimensional vector of the excitation currents at the feeding points of the n_A antennas can be represented in the form:

$$\mathbf{i}_A(t) = Re\left[\sum_{k=1}^{n_H} \mathbf{I}_{A,k} e^{jk\omega_0 t}\right] \tag{3.2}$$

It is known [24] that for an array with n_A antennas, the **E** far-field value at a point (r,θ,φ) given that the excitation signal has frequency ω_0 can be given by (3.3). This result comes from the exploitation of both the linear behavior of the array and the time-invariant vector $\mathbf{I}_{A,1}$ representing the currents at the fundamental frequency ω_0 ($k = 1$) entering the antennas ports.

$$\mathbf{E}(r, \theta, \phi, \omega_0) = \frac{e^{-j\beta r}}{r} \sum_{i=1}^{n_A} [\hat{a}_\theta A_\theta^i(\theta, \phi, \omega_0) + \hat{a}_\phi A_\phi^i(\theta, \phi, \omega_0)] \mathbf{I}_{A,1}^i \tag{3.3}$$

where β is the free-space phase constant, $I_{A,1}^i$ the excitation current at the generic ith antenna port, A_θ^i and A_ϕ^i are the scalar components of the normalized field. Such components are generated by full-wave simulation of the whole array with only the ith antenna excited by a unit-current sinusoidal source of angular frequency ω_0. In case of a TMA radiating nonlinear system, the regime is given by a modulated RF drive, because there is the superposition of the sinusoidal carrier (ω_0) and the slow (with respect to the RF) modulation law of the switches, that are periodically (with period $T_M = 2\pi/\omega_M$) biased. In this case, the currents feeding the antennas become time-dependent currents $I_{A,1}^i(t_M)$ that can be given by (3.4)

$$\mathbf{I}_{A,k}(t_M) = \sum_{h=-n_M}^{n_M} \mathbf{I}_{A,kh} \exp(jh\omega_M t_M) \tag{3.4}$$

Where k is the harmonic order with respect to the carrier oscillation frequency, h is the harmonic order of the modulated excitation current. The input current vector thus becomes dependent on two time-bases, the fast carrier time t and the slow modulation time t_M, and assumes the expression given in (3.5).

$$\mathbf{i}_A(t, t_M) = Re\left[\sum_{k=1}^{n_H} \mathbf{I}_{A,k}(t_M) \exp(jk\omega_0 t) \right] \tag{3.5}$$

The modulation is obviously transferred from the driving currents to the radiated far field. The time-dependent far-field envelope may be now computed for any modulation speed by the general convolution algorithms [25]. However, based on the assumption of slow variation of the excitation, the resulting **E** field radiated at the fundamental frequency ($k = 1$) can be expressed as follows (3.6) [24].

$$E(r, \theta, \phi, \omega_0) = \sum_{i=1}^{n_A} \varepsilon^i \cdot \mathbf{I}_{A,1}^i(t_M) - j \sum_{i=1}^{n_A} \left.\frac{\partial \varepsilon^i}{\partial \omega}\right|_{\omega=\omega_0} \cdot \frac{d\mathbf{I}_{A,1}^i(t_M)}{dt_M}$$

where $\varepsilon^i = \hat{a}_\theta A_\theta^i(\theta, \phi, \omega_0) + \hat{a}_\phi A_\phi^i(\theta, \phi, \omega_0)$ $\hspace{2em}$ (3.6)

The field intensity shown in (3.6) represents the time waveform, which can be also introduced using the array factor representation. By referring again to the TMA architecture of Figure 3.3 where the elements of the array are aligned along the direction $\hat{a}(\hat{a}\hat{r} = \cos\chi)$, the far-field can be represented as in (3.7):

$$\mathbf{E}(r, \theta, \varphi, t_M) = \mathbf{E}_0(r, \theta, \varphi) \sum_{i=1}^{n_A} \Lambda_i U_i(t_M) e^{j(i-1)\beta L \cos\chi} = \mathbf{E}_0(r, \theta, \varphi) AF(\theta, \varphi, t_M) \tag{3.7}$$

where E_0 represents the far-field radiated at the carrier frequency f_0 by the base-element of the array, and the RF switch at the ith antenna port is driven by a periodical sequence of rectangular pulses of period $T_M = 1/f_M$ and normalized amplitude $U_i(t_M)$. In this way, the standard constant excitation coefficient A_i of standard arrays is replaced with the corresponding time-dependent version $\Lambda_i \cdot U_i(t)$.

This leads to a time-dependent array factor whose frequency-domain representation using Fourier transformation is shown in (3.8).

$$AF(\theta, \varphi, t_M) = \sum_{h=-\infty}^{\infty} e^{j2\pi(f_0 + hf_M)t_M} \sum_{i=1}^{n_A} \Lambda_i u_{hi} e^{j(i-1)\beta L \cos\chi} \tag{3.8}$$

where u_{hi} is the hth Fourier coefficient of the $U_i(t_M)$ pulse.

The direct consequence of the replacement of (3.8) in (3.7) is the multifrequency radiation contribution: indeed, the TMA is able to radiate at the usual fundamental frequency f_0 ($h = 0$), but also at the near-carrier sideband harmonics $f_0 + h f_M$ with ($h = \pm 1, \pm 2, \ldots$). This sideband radiation is efficiently transmitted/received due to the low value of the modulation frequency f_M with respect to the carrier f_0 (MHz versus GHz), which allows the array elements to be still almost resonant.

3.3 Nonlinear/Full-Wave Co-simulation of TMAS

The TMA simulation approaches available in literature typically focus on the goal of the control sequences optimization but limit the analysis to the case of both ideal radiators and ideal switches. As described in the previous Section 3.2, the non-linear behavior of TMA systems is quite complex, because of the bidimensional regime, the unavoidable presence of electromagnetic (EM) couplings among the radiating elements, and the nonlinear nature of the driving switches: neglecting these effects can lead to highly inaccurate results. For this reason, a rigorous approach for the accurate modeling of the TMA regime accounting for all the non-idealities and the actual system dynamic is needed. In this section, we adopt the simulation tool described in [24], here briefly recalled for the sake of completeness.

The approach consists of the nonlinear/EM analysis of the whole radiating system through the combination of harmonic balance (HB) technique [25], for the time-based description of the nonlinear switches, and the full-wave simulation of the array and its feeding network in the frequency domain. By referring again to the array with n_A ports of Figure 3.3, with an equal number of switches, the EM-based description of the system consists of a n_{A+1}-port network (including the RF input port), with n_A internal ports for the inclusion of the switches circuit-model: each internal port consists of a couple of floating nodes directly connected to the mesh of the EM simulator.

Figure 3.4 shows three possible periodic switch controls: the superposition of the slow switch modulation to the fast carrier frequency f_0 to be radiated, allows to resort to the modulation-oriented HB method [25] leading to (3.6, 3.7) for the feeding currents and the radiated field, respectively. One can note that the condition $f_0 \gg f_M$ is always valid in practical TMA applications (GHz versus a few MHz): hence, the modulated regime can always be seen as a sequence of slowly changing unmodulated ones.

The direct application of (3.6) provides the field envelope at the desired harmonic (typically the fundamental, $k = 1$), in any direction of radiation. Hence, from (3.6), the far-field harmonics of the bidimensional regime due to the intermodulation between f_0 and f_M are available, too, and the far-field can be also represented by their superposition as in (3.9).

$$\mathbf{E}(r, \theta, \varphi, t_M) = \sum_{h=-\infty}^{\infty} \mathbf{E}_h(r, \theta, \varphi) e^{j h \omega_M t_M} \tag{3.9}$$

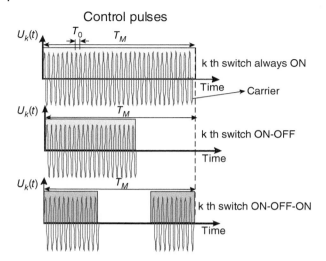

Figure 3.4 Examples of periodical switches excitation sequences, modulating the RF carrier waveforms, for three different switching patterns. Source: Masotti et al. 2016 [22]. Reproduced with pemrission of IEEE.

This approach allows to have any kind of result in terms of radiated far-field: by varying the harmonic number h in (3.9) and plotting the corresponding $\mathbf{E}_h$ magnitude, the far-field spectrum around the carrier harmonic $k = 1$ is obtained, for a given distance/direction. Alternatively, by varying (θ, φ) and plotting again the $\mathbf{E}_h$ magnitude, the radiation surface/pattern at a given distance r and for a given harmonic is available.

It is worth noting that this approach directly investigates the nonlinear nature of the TMA system while radiating.

3.4 Two-Step Agile WPT Strategy

The proposed scenario for the full exploitation of TMAs potentialities is an indoor room with N_{tag} tag/sensors, equipped with rectennas, randomly distributed. Figure 3.5 describes the WPT two-step procedure based on the dynamic reconfigurability of a linear TMA: (i) in the first step only, two elements of the TMA are operating to localize the N_{tag} tags; (ii) the second step uses the entire n_A-element array to send the power to the previously detected tags, only. Both the phases take advantage from the TMA multiple radiation capability, at the fundamental f_0 and the first sideband harmonics $f_0 \pm f_M$.

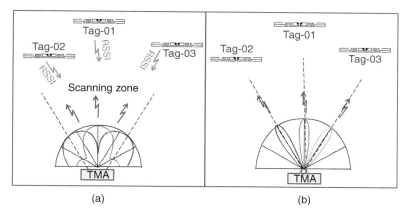

Figure 3.5 Two-step WPT procedure exploiting a linear TMA. (a) First step: tags localization; (b) Second step: power transmission to the previously detected tags. Source: Masotti et al. 2016 [22]. Reproduced with pemrission of IEEE.

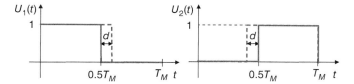

Figure 3.6 Switches driving sequences of a two-element array for localization purposes: d is the pulse shape design parameter, for Δ pattern steering. Source: Masotti et al. 2016 [22]. Reproduced with pemrission of IEEE.

3.4.1 Localization Step

An interesting direction finding functionality of a two-element TMA is achievable through properly biasing the two switches [18]. A control sequence of the kind reported with solid lines in Figure 3.6 (i.e. duty cycle of the two anti-symmetric driving sequences equal to 50%) leads the two-element array to a symmetric radiation: of the sum-type (Σ) at f_0 (two elements in phase), of the difference-type (Δ) at $f_0 \pm f_M$ (two elements out of phase). Through the tuning parameter d (shown in Figure 3.6), one can symmetrically vary the duty cycle of the control sequences: in this way, the Δ pattern can be steered. If we imagine to have the radiating elements aligned in the x-axis (z-axis coming out from the page), the xz-plane can be considered as the scanning plane ($-90° \leq \theta \leq 90$): increasing d brings to a shift increase of the Δ null in the right half-plane ($0° \leq \theta \leq 90$) at $f_0 - f_M$; a symmetric result in the left half-plane ($-90° \leq \theta \leq 0$) is observed at $f_0 + f_M$.

While the Δ-pattern is steering (because the complex nature of the corresponding Fourier coefficient $u_{\pm 1i}$ (3.8)), the Σ-pattern remains fixed (because of the real coefficient u_{0i}). This implies that in order to also have a scanning Σ-pattern, something additional must be done, as will be described in Section 3.7. Figure 3.7a shows the corresponding plots of the array factor (3.8) for a two-element array with $L = \lambda/2$ and $U_i(t_M)(i = 1, 2, \ldots)$ given by the waveforms of Figure 3.6(i.e. the radiation patterns of an ideal array of two isotropic radiators). With an increase of d up to 20%, a scanning region of about $\pm 60°$ is achieved.

These radiation capabilities can be fruitfully exploited in the localization phase. If the passive tags/sensors identification (ID) acquisition has been already performed through a standard radio frequency identification (RFID) reading operation, the aforementioned scanning phase takes place, as described by the patterns of Figure 3.6, mainly thanks to the sharp nulls of the steered Δ patterns: the backscattered received signal strength indicators (RSSIs) due to the Σ and Δ patterns provide the maximum power ratio (MPR) [7] according to the following formula:

$$\text{MPR}(\theta_j) = \Sigma_{\text{RSSI}}^{\text{dB}}(\theta_j) - \Delta_{\text{RSSI}}^{\text{dB}}(\theta_j); \quad j = 1, 2, \ldots, N_{\text{tag}} \tag{3.10}$$

The combination of these patterns (obeying to the radar-monopulse principle) with the scanning capability is highly effective in indoor tags localization (resolution up to few cm at 2.45 GHz [7]).

In this first step, [18] suggests a reduced element spacing L, because a flatter Σ-pattern is obtained, and consequently almost unchanged localization properties are guaranteed by (3.10), while steering. For this purpose, Figure 3.7b shows the same ideal patterns as in Figure 3.6a closer antennas ($L = \lambda/8$): theoretically, superior scanning performance pertain to this case, i.e. a scanning region of about $\pm 70°$ with lower values of the tuning parameter d.

At the end of the scanning activity, an on-board microprocessor records the vector containing the θ values with the N_{tag} maxima of (3.10) (θ_{max}).

3.4.2 Power Transfer Step

In the second step, starting from the known position of the tags, the entire array has to send the power in all the θ_{max} directions. A proper control sequence involving all the switches of the array needs to be synthesized: this can be done by choosing one of the optimization strategies available in the literature, as in [26, 27]. The sideband radiation phenomenon of TMAs can be exploited in this step, too, in order to point the pattern maxima at the sideband harmonics $f_0 \pm h f_M$ ($h \neq 0$) in the desired directions, and this can be done at several near-carrier frequencies ($h \geq 1$) while simultaneously sending power with the fundamental pattern ($h = 0$).

Figure 3.8 shows how to perform this transmission on power: the scanning region of an array of 16 dipoles can be split into sectors of amplitude equal to the half power beam width (HPBW = 7° in this case), centered around θ_{HPBW}.

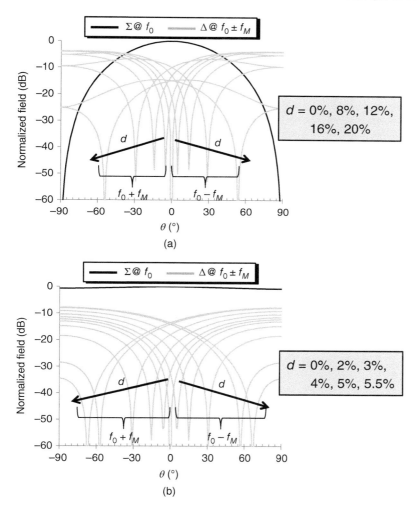

Figure 3.7 Fixed Σ and steerable Δ patterns of an ideal array of two isotropic elements (a) with spacing $\lambda/2$ and (b) with spacing $\lambda/8$, as a function of d. Source: Masotti et al. 2016 [22]. Reproduced with pemrission of IEEE.

In the figure, the exploitation of the first sixth harmonics is foreseen. For pre-recorded θ_{max} directions falling in some θ_{HPBW}-centered sectors, one of the preloaded control sequences pointing the desired harmonic patterns (at $f_0 \pm hf_M$) to the θ_{HPBW} direction is used.

A limitation of the described radiating system is that the fixed fundamental harmonic beam can be used to energize a sensor falling in the sector centered around $\theta_{HPBW} = 0°$, whereas the sideband harmonics can simultaneously transfer the power to other couples of almost symmetrically placed tags.

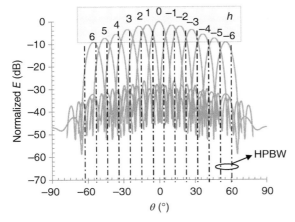

Figure 3.8 Ideal full exploitation of a 16-dipole TMA sideband radiation, when considering up to the sixth harmonic, for multiple tags powering. Source: Masotti et al. [22]. Reproduced with pemrission of IEEE.

3.5 Simulation Results

The computer-aided design (CAD) approach described in Section 3.2 and 3.3 can be used to select the proper antenna topology for the two-step WPT procedure, by evaluating plots of the kind of Figures 3.7 and 3.8, now obtained by taking into account the actual layouts and radiation patterns.

3.5.1 Localization Step

For localization purposes, we examine the planar, two-monopole array topologies with $L = \lambda/2$ and $L = \lambda/8$, as described in Figure 3.9a,b, in order to compare their Σ and Δ patterns with those of Figure 3.7. The arrays are realized on a Taconic RF60A substrate ($\varepsilon_r = 6.15$, thickness = 0.635 mm) and operate at $f_0 = 2.45$ GHz. Medium-power microwave Schottky diodes (Skyworks SMS7630-079) are adopted as switching elements, and the modulation frequency is $f_M = 25$ kHz. A low power level of -10 dBm RF power at each antenna port is used in the following analyzes.

The two elements of Figure 3.9 correspond to the inner elements of larger arrays (in the following step of the WPT procedure, the arrays will have $n_A = 16$ ports). Of course, the EM-database ($A_\theta^{(i)}$, $A_\varphi^{(i)}$) used in (3.6) is obtained by full-wave analysis of the whole large array, thus taking into account all the unavoidable EM-couplings between the nearby elements. Note that the same database will be exploited in the analyzes involving all the n_A switches of the second phase.

Figure 3.9c,d shows the results in terms of Σ and Δ radiation patterns, for the arrays of Figure 3.9a,b, respectively, in the scanning θ-plane, for different d-values: it can be noticed that both the monopole arrays have very wide steering of the sideband patterns: the scanning region is about $-50° \leq \theta \leq 50°$ in both cases. It is also worth noting the higher needed d parameter value with respect to the ideal plots of Figure 3.7: $d = 32\%$ determines Δ peaks (at $f_0 \pm f_M$) at $\pm 45°$, and the corresponding

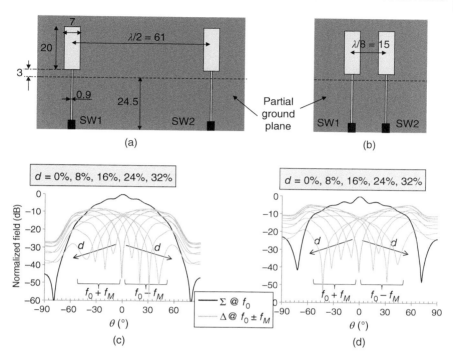

Figure 3.9 Layouts and dimensions (in mm) of planar two-element TMAs with: (a) $\lambda/2$-spaced monopoles, (b) $\lambda/8$-spaced monopoles. (c,d) Corresponding Σ and Δ radiation patterns for different d values. Source: Masotti et al. 2016 [22] Reproduced with pemrission of IEEE.

MPR plots are given in Figure 3.10 for some of patterns of Figure 3.9c. The higher d values of these plots are due to not only to the nonideal antenna radiating characteristics but also to the real control switch behavior.

These first results indicate that there is not a significant difference between the two topologies of Figure 3.9a,b in localizing the tags, and a further comparison between the corresponding 16-element arrays will be given soon. Before this, an additional trial is performed with a more directive multilayer patch antenna, again arranged in a two-element topology, as reported in Figure 3.11a. The same substrate, switches, and frequencies of the previous example are adopted. The simulation results of Figure 3.11b demonstrate that the more directive patch antenna behavior is a drawback for localization purposes: the scanning region is reduced and the negative peaks are less neat.

3.5.2 Power Transfer Step

In the power transfer step, a highly directive array is needed for a precise energization of the previously detected tags. Higher directivity is guaranteed by larger

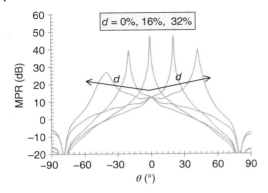

Figure 3.10 MPR plots of the planar two-element TMAs with $\lambda/2$-spaced monopoles by varying d [22].

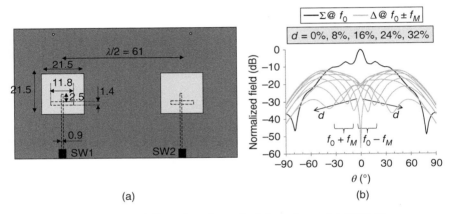

(a)   (b)

Figure 3.11 (a) Layout and dimensions (in mm) of planar two-patch TMA; (b) corresponding Σ and Δ radiation patterns for different d values. Source: Masotti et al. 2016 [22]. Reproduced with pemrission of IEEE.

interelement spacing (i.e. standard choice of $L = \lambda/2$). However, in order to show the flexibility of the developed CAD tool, in the following, we compare the WPT performance of the two 16-monopole arrays of Figure 3.12a,b, having almost identical directivity of about 14.5 dBi, but a standard equally spaced topology ($L = \lambda/2$) (Figure 3.11a) and an unequally spaced one, with the inner couple of monopoles at a distance of $\lambda/8$, with the remaining ones at a distance of $\lambda/2$.

As regards the switches excitation, the sequence proposed in [21] for a 16-ideal-dipole array is adopted and reported in Figure 3.13, for the sake of clarity: the ith blue rectangles correspond to normalized on-time of the $U_i(t)$ pulse of the ith switch.

Figure 3.14a,b shows the radiation patterns simultaneously available at the harmonics corresponding to $h = 0, 1, -1$, for the two arrays of Figure 3.12a,b,

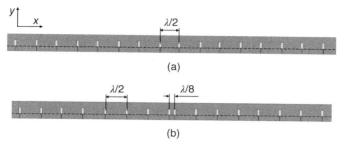

Figure 3.12 (a) Equally and (b) unequally spaced arrays of 16 planar monopoles. Source: Masotti et al. 2016 [22]. Reproduced with pemrission of IEEE.

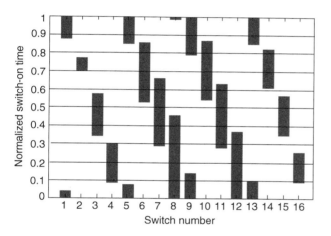

Figure 3.13 Switches control pattern for the 16-dipole arrays of Figure 4.12. Source: Poli et al. 2011 [21].

respectively: the patterns show that the two TMA topologies can energize three tags, within sectors centered around $\theta_{\mathrm{HPBW}} = -30°$, $0°$, $30°$, without a significant difference in terms of side-lobe level (SLL). Note that the adopted control sequence has been optimized for ideal, uniformly distributed isotropic antennas [21]: despite this, the performance are quite satisfactory for the two array arrangements of Figure 3.12, even if a direct optimization of the whole assembly with the proposed simulation tool could lead to higher performance [28].

Through the exploitation of (3.6), applied to the array of Figure 3.12a, the plots of Figure 3.15 can be straightforwardly obtained: they are the far-field envelopes at a distance $r = 1$ m, at the fundamental harmonic ($k = 1$) in the three directions of the power transmission of Figure 3.13 ($\theta = -30°$, $0°$, $30°$). The field intensity oscillation of the waveforms is directly related to the number of active switches at each sampling instant, which is typical of TMAs radiating behavior.

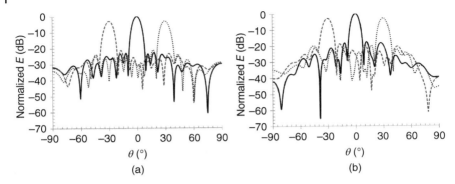

Figure 3.14 Radiation patterns at the fundamental (solid line) and at the two first symmetrical sideband harmonics (dashed lines) due to the excitation of Figure 3.13: (a) for the equally spaced and (b) unequally spaced arrays of Figure 3.12. Source: Masotti et al. 2016 [22]. Reproduced with pemrission of IEEE.

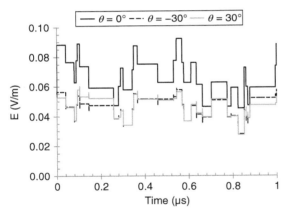

Figure 3.15 Far-field envelopes at 1 m-distance, in correspondence of the three radiation maxima of Figure 3.14a. Source: Masotti et al. 2016 [22]. Reproduced with pemrission of IEEE.

In presence of distant and/or power-hungry tags, the power transfer phase needs for a transmission of high power levels. For this reason, an investigation of the effects of the input RF power level on the TMA nonlinear performance is carried on. Figure 3.16 shows the radiation patterns (of the equally spaced array under the excitation of Figure 3.13) at the carrier frequency ($h = 0$) for different values of the RF power at each antenna port. An evident nonlinear phenomenon takes place in the TMA circuit: when the RF input signal increases, the diodes start rectifying a significant portion of it; as a result, the bias control sequence can be completely overrun by this strong additional dc contribution. The macroscopic effect of this undesired rectification is evinced from Figure 3.16, where the progressive increase of the SLL in the patterns is a consequence of the weak control of the diodes. For the adopted Skyworks diode, this degradation is reached at $P_{IN} = 10$ dBm: all the

Figure 3.16 Effects of the input RF power level on the radiation pattern at the fundamental frequency of the array of Figure 3.12a under the excitation of Figure 3.13.

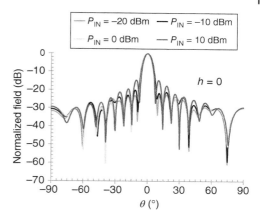

switches are almost forced to the on state for the whole modulation period T_M by the rectification phenomenon, as demonstrated by the corresponding radiation pattern which is identical to the one of a standard 16-element static array. This investigation suggests that the Skyworks SMS7630-079 has 0 dBm as input power limit. For higher power needs, the use of different devices (such as the $p–i–n$ diode Infineon BAR64–02V) can release this limit.

3.6 Measured Results

An experimental proof of the TMA capability of accurate tag detection is carried out with a prototype consisting of the two-element array of Figure 3.9a, driven by the control sequence used in the first step of operation to generate the Σ and Δ patterns. The layout of the realized prototype is shown in Figure 3.17 together with the component values, according to the scheme of Figure 3.3. A microprocessor TI MSP430 is used as switch controller: Figure 3.18 shows the corresponding real waveforms (with $f_M = 25\,\text{kHz}$ and period $T_M = 1/f_M = 40\,\mu\text{s}$), measured at the microprocessor ports, for $d = 0\%$ and $d = 32\%$. With respect to the control sequences reported in Figure 3.6, here positive (3.5 V) and negative (−3.5 V) waveform levels are adopted, thus ensuring the proper bias of the diodes, and hence the correct control of the antenna ports. The waveforms show slight oscillations around the ±3.5 V values, but rapid transition between the two states ($t_{\text{rise}} = t_{\text{fall}} \approx 60\,\text{ns}$). A 3.5-mA current flows through the diodes in the on-state.

The first test of the prototype consists of spectra measurements to verify the complex radiation mechanism of the system under examination: Figure 3.19a,b shows the measured spectra for the broadside direction ($\theta = 0°$), and for $\theta = 45°$, respectively. The sideband radiation phenomenon at 2.499 975 and

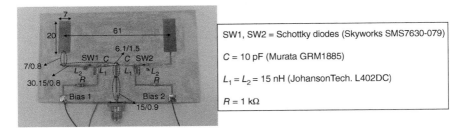

Figure 3.17 2.45 GHz prototype of a two-monopole TMA, with lines dimensions (length/width) in mm. Source: Masotti et al. 2016 [22]. Reproduced with pemrission of IEEE.

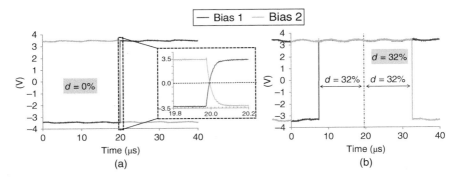

Figure 3.18 Pulse waveforms measured at the microprocessor output ports, for two *d* values: (a) *d* = 0% and (b) *d* = 32%. Source: Masotti et al. 2016 [22]. Reproduced with pemrission of IEEE.

2.450 025 GHz is clearly visible in the figure: when $d = 0\%$, the Δ field strength, at the sideband harmonics, increases with increasing θ value, whereas the Σ field at the fundamental remains almost unchanged. This is also confirmed by the simulated results reported again in Figure 3.19c,d for the two pointing directions, for the sake of clarity: the circles indicate the Σ predicted values for the two values of d (same color of the spectra), whereas the squares do the same for the Δ. When $d = 32\%$, the Δ field decreases with increasing θ values. These preliminary results also demonstrate that the Δ minimum in the $d = 32\%$ case is not deep as in simulation (see the dark grey square of Figure 3.19d), since the sideband harmonic spectral lines are not much lower than the fundamental one (see Figure 3.19b).

The resulting measured radiation patterns shown in Figure 3.20 for $d = 0\%$, 32% confirm the last sentence, i.e. the Δ peaks are -25 dB, only, for $d = 32\%$, whereas a very deep peak (-40 dB) of the Δ pattern with $d = 0\%$ is reached. However, these

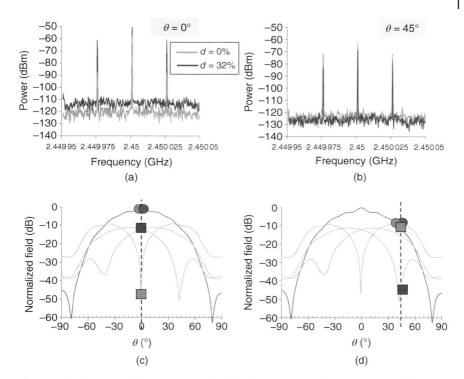

Figure 3.19 Measured fundamental and first sideband harmonics spectra, radiated in two different directions: (a) $\theta = 0°$, (b) $\theta = 45°$. Corresponding predicted patterns with highlighted Σ and Δ values for: (c) $\theta = 0°$, (d) $\theta = 45°$, for $d = 0\%$, 32%. Source: Masotti et al. 2016 [22]. Reproduced with pemrission of IEEE.

measurements confirm that the TMA operation is reached with good steering capabilities. Some discrepancies with the simulated results are evident: there is a slight misalignment of the measured Δ patterns of about 10° when $d = 0\%$ (the negative peak is at $\theta \approx -10°$ instead of 0°) and of 5° when $d = 32\%$ (the Δ peaks are at $\theta \approx -50°$ (expected −45°) and $\theta \approx 40°$ (expected 45°), for the higher and lower sideband, respectively). This can be due to slight asymmetries of the in-house prototype, which could cause currents phase unbalance at the antenna ports.

Furthermore, a lower amplitude of the measured Δ-field with respect to the corresponding modeled one is observed. This is probably due to the uncertainties of the available diode package model used in the nonlinear simulation of the entire system. Indeed, we experimentally observed that in real operation, an alternative path for the RF signal is created by the diode capacitance parasitic, which allows a small amount of the RF signal to reach the antenna ports, thus perturbing the delicate equilibrium which the sideband radiation is based on.

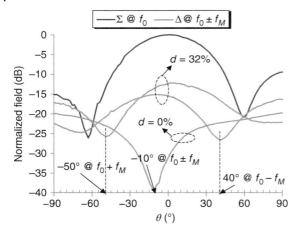

Figure 3.20 Measured Σ (at fundamental) and Δ radiation patterns (at first sideband harmonics) for $d = 0\%$, 32%. Source: Masotti et al. [22]. Reproduced with pemrission of IEEE.

3.7 TMA Architecture for Fundamental Pattern Steering

As described in the previous WPT procedure, the radiation patterns at the sideband harmonics of TMAs become auxiliary efficient radiating tools. Indeed, they are more agile design options than the fundamental one because of the complex nature of the Fourier coefficients u_{hi} $(h \neq 0)$: no bulky and/or complex phase shifters are needed to steer them. The real nature of the coefficient at the fundamental $u_{0i} = \tau_i$ (i.e. the on-duration of the rectangular pulse $U_i(t)$) implies a fixed broadside fundamental lobe, as demonstrated by the fundamental $(h = 0)$ harmonic of array factor derived from (3.8), in the case of elements aligned along the x-axis and radiation patterns detected in the xz-plane (hence, $\cos \chi = \sin \theta$):

$$e^{j2\pi f_0 t} AF_0(\theta) = e^{j2\pi f_0 t} \sum_{i=1}^{n_A} \tau_i \, e^{j(i-1)\beta L \sin \theta} \tag{3.11}$$

Figure 3.21a shows a time-modulated multielement architecture [29] able to steer the fundamental pattern, too: four patch antennas are realized on the same Taconic RF-60A substrate used before, and resonate at 5.8 GHz (f_0). The antennas/feeding network are geometrically defined in such a way that (i) all the elements are vertically polarized; (ii) the field radiated by the reference antenna (no. 1) has a phase arbitrarily fixed to 0°; and (iii) the antennas radiate a field with the same amplitude and phase shift equal to $i \cdot 90°$ (with $i = 0, 1, 2, 3$ starting from the reference one, in a counterclockwise direction). The switch (SWi+1 in Figure 3.21a) at each antenna port allows to radiate an **E** phasor with arbitrary phase Ψ, as described in the following with the help of the complex plane of Figure 3.21b: the four light grey vectors on the axes correspond to the **E** field

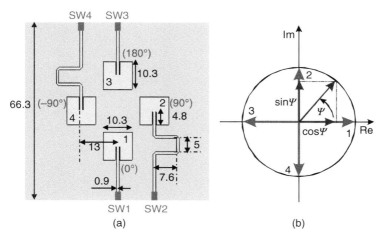

Figure 3.21 (a) 4-patch TMA architecture @ 5.8 GHz able to control the phase of the radiated far-field at the carrier frequency (dimension in mm); (b) vectorial combination of two neighboring antennas far-field to obtain the desired phase ψ. Source: Masotti 2017 [29]. Reproduced with pemrission of IEEE.

radiated by the assembly of Figure 3.21a (the vector number is the same as the corresponding antenna). The resulting radiated field can have a phase equal to Ψ if the antenna excitation amplitudes are controlled in such a way as to assign at two neighboring antennas the (normalized) amplitudes $\cos\psi$ and $\sin\psi$. If $\cos_\psi \cdot T_M$ and $\sin_\psi \cdot T_M$ are adopted as pulses duration of the selected neighboring antennas (where $T_M = 1/f_M$ is the switches modulation period), the desired result is reached.

The radiating assembly of Figure 3.21a can now be adopted as one of the two elements of the TMA architecture described in Figure 3.22 in order to overcome the limitation of fixed fundamental (at f_0) beam of this family of arrays. The new antenna no. 5 is placed at a distance D from the reference one (no. 1) and has the same **E** field phase.

According to the switches on period representation previously adopted, Figure 3.23a–c shows how to excite the assembly of Figure 3.22 in order to guarantee a phase shift Ψ between the fields radiated by the multiantenna architecture (antennas nos. 1, 2, 3, 4) and the standard antenna (no. 5) equal to 30°, 60°, and 120°, respectively (the on-period rectangles are light grey for the four-antenna subsystem, and dark grey for the antenna no. 5).

By referring to the fundamental harmonic array factor formulation (3.11), the steering of the corresponding main beam can be easily evaluated with the following formula:

$$\theta_{\max} = \arcsin\left(\frac{\lambda\psi}{2\pi D}\right) \tag{3.12}$$

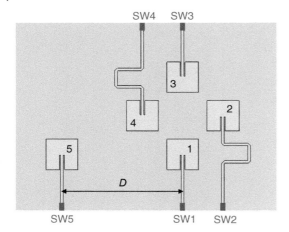

Figure 3.22 TMA layout for precise localization purposes: a standard antenna (no. 5) forms a two-element array with the multiantenna architecture of Figure 3.21a. Source: Masotti 2017 [29]. Reproduced with pemrission of IEEE.

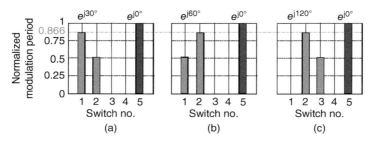

Figure 3.23 Switches on-sequences for a phase-shift ψ equal to (a) 30°; (b) 60°; and (c) 120° in the far-field radiated by the two-element array of Figure 4.22. Source: Masotti 2017 [29]. Reproduced with pemrission of IEEE.

where θ_{max} indicates the maximum radiation direction in the scanning (elevation) plane. For $D = 0.75\lambda$, this leads to a theoretical discrete steering step of about 6° for each ψ increase by 30°.

The circuit-based simulation of the nonlinear radiating system of Figure 3.22, with $D = \lambda$, is carried out by resorting to the technique described in Section 3.3 [22], through the combination of the HB technique for the accurate description of the nonlinear devices (the same Schottky diodes as before) with the full-wave description of the five-port array of Figure 3.22. As modulation frequency, $f_M = 1$ MHz is adopted: the difference of several orders of magnitude with respect to the fundamental carrier ($f_0 = 5.8$ GHz) again allows to exploit the efficient envelope-oriented HB analysis approach [25]. One can note that the proposed approach, being based on the aforementioned co-simulation strategy, does not exactly obey to the theory of time-based linear arrays given by (3.11), (3.12), thus

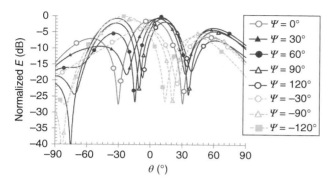

Figure 3.24 Steering performance of the fundamental radiation pattern by changing the phase-shift between the two-element TMA of Figure 3.22 when $D = \lambda$. Source: Masotti 2017 [29]. Reproduced with pemrission of IEEE.

providing the actual performance of the misaligned and unequally spaced layout under exam (Figure 3.22).

The results, in terms of the radiation pattern steering at f_0, are shown in Figure 3.24: the blue solid lines, obtained for $\psi = 30°, 60°, 90°, 120°$, show maxima on the right side of the scanning plane at θ values close to the theoretical ones given by (3.12) (i.e. 5°, 10°, 15°, 20°). For the right steering, antennas # 1, 2, 3, and 5 are involved. The same figure reports (in dashed green line) some radiation patterns for $\psi < 0$: in this case, antennas # 1, 3, 4, and 5 are used and a less neat behavior is visible, due to the higher EM-coupling between antennas 4 and 5. One can note the asymmetry of the patterns (for $\psi < 0$ and $\psi > 0$) due to the different involved radiating structures.

This radiating mechanism can be favorably exploited for accurate localization of tagged objects, in a way similar to that one described in Section 3.4.

In practical realizations, a flat Σ pattern is useful because it provides a more effective MPR (according to (3.10)), but it means very small interelement distance (e.g. $\lambda/8$ as in [18]), hence, strong interelement couplings and Δ pattern distortion [30]. The architecture of Figure 3.22 allows to circumvent this problem because the Δ pattern can steer together with the Σ one, thus guaranteeing the maximization of the MPR while steering. In fact, the out of phase excitations with duty cycle = 50% of Figure 3.6 is now given by the control sequences of Figure 3.25a. The Σ and Δ patterns obtained by exciting in this way the switches 1 and 5 of the TMA of Figure 3.22 (with $D = \lambda$) are the solid and dashed black lines of Figure 3.26, respectively: as expected, the maximum of Σ and the minimum of Δ are both in the broadside direction ($\theta = 0°$).

Less intuitive control sequences provide an identical Σ and Δ steering: imagine to aim at a phase shift $\psi = 30°$ for both the patterns, following the method

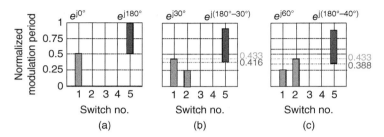

Figure 3.25 Switch control sequences for the two-element TMA of Figure 3.22 for identical steering of both the Σ and Δ patterns at the fundamental and first sideband harmonic, respectively, for ψ: (a) 0°; (b) 30°; and (c) 60°. Source: Masotti 2017 [29]. Reproduced with pemrission of IEEE.

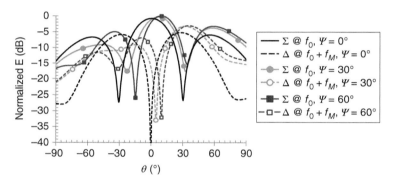

Figure 3.26 Σ (solid line) and Δ (dashed line) patterns identical steering for the two-element TMA of Figure 3.22 (with $D = \lambda$) at the fundamental and first sideband harmonic, respectively. Source: Masotti 2017 [29]. Reproduced with pemrission of IEEE.

previously described. Halving the period of the sequences of Figure 3.23a is not enough: the out-of-phase condition of antenna no. 5 must be added, too, by driving it with a sequence shifted by 180°-ψ, corresponding to 150° in this case (Figure 3.25b). The solid and dashed curves with circular marker of Figure 3.26 demonstrate the corresponding steered Σ and Δ patterns. If the ψ value is 60°, the previous rule is not exactly applicable: in fact, the phase shift of the dark grey sequence of Figure 3.25c (for the antenna no. 5) reproducing the same Σ and Δ steering is not the theoretical 180° to −60° = 120°, but a lower one (180° to −40° = 140°) (as demonstrated by the lines [solid and dashed] with square markers of Figure 3.26). The latter result underlines the importance of a rigorous simulation tool for TMA analysis: the interactions between the radiating elements and the actual behavior of the switches (responsible for unwanted phase delay) must be taken into account to predict this complex nonlinear radiating mechanism.

Figure 3.27 MPR peaks provided by the two-element TMA of Figure 3.22 (with $D = \lambda$). Source: Masotti 2017 [29]. Reproduced with pemrission of IEEE.

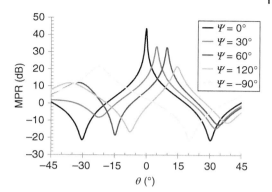

Figure 3.28 Layout of an nine-element linear TMA for energy-aware WPT with the inner one given by the layout of Figure 3.21a.

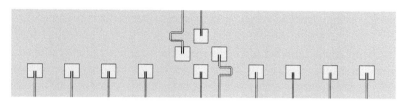

The perfect alignment between Σ maxima and Δ minima of the proposed architecture is thus responsible for sharp MPR peaks, as demonstrated in Figure 3.27, within a 90° scanning window, where additional ψ values are added with respect to Figure 3.26. As foreseen by the results of Figure 3.24, an asymmetric scanning is achieved. Despite this, the proposed TMA architecture can be useful in precise selection of tags, even in harsh EM environments both from the channel and the tags' crowd point of view because of the almost constant pointing capability while steering.

Of course, the second step of the WPT procedure described in Section 3.4.2, i.e. the precise power transmission through the exploitation of a highly directive radiating system can be performed by the quasilinear array of patches reported in Figure 3.28. The proper control sequences must be obtained through the exploitation of the nonlinear/EM co-simulation strategy described in Section 3.3, thus taking into account all the linear and nonlinear dynamics of the system.

3.8 Conclusion

In this chapter, the unrivaled level of configurability of TMAs has been deeply investigated with special emphasis on a dynamic two-step intentional WPT operation. By simultaneously controlling in real-time the bias waveforms of nonlinear switches feeding the array elements, the system is able to first precisely localize and then energize randomly placed sensors, without wasting of power and with no need of bulky phase shifters. The actual nonlinear switch dynamics and the linear EM couplings between different portions of the WPT system need to be rigorously considered during the analysis: for this purpose, a simulation tool based on the combined use of the HB technique and full-wave analysis has been effectively adopted. This software tool must be the inner loop of any optimization strategy adopted for the synthesis of the needed control sequence of the time-based radiating system, as some attempts in the literature start to demonstrate.

The effectiveness of TMA radiation capability is demonstrated by some preliminary measurements of a realized two-monopole array: the obtained behavior demonstrates multiband radiation mechanism as well as the feasibility of the localization function. These first promising results indicate TMAs as potential candidates for agile and reconfigurable WPT systems, to be exploited in many civil and industrial energy-aware applications. Future work will be devoted to the investigation of TMAs as radiating systems able to focus the power in precise locations.

References

1 Dolgov, A., Zane, R., and Popovic, Z. (2010). Power management system for online low power RF energy harvesting optimization. *IEEE Transactions on Circuits and Systems I: Regular Papers* 57 (7): 1802–1811.

2 Costanzo, A., Romani, A., Masotti, D. et al. (2012). RF/baseband co-design of switching receivers for multiband microwave energy harvesting. *Sensors and Actuators A: Physical* 179 (1): 158–168.

3 Essel, J., Brenk, D., Heidrich, J., and Weigel, R. (2009). A highly efficient UHF RFID frontend approach. *Proceedings of the IEEE MTT-S International Microwave Workshop on Wireless Sensing, Local Positioning, and RFID (IMWS 2009)*, Cavtat, Croatia (24–25 September 2009), pp. 1–4.

4 Cost Action IC1301 Team (2017). Europe and the future for WPT: European contributions to wireless power transfer technology. *IEEE Microwave Magazine* 18 (4): 56–87.

5 Costanzo, A., Masotti, D., Fantuzzi, M., and Del Prete, M. (2017). Co-design strategies for energy-efficient UWB and UHF wireless systems. *IEEE Transactions on Microwave Theory and Techniques* 65 (5): 1852–1863.

6 Masotti, D., Costanzo, A., Francia, P. et al. (2014). A load-modulated rectifier for RF micropower harvesting with start-up strategies. *IEEE Transactions on Microwave Theory and Techniques* 62 (4): 994–1004.

7 Del Prete, M., Masotti, D., Arbizzani, N., and Costanzo, A. (2013). Remotely identify and detect by a compact reader with mono-pulse scanning capabilities. *IEEE Transactions on Microwave Theory and Techniques* 61 (1): 641–650.

8 Sharp, E. and Diab, M. (1960). Van Atta reflector array. *IRE Transactions on Antennas and Propagation* 8 (4): 436–438.

9 Toh, B.Y., Fusco, V.F., and Buchanan, N.B. (2002). Assessment of performance limitations of Pon retrodirective arrays. *IEEE Transactions on Antennas and Propagation* 50 (10): 1425–1432.

10 Ijaz, B., Roy, S., and Masud, M.M. et al. (2013). A series-fed microstrip patch array with interconnecting CRLH transmission lines for WLAN applications. *2013 7th European Conferrence on Antennas and Propagation (EuCAP)*, Gothenburg, Sweden (8–12 April 2013), pp. 2088–2091.

11 Gomez-Tornero, J.L. (2011). Analysis and design of conformal tapered leaky-wave antennas. *IEEE Antennas and Wireless Propagation Letters* 10: 1068–1071.

12 Shan, L. and Geyi, W. (2014). Optimal design of focused antenna arrays. *IEEE Transactions on Antennas and Propagation* 62 (11): 5565–5571.

13 Xiong, J., Wang, W., Shao, H., and Chen, H. (2017). Frequency diverse array transmit beampattern optimization with genetic algorithm. *IEEE Antennas and Wireless Propagation Letters* 16: 469–472.

14 Wang, W. (2019). Retrodirective frequency diverse array focusing for wireless information and power transfer. *IEEE Journal on Selected Areas in Communications* 37 (1): 61–73.

15 Matsumuro, T., Ishikawa, Y., Mitani, T. et al. (2017). Study of a single-frequency retrodirective system with a beam pilot signal using dual-mode dielectric resonator antenna elements. *Wireless Power Transfer* 4 (2): 132–145.

16 Mitani, T., Kawashima, S., and Shinohara, N. (2018). Direction-of-arrival estimation by utilizing harmonic reradiation from rectenna. *Proceedings of the 2018 IEEE MTT-S Wireless Power Transfer Conference*, Montreal, Canada (3–7 June 2018).

17 Shanks, H. and Bickmore, R. (1959). Four-dimensional electromagnetic radiators. *Canadian Journal of Physics* 37: 263–275.

18 Tennant, A. and Chambers, B. (2007). A two-element time-modulated array with direction-finding properties. *IEEE Antennas and Wireless Propagation Letters* 6: 64–65.

19 Chong, H., Xianling, L., Zhaojin, L. et al. (2015). Direction finding by time-modulated array with harmonic characteristic analysis. *IEEE Antennas and Wireless Propagation Letters* 14: 642–645.

20 Li, G., Yang, S., Chen, Y., and Nie, Z. (2008). An adaptive beamforming in time modulated antenna arrays. *Proceedings of the ISAPE 2008*, Kunming, China (2–6 November 2008), pp. 166–169.

21 Poli, L., Rocca, P., Oliveri, G., and Massa, A. (2011). Harmonic beamforming in time-modulated linear arrays. *IEEE Transactions on Antennas and Propagation* 59 (7): 2538–2545.

22 Masotti, D., Costanzo, A., Del Prete, M., and Rizzoli, V. (2016). Time-modulation of linear arrays for real-time reconfigurable wireless power transmission. *IEEE Transactions on Microwave Theory and Techniques* 64 (2): 331–342.

23 Masotti, D. and Costanzo, A. (2017). Time-based RF showers for energy-aware power transmission. *2017 11th European Conference on Antennas and Propagation (EUCAP)*, Paris (19–24 March 2017), pp. 783–787. https://doi.org/10.23919/EuCAP.2017.7928158.

24 Masotti, D., Francia, P., Costanzo, A., and Rizzoli, V. (2013). Rigorous electromagnetic/circuit-level analysis of time-modulated linear arrays. *IEEE Transactions on Antennas and Propagation* 61: 5465–5474.

25 Rizzoli, V., Masotti, D., Mastri, F., and Montanari, E. (2011). System-oriented harmonic-balance algorithms for circuit-level simulation. *IEEE Transactions on Computer-Aided Design of Integrated Circuits and Systems* 30 (2): 256–269.

26 Poli, L., Rocca, P., Manica, L., and Massa, A. (2010). Pattern synthesis in time-modulated linear arrays through pulse shifting. *IET Microwaves, Antennas & Propagation* 4: 1157–1164.

27 Zhu, Q., Yang, S., Zheng, L., and Nie, Z. (2012). Design of a low sidelobe time modulated linear array with uniform amplitude and sub-sectional optimized time steps. *IEEE Transactions on Antennas and Propagation* 60: 4436–4439.

28 M. Salucci, L. Poli, D. Masotti et al. (2018). PSO-driven synthesis of realistic time modulated arrays with optimal instantaneous directivity through a system-by-design implementation. *2018 IEEE International Symposium on Antennas and Propagation and USNC-URSI Radio Science Meeting*, Boston (8–13 July 2018).

29 Masotti, D. (2017). A novel time-based beamforming strategy for enhanced localization capability. *IEEE Antennas and Wireless Propagation Letters* 16: 2428–2431.

30 Zhu, Q., Yang, S., Yao, R., and Nie, Z. (2014). Directional modulation based on 4-D antenna arrays. *IEEE Transactions on Antennas and Propagation* 62 (2): 621–628.

4

Backscatter a Solution for IoT Devices

Daniel Belo[1], Ricardo Correia[1], Marina Jordao[1], Pedro Pinho[1,2], and Nuno B. Carvalho[1]

[1]*Departamento de Electrónica, Telecomunicações e Informática, Instituto de Telecomunicacoes, Universidade de Aveiro, Aveiro, Portugal*
[2]*Instituto Superior de Engenharia de Lisboa - ISEL, Lisboa, Portugal*

4.1 Backscatter Basics

Radio frequency identification (RFID) is a kind of the contactless automatic identification technology via radio frequency (RF) signal, and it is one of the most developed rapid technologies in automatic identification technology field. It uses RF to identify the target and collect data via noncontact radio communication. With the operating frequency of RFID system increasing, for passive RFID systems working at the ultra-high frequency (UHF) and microwave bands and on the basis of backscattering modulation principle to work, it becomes very important to research on response model of electromagnetic field and calculation of the read distance.

To modulate the backscattered signal, the RFID chip switches its input impedance between two states (amplitude modulation). In order to improve the reading range, the tag's antenna is generally matched to the impedance of the chip. The matching is obtained in special conditions, at a given frequency and usually for tag placed in free space. The impedance of the chip is a function of frequency and received power, and the impedance of the antenna is highly dependent on the support on which the tag is placed.

Figure 4.1 illustrates the operation of a modern passive RFID system that includes an RFID reader and an RFID tag, composed of an antenna and an integrated circuit (IC) chip. The reader signal alternates between a continuous wave (CW) and modulated transmissions. The tag sends data during one of the CW periods by switching its input impedance between two states, effectively changing its radar cross section (RCS) and thus modulating the backscattered

Wireless Power Transmission for Sustainable Electronics: COST WiPE - IC1301,
First Edition. Edited by Nuno Borges Carvalho and Apostolos Georgiadis.
© 2020 John Wiley & Sons, Inc. Published 2020 by John Wiley & Sons, Inc.

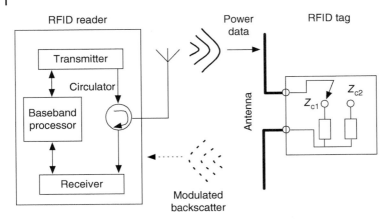

Figure 4.1 Passive RFID system overview. Source: Adapted from Skolnik 2001 [1].

field. The RCS of a scattering target is defined by the equivalent area of the target based on the target scattering the incident power isotropically [1].

One of the impedance states is usually high and another is low to provide a significant difference in the backscattered signal. Data exchange between RFID reader and tag can employ various modulation and coding schemes (amplitude modulation and Manchester coding).

Passive RFID tags use the RF power, from the reader, to energize the digital part of the tag, which is responsible for the modulation of the incoming wave. To enable the possibility of having a totally passive RFID tag with data logging and advanced computing, a more careful and detailed WPTs should be designed and optimized to power the intelligent tags. This integration of passive RFID, passive sensing and increased computing capabilities have enabled the interest in the concept of passive wireless sensors. The concept of having passive wireless sensors with data logging and advanced computing capabilities will play an important role in the IoT context, where a lot of sensors can be connected, deployed anywhere, and give information about environment without the need of batteries. Nonetheless, the increase of IoT sensors will imply the height of batteries to be deployed, which will have a negative ambient impact.

As it was mentioned previously, battery-powered tags can improve the distance of communication but have some limitations when referring to the battery cost and its replacement. Thus, the alternatives to the battery systems are based on energy harvesting (EH) technology or other different sources (solar [2], motion or vibration [3], ambient RF [4]).

To overcome the drawbacks employed from the EH and batteries, the concept of WPT was explored to supply the tags with power.

As it was previously mentioned, in most RFID systems and passive sensors, the reader to tag communication is an amplitude shift keying (ASK) or phase-shift keying (PSK) that modulates either the amplitude or both the amplitude and phase of the reader's transmitted RF carrier. The use of this technology entails a number of advantages over barcode technologies such as tracking people, items, and equipment in real time, nonline of sight requirement, long reading range, and standing harsh environment.

However, the work [5] has shown that modulated backscatter can be extended to include higher order modulation schemes, such as 4-QAM. While ASK and PSK transmit 1 bit of data per symbol period, 4-QAM based can transmit 2 bits per symbol period, thus increasing the data rate and leading to reduced on-chip power consumption and extended read range. The work presented in [5, 6] refers to a 4-QAM backscatter in semipassive systems, by using a coin cell battery as a power source for the modulator and a microcontroller that needs 3 V of supply. This way, the authors proved the quadrature phase shift keying (QPSK) modulator and battery powered system, by using an approach with four lumped impedances connected to an RF switch that is controlled by a microcontroller.

The same authors developed a 16-QAM modulator for UHF backscatter communication with a consumption of 1.49 mW at a rate of 96 Mbps only in the modulator (not the overall system with data generation logic feeding the modulator) [7]. This modulator was implemented with five switches with lumped terminations as a 16-to-1 multiplexer to modulate the load between 16 different states.

4.1.1 Different Backscatter Sensors Development

Nowadays, the wireless sensing devices are growing at a phenomenal rate, with billions of wireless sensors reaching a much larger proportion than the world's population. However, this rapid increase presents two main issues: increased use of batteries and energy consumption. In the IoT context, passive backscatter radios will certainly play a crucial role due to their low cost, low complexity, and battery-free operation.

Therefore, there is a significant need to design novel wireless communication techniques to achieve higher data rates while simultaneously minimizing energy consumption. In the backscatter communication, the tag reflects a radio signal transmitted by the reader, and modulates the reflection by controlling its own reflection coefficient.

Consequently, it will be presented some designs that can potentiate the use of backscatter radio combined with WPT and with high-order modulation.

4.1.2 Backscatter with WPT Capabilities

The first proposed system is based on Figure 4.2, and it was evaluated in [9]. It is composed by two matching networks, a backscatter modulator and a dual band

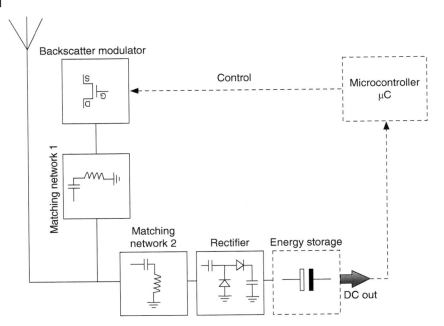

Figure 4.2 Block diagram of the system based on backscattering with WPT. Source: Correia et al. 2016 [8]. Reproduced with permission of IEEE.

rectifier. The goal is to harvest electrical energy with one tone (1.8 GHz) and with the other tone (2.45 GHz) transfer data by backscatter means. The RF power harvester employs a receiving antenna, an impedance matching network, DC power conditioning, and the sensor to be powered. The backscatter modulator employs the receiving antenna, an impedance matching network, and a semiconductor device to control the reflection coefficient.

The RF circuit is shown in Figure 4.3. It is divided into three main sections: the backscatter modulator, which is composed of a switch transistor that modulates the impedance of the antenna; the matching network, which was designed to provide backscatter load modulation at one frequency and continuous flow of WPT at other frequencies; and the five-stage Dickson multiplier, which allows for RF-DC conversion to provide sufficient DC power to supply the microcontroller.

4.1.3 High-Order Backscatter Modulation

The second design proposed consists of a backscatter modulator with high-order backscatter modulation format, by proposing an RF circuit design that is able to be replicated to include more modulation levels [11].

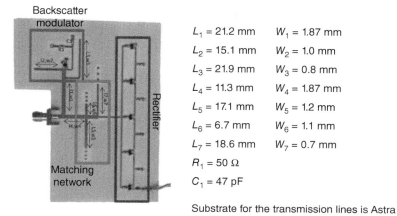

Backscatter modulator

L_1 = 21.2 mm W_1 = 1.87 mm

L_2 = 15.1 mm W_2 = 1.0 mm

L_3 = 21.9 mm W_3 = 0.8 mm

L_4 = 11.3 mm W_4 = 1.87 mm

L_5 = 17.1 mm W_5 = 1.2 mm

L_6 = 6.7 mm W_6 = 1.1 mm

L_7 = 18.6 mm W_7 = 0.7 mm

R_1 = 50 Ω

C_1 = 47 pF

Matching network

Rectifier

Substrate for the transmission lines is Astra MT77, thickness = 0.762 mm, ε_r = 3.0, tan δ = 0.0017.

Figure 4.3 Photograph of implemented system with backscatter modulator combined with wireless power transmission (WPT). Source: Pereira et al. 2017 [10]. https://www.mdpi.com/1424-8220/17/10/2268. Licensed under CC BY 4.0.

Figure 4.4a shows the design of the proposed 16-QAM scheme, and Figure 4.4b the prototype implemented that includes the Wilkinson power divider, two matching networks, and two transistors. This prototype was optimized for 0 dBm at 2.45 GHz, and it was simulated in advanced design system (ADS). In Figure 4.4b, it is possible to view the difference of line length in each branch, and it is related with the 45° phase shift.

Using two transistors and changing the voltage at the gate of each transistor, it is possible to create a multitude of impedance arrangements, and with this enable high-order backscatter modulation.

4.1.4 Modulated High-Bandwidth Backscatter with WPT Capabilities

The third system, in addition to the high-order modulator, has a rectifier that has the capability of harvesting dc power through the RF CW from the transmitter. It was developed a 16-QAM backscatter modulator with 59 μW of power consumption and a remarkable data rate of 960 Mb/s, and it presents wireless power transfer (WPT) capabilities [12]. The modulator was implemented at 2.45 GHz and 58 presents 61.5 fJ/b consumption. The circuit has a rectifier with a tuned matching network to maximize the conversion efficiency at 1.7 GHz and provide the backscatter communication at 2.45 GHz (Figure 4.5).

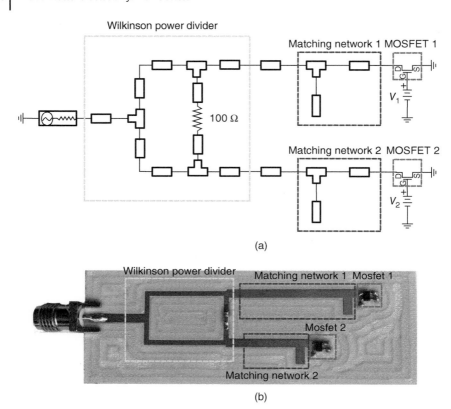

Figure 4.4 16-QAM backscatter modulation scheme. (a) Design of 16-QAM backscatter modulator. (b) Photograph of the 16-QAM backscatter circuit. Source: Correia et al. 2017 [11]. Reproduced with permission of IEEE.

4.2 An IoT-Complete Sensor with Backscatter Capabilities

At some years ago, concepts like IoT or wireless sensors networks were not more than a vision from the future. Nowadays, these concepts are already a reality, and the number of IoT devices connected by 2020 is expected to be 50.1 billion, near twice the number of IoT devices in 2017. This growth is only possible due advances in semiconductor, networking, and material science technologies [10].

The most recent IoT devices have sensing, computational, and communication capabilities allowing them to create an autonomous network. This new reality allows applications that can range from environmental sensing and wearable biometric monitoring to security and structural monitoring [13]. The benefits

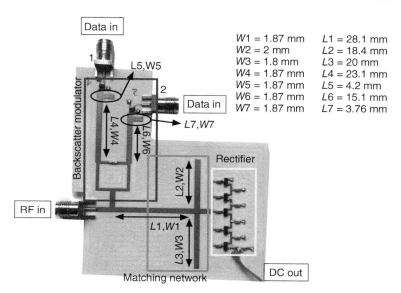

Figure 4.5 Photograph of the proposed system, composed by a 16-QAM modulator and a rectifier. Substrate for the transmission lines is Astra MT77, thickness = 0.762 mm, $\varepsilon_r = 3.0$, and tan $\delta = 0.0017$. Source: Correia and Carvalho 2017 [12]. Reproduced with permission of IEEE.

that come from the use of this type of technology cannot be denied; however, it is important to look to the other side. This IoT devices need to have a power source, and many times, the solution is the use of batteries. The use of batteries can introduce some limitations, application where battery replacement is a challenge, where extending battery life is important or where there is no space to use a battery [14].

An alternative to the use of batteries is the work presented in [8, 15, 16], where is developed a passive sensor based on wireless power transmission (WPT) and backscatter techniques. This work is developed with the intention of use of the cited research and complete it with the integration of a digital structure that allows it to become a real passive sensor. Besides the digital part, in this work is also presented the results obtained using the final passive sensor.

4.2.1 System Description

The development of a passive sensor has two crucial questions, how to power the sensor? and how is it going to communicate?

In this work, the first problem is solved using microwave power transmission (MPT). The MPT technique uses a device to create CWs, in this case the Rohde & Schwarz SMW200A Vector Signal Generator, and a patch antenna. From the

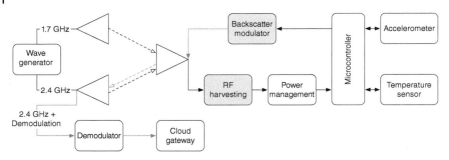

Figure 4.6 Block diagram of the implemented system. Source: Pereira et al. 2017 [10]. https://www.mdpi.com/1424-8220/17/10/2268. Licensed under CC BY 4.0.

sensor side, there is also patch antennas and a RF-DC circuit in order to transform the microwave power into DC. Two frequencies are transmitted at 30 dBm, 1.7 GHz, and 2.4 GHz.

The communication problem is solved using a RF backscatter. The backscatter works at 2.4 GHz and has different impedances according with the voltage level at the MOSFET gate. When there is no voltage, the backscatter does not reflect the CW, when there is 600 mV at its gate, it reflects the CW. This voltage is defined by a microcontroller that controls the voltage according with the information that wants to send. The reflected CW is received by a reader, in this case, the Rohde & Schwarz FSP Spectrum Analyzer.

The microcontroller is the processing unit, besides controlling the backscatter voltage, it is responsible for acquiring the information from the sensors. The work uses two sensors, one accelerometer and a temperature sensor.

Figure 4.6 shows a block diagram that represents the proposed system.

4.2.2 Digital Component

The amount of energy that arrives to the sensor is only a small part compared with the 30 dBm delivered to the system. In order to calculate the amount of energy that arrives to the sensor can be used the Friis Equation, presented in Eq (4.1).

$$\frac{P_r}{P_t} = G_t G_r \left(\frac{\lambda}{4\pi r} \right)^2 \tag{4.1}$$

Considering:

$$P_t = 30 \text{ dBm}; \quad P_r = -6 \text{ dBm}; \quad G_t = G_r = 8 \text{ dB}; \quad f = 2.4 \text{ GHz}; \quad \lambda = 0.125 \text{ m}$$

Comes:

$$r \approx 4 \text{ m}$$

This demonstration shows that if there is 4 m of separation between sensor and the power transmitter, then the power delivery to the sensor is −6 dBm. Besides

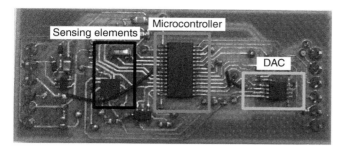

Figure 4.7 Developed PCB. Source: Pereira et al. 2017 [10]. https://www.mdpi.com/
1424-8220/17/10/2268. Licensed under CC BY 4.0.

that the efficiency of the RF-DC need to be considered, according with [5] it goes
around 30%.

All these factors are crucial to understand the key aspect of the digital component, the low power consumption. The distance at which the sensor can be from
the transmitter is going to be defined mainly by the consumption of the digital
component. Having this in mind, it was selected the MSP430F2132 from Texas
Instruments as microcontroller, the LM94021 also from Texas Instruments as
temperature sensor, and the ADXL362 from Analog Devices as accelerometer. As
the microcontroller does not have an internal digital-analog converter (DAC), it
was necessary to have an external, it was chosen the MAX555 from MAXIM. The
DAC is an indispensable component in the system once it controls the 600 mV
needed by the backscatter.

Figure 4.7 shows the printed circuit board (PCB) with sensing and processing
elements.

Besides the components choice, it is also important to ensure an efficient use of
the processing capabilities. The MSP430F2132 from Texas Instruments has con-
sumptions of 220 µA when active and 1.2 µA when in low-power mode [17]. Due
to that it is important that it is only active when needed.

After being supplied, the sensor does the following routine:

- Configure all the components (clock, ADC, I/O ports, sleep-mode, etc.);
- The sensors perform measurements and communicate the acquired values to
  the microcontroller;
- Convert the acquired values to binary, format that corresponds to 0 and 600 mV
  required by the backscatter;
- Send the information to the backscatter using the DAC;
- Enter in sleep-mode;
- Wake-up from sleep mode and go to the step 2;

The routine diagram is showed in Figure 4.8.

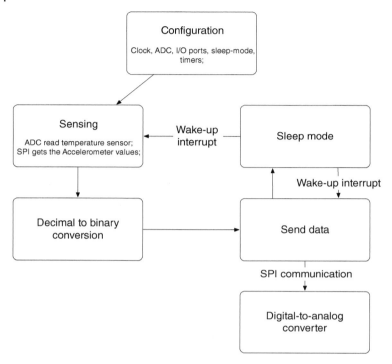

Figure 4.8 Code diagram. Source: Pereira et al. 2017 [10]. https://www.mdpi.com/1424-8220/17/10/2268. Licensed under CC BY 4.0.

The message frame is composed of a total of 11 bits:

- 3 bits to identify the sensor node;
- 2 bits to identify the measured parameter;
- 4 bits for the measurement;
- 2 bits to detect errors;

4.2.3 Measurements

The first measurements were to ensure the correct operation of the digital logic. Figure 4.9 shows the module output, where it is possible to see that it is performing correctly.

Following the digital component as the set of antennas and RF-DC converter. Both elements should be adapted at the desired frequencies, Figure 4.10 shows exactly that.

After tested each component individually, the next step as to join the elements. Figure 4.11 shows the setup with all the components highlighted.

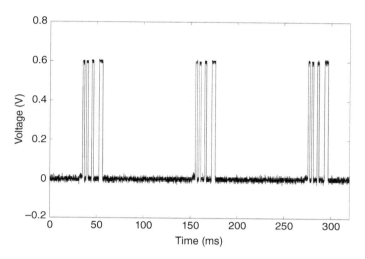

Figure 4.9 Digital output. Source: Pereira et al. 2017 [10]. https://www.mdpi.com/1424-8220/17/10/2268. Licensed under CC BY 4.0.

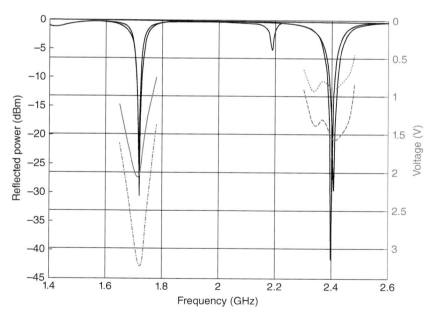

Figure 4.10 Antennas and RF-DC converter adaptation. Source: Pereira et al. 2017 [10]. https://www.mdpi.com/1424-8220/17/10/2268. Licensed under CC BY 4.0.

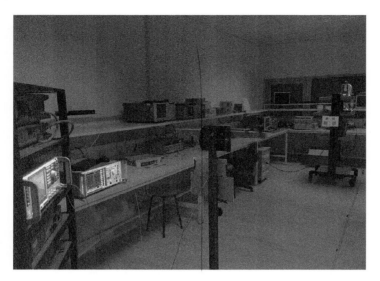

Figure 4.11   Final setup. Source: Pereira et al. 2017 [10]. https://www.mdpi.com/1424-8220/17/10/2268. Licensed under CC BY 4.0.

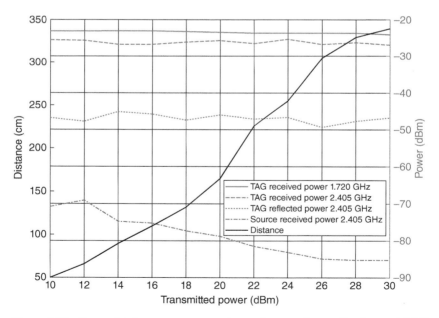

Figure 4.12   Transmitted power versus maximum distance. Source: Pereira et al. 2017 [10]. https://www.mdpi.com/1424-8220/17/10/2268. Licensed under CC BY 4.0.

From this, it was possible to do multiple tests to understand the maximum distance at which the system works. Having a consumption of 45 µA, the maximum distance between the MPT transmitter and the sensor is 4.4 m.

Figure 4.12 presents multiple results obtained. As expected, the power received by the sensor does not suffer large variation, which is natural once the minimum received power to turn it on is constant. Being the power received constant, and having no variation on the reflection method, the reflected power is also constant. As the distance between the transmitter and the sensor increases, the power received by the reader decreases. Figure 4.12 is limited at the distance of 3.4 m because at higher distances and with the available equipment, it was not possible to clearly identify the reflected wave. However, the values of −85 dBm received show that there is a big margin to receive the signal if used a system with lower sensitivity. Having this, it is possible to conclude that the limitation is imposed by the amount of energy that is acquired by the sensor, which points to a distance of 4.4 m.

4.3 The Power Availability for These Sensors

The goal of this section is to present a complete system able to detect, track, and power a battery-less receiving device [18]. A general view of such system is shown in Figure 4.13. The transmission is composed by a dedicated wireless power transmitter and an additional transceiver module that works in a different frequency used to process a pilot signal. The wireless power transmitter is an electronically controlled phased array, and in this application, it may change

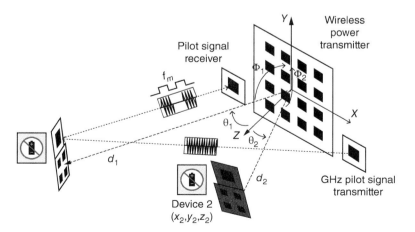

Figure 4.13 Representation of the proposed wireless power transfer system. Source: Belo et al. 2018 [19]. Reproduced with permission of IEEE.

between some predefined states, where each of them transmits a specific amount of power [20]. The pilot signal is intended to be backscattered by the receiver with information about its received signal strength (RSS), and it will be used by the wireless power transmitter to track remote devices. The receiver is a general energy receiving device that converts the received RF energy into usable dc energy and is supposed to be attachable to any mobile device [19].

The system shall operate as follows: the transmitter starts by scanning its covering range using full power. At each beam direction, the pilot signal is monitored, and if a receiver is present, it will convert the collected energy into dc energy and will activate the backscatter system, indicating the transmitter that on that direction a receiver is present. Furthermore, by decoding the backscattered information, the transmitter may adapt the transmitted power.

4.3.1 Electronically Steerable Phased Array for Wireless Power Transfer Applications

A block diagram for a possible CW steerable phased array is shown in Figure 4.14. This antenna array is composed by sixteen active elements in a planar 4×4

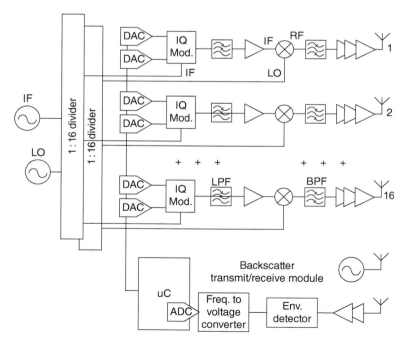

Figure 4.14 Transmitter block diagram including the backscattering transceiver module. Source: Belo et al. 2019 [18]. Reproduced with permission of IEEE.

distribution. Each of them is composed by a microstrip patch antenna fed with a coaxial probe and linear polarization.

The interelement spacing is $0.58\lambda0$. It is well known that the array factor for a planar array with N elements in each row and M elements in each column and a certain phase progression is given by, [21]:

$$AF(\theta, \varphi) = \sum_{n=1}^{n=N} \sum_{m=1}^{m=M} I_{n,m} e^{j(\Delta(\psi_n+\psi_m)+kdn \sin\theta \cos\varphi+kdm \sin\theta \sin\varphi)} \tag{4.2}$$

where k is the wave number, d the interelement spacing, $I_{(n,m)}$ the excitation amplitude for each element, and $\Delta(\psi_n+\psi_m)$ is the progressive phase increment. It can be shown that (4.2) has its maximum for $\theta = 0$ and $\varphi = 0$, if the elements are fed in phase, that is, $\Delta(\psi_n+\psi_m) = 0$. If it is desired to point this maximum to a specific direction (θ_p, φ_p), then:

$$k\,d\,n\sin(\theta_p)\cos(\varphi_p) + k\,d\,m\sin(\theta_p)\sin(\varphi_p) + (\psi_n + \psi_m) = 0 \tag{4.3}$$

and the phase progression that must be applied to each element is given by (4.4)

$$\psi = \psi_n + \psi_m \tag{4.4}$$

$$\psi = -k\,d\,n\sin(\theta_p)\cos(\varphi_p) - k\,d\,m\sin(\theta_p)\sin(\varphi_p) \tag{4.5}$$

In theory, an active antenna array has complete flexibility to change the amplitude and phase for each of the elements that comprises the antenna array. In this application, instead of having a discrete phase shifter and attenuator, an IQ modulator is used on every element to provide a similar flexibility. To produce a CW signal (s) with a certain amplitude (A) and any desired phase (ψ), as shown in Figure 4.15, it follows that

$$i(t) = i_{dc}(t)\cos(\omega_0 t) \tag{4.6}$$

$$q(t) = q_{dc}(t)\cos(\omega_0 t + 90°) \tag{4.7}$$

$$q(t) = q_{dc}(t)\sin(\omega_0 t) \tag{4.8}$$

$$q(t) = q_{dc}(t)\sin(\omega_0 t) \tag{4.9}$$

and

$$s(t) = i(t) + q(t) \tag{4.10}$$

$$s(t) = \left(\sqrt{i_{dc}^2(t) + q_{dc}^2(t)}\right)\cos(\omega_0 t + \psi(t)) \tag{4.11}$$

Therefore, the phase shift and amplitude are given by

$$\psi = \tan^{-1}\left(\frac{i_{dc}}{q_{dc}}\right); \quad A = \sqrt{i_{dc}^2(t) + q_{dc}^2(t)} \tag{4.12}$$

Thus, each IQ modulator will generate a signal whose phase and amplitude can be adjusted with only two control signals. Each of those IQ modulators will

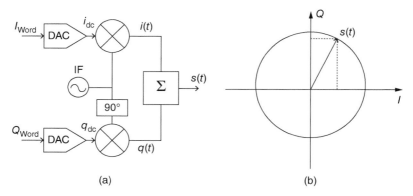

Figure 4.15 Basic beamforming block. (a) IQ modulator. (b) Representation in the IQ plane.

modulate an intermediate frequency (IF) signal that was previously splitted into 16 equal signals and mixed with an local oscillator (LO) signal, previously splitted by 16 as well, to upconvert the IF signal to an RF signal. An important prerequisite for any beamforming architecture is a phase coherent signal generation, which is, in this case, achieved by splitting both IF and LO signals. Note that all source signals do not need to be in phase on every element to be able to apply beamforming, they just need to have a constant phase relation. In order to track a receiving device, the backscattering transceiver module operates at a different frequency. This subsystem is responsible to radiate a pilot signal and to monitor a backscattered version of it with information about the receiver's RSS. It will be shown that this information will be modulated in frequency; thus, a conventional envelope detector combined with a frequency-to-voltage converter circuit is used to demodulate the information backscattered by the receiver.

Besides the ability to point the main beam to any desired direction, the proposed architecture is also able to switch between states the states represented in Figure 4.16. Each state is composed by a specific number of active adjacent elements that forms subarrays (groups of rows). Assuming that the equivalent isotropic radiated power (EIRP) achieved by state 1 is $\text{EIRP}_{\text{State1}}$, then:

$$\text{EIRP}_{\text{State } n} = \text{EIRP}_{\text{State } n-1} + \Delta G_n, \quad n = 2, 3 \tag{4.13}$$

where n is the transmitter state and ΔG_n is the incremental EIRP gain given by:

$$\Delta G_n = (G_{P,n} - G_{P,n-1}) + (G_{A,n} - G_{A,n-1}), \quad n = 2, 3 \tag{4.14}$$

where G_P and G_A are the power and antenna gains, respectively. It is assumed that $G_{(P,1)}$ and $G_{(A,1)}$ are gains relative to a single antenna element. Since the number of active elements doubles with respect to the previous state, the power fed to the

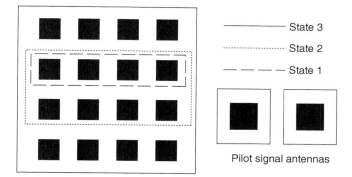

Figure 4.16 Representation of the planar 4×4 wireless power microstrip antenna array with its configurable states and the pilot signal transmit/receive antennas. Source: Belo et al. 2019 [18]. Reproduced with permission of IEEE.

antenna doubles as well as the antenna gain, therefore $\Delta G_n = 6\,\mathrm{dB}$. This feature will allow the transmitter to adapt the transmitted power by simply turning on a specific number of antenna elements. It should be highlighted that each element will either be off or operating at saturation maximizing the transmitter efficiency in all states. If the dc power consumed at state 1 is $P_{(\mathrm{dc},1)}$, then

$$P_{\mathrm{dc},n} = nP_{\mathrm{dc},1}, \quad n = 2, 3 \tag{4.15}$$

The idea behind this concept is that the amount of transmitted power as well as the dc power consumed can be controlled based on the power that a receiving device is collecting. For example, if a receiver is too close to the transmitting antenna, state 1 may be sufficient to deliver enough power to it. On the other hand, if it is too far away or blocked by an object, higher states can be used. Note that every remote electronic device needs a certain amount of power for its proper operation, not less and not more. After proper calibration, the radiation patterns of such array are depicted in Figure 4.17.

4.3.2 Wireless Energy Receiving Device

In order to cooperate with the previous wireless power transmitter, an energy receiving device is presented and discussed. Due to the second frequency backscattering subsystem, the receiving device will be able to continuously receive the energy that is being transmitted to it, while being able to backscatter the pilot signal with useful information modulated in frequency. With this implementation, strong self-jamming due to the high transmitted power levels at the wireless power frequency are avoided [22]. A schematic of such device is depicted in Figure 4.18. The main components of this circuit are two antennas,

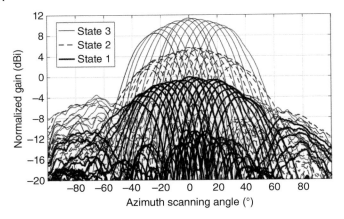

Figure 4.17 Radiation patterns in the azimuth plane measured after calibration. All radiation patterns are normalized to the maximum of the broadside direction of state 1, [1].

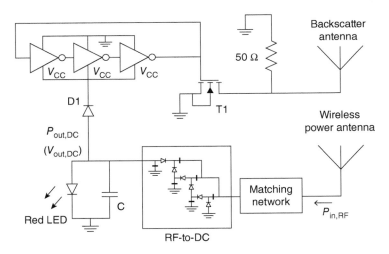

Figure 4.18 Receiving device schematic composed by two antennas, an RF to dc converter, a ring oscillator, and a switching transistor [1].

one to receive energy and one to deal with the backscatter system, a conventional RF to dc converter circuit with a matching network, a low cost ring oscillator and a switching transistor. In order to show the concept, the load presented to this circuit is a red LED. Note that this device is intended to be a general energy receiving unit, and with proper optimization, it may be attached to any portable electronic device. In order to modulate the pilot signal with information about its

RSS, a low complexity coding circuit was developed. By connecting in series, an odd number of inverter gates and feeding back the output of the last inverter to the first result in a so-called ring oscillator [23]. Here, the ring oscillator operates as a voltage-controlled oscillator (VCO) whose oscillation frequency (f_m) depends on the device's input power. This behavior can be achieved by connecting the inverters' power rails directly to the output of the RF to dc converter. This relation can be understood in two steps. Firstly, it is well known that the efficiency achieved by an RF to dc converter is a function of the input power and load. It may be measured and given by (4.16)

$$\eta = \frac{P_{out,DC}}{P_{in,RF}} = \frac{V_{out,dc}^2}{R_{Load} \cdot P_{in,RF}} \tag{4.16}$$

where $V_{out,dc}$ is the output dc voltage, P_{dc} is the output dc power, $P_{in,RF}$ the collected RF power, and R_{Load} the load presented to the converter circuit, in this case a red LED. Thus, by characterizing the device's power conversion efficiency (η) given by (4.16), the dc voltage produced is shown to be variable and dependent on the available input power, as in (4.17)

$$V_{out,dc} = \sqrt{\eta \cdot R_{Load} \cdot P_{in,RF}} \tag{4.17}$$

Secondly, there is also a direct relation between the ring oscillator's supply voltage ($V_{out,dc} = V_{cc}$) and the delay associated with the inverters, which defines the oscillating frequency. Since the power rails are directly connected to the output of the RF to dc converter circuit, the delay becomes a function of $V_{out,dc}$. Therefore, the total delay (d) can be represented by (4.18)

$$d = f(V_{out,dc}) \tag{4.18}$$

and since $V_{out,dc}$ is a function of the input power

$$d = g(P_{in,RF}) \tag{4.19}$$

where g is the relationship between the total delay and the input power. Please note that g can also be obtained from measurements. Finally, the oscillation frequency produced by the ring oscillator is given by (4.20)

$$f_m = \frac{1}{2 \cdot d} = \frac{1}{2 \cdot g(P_{in,RF})} \tag{4.20}$$

Having previously measured the g function, the backscattered pilot signal can be monitored/decoded at the transmitter side, and an accurate estimation of the device's available input power can be taken by inverting (4.20).

For the practical implementation of this receiver, the RF to dc converter circuit is a conventional three-stage Dickson charge pump built using Schottky diodes and optimized to operate at the working frequency. Since an LED behaves as a variable load, it was found through measurements that a reasonable LED

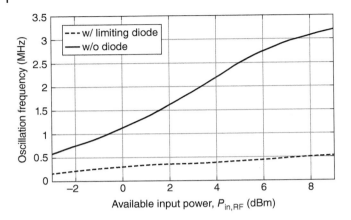

Figure 4.19 Measured relationship between f_m and the available RF input power $P_{(in,RF)}$ for the prototyped device (*g* function), [1].

brightness is achieved with 2.5 mW of dc power (1.78 V and 1.3 mA), which corresponds to a resistive load of approximately 1.3 kΩ. This load value was used for simulation/optimization purposes. The ring oscillator is composed by three common inverter gates that will switch the transistor T1 with a square wave, producing an absorption (transistor off) or reflection of the pilot signal (transistor on). With the aid of a cabled and calibrated measuring laboratorial setup, the operation of this receiving device was characterized. The input power presented to it was swept, and the modulation frequency (f_m) as well as the dc power delivered to the load (LED) were measured. The results are presented in Figures 4.19 and 4.20, respectively. It can be seen that the modulation frequency increased with the increase of the input power, up to a few MHz. It was found that without diode D1, the ring oscillator was consuming most of the dc power produced by the converter circuit. In order to reduce the power consumed by it, diode D1 was placed in series with the power rails as a limiting device. By doing so, the ring oscillator will turn on only when the output of the RF to dc converter reaches the threshold voltage of the set composed by the diode plus ring oscillator. For the optimum chosen dc power delivered to the LED (2.5 mW), the backscatter system consumes 4% of the total available input power, $P_{in,RF}$.

4.3.3 Experimental Results

In order to validate the proposed system design working at 5.8 GHz for wireless power and 3.6 GHz for the pilot signal, three similar receivers were built. These devices were positioned in three different locations within a laboratorial environment, at a certain angle and distance from the transmitting antenna. The

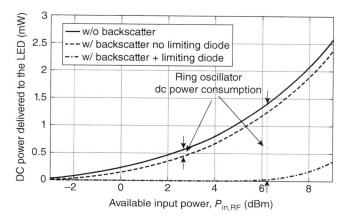

Figure 4.20 Measured LED dc power consumption when the backscattering circuit is not attached and when it is attached with and without limiting diode, [1].

prototyped transmitting antenna scans its covering sector with the beams that were previously stored in the look-up table (LUT), from −45° to +45° in steps of 5°. For target detection, the transmitting antenna is initially configured with state 3 (maximum power). When the power that reaches the receiver is enough to turn it on, the pilot signal is backscattered and monitored at the transmitter side. Note that at the detection stage, it was found that at certain angles, it was monitored more than one oscillation frequency. That is, at some directions, it was possible to turn on more than one receiver. For this experience, the higher oscillation frequency was monitored and all the others were ignored. With the aid of a spectrum analyzer, the higher oscillation frequency was monitored while the transmitter was scanning, and the results are in Figure 4.21. It can be seen that at each receiver position, a local maximum occurs indicating the presence of a receiver. As expected, the closer the receiver is to the transmitting antenna, the higher is the modulation frequency. After detection, the transmitter sequentially points the radiation pattern to each receiver, and the state is decreased (or kept) until an oscillation frequency of at least 0.4 MHz is monitored, corresponding to the power that is necessary to keep the LED to light up with a reasonable brightness, as indicated previously. The final radiation patterns used by the transmitter to power up those devices are shown in Figure 4.22. It is important to highlight that on weak multipath environments, it is possible to have an accurate estimation of the receiving device position. However, on indoor scenarios where the multipath effect is severe, instead of having knowledge about the device location, information about the optimum channel/direction for energy transmission is given. A higher beam steering resolution and a lower beam width would lead to a better estimation of the optimum direction for energy transmission.

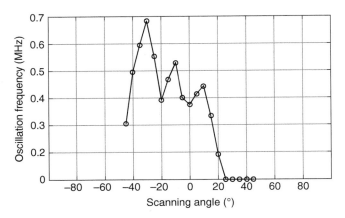

Figure 4.21 Measured backscattered frequency modulation (f_m) for each beam direction when there are three receivers present ready to be powered up, [1].

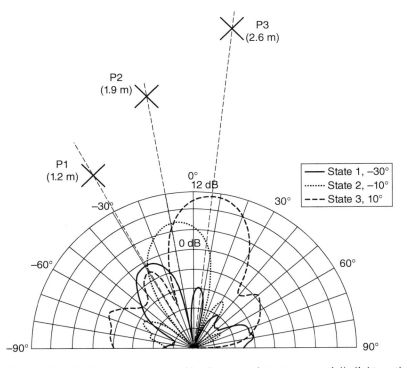

Figure 4.22 Radiation patterns used by the transmitter to sequentially light up the LEDs with the power required to keep them with a reasonable brightness, [1]. Source: Adapted from Skolnik 2001 [1].

4.4 Characterization of High-Order Modulation Backscatter Systems

Characterization and measurements have an important role in circuits and system verification, since they provide information to improve and to optimize their operation. In Internet-of-Things (IoT) and wireless power transmission (WPT), system characterization and measurements procedures work as a useful tool for their validation. Nowadays, the use of backscatter RFID sensors is increasing [6] because they have a diversified application, for instance in 5G scenarios. There are several characteristics that highlight their use, since they are small devices with lower consumption.

In terms of characterization in RFID tags [24] in IoT, to guarantee a proper RFID performance, the impedance match between antenna and chip is the main focus, since the tag impedance variation affects frequency and power performance. In addition, in IoT, backscatter communication [7] is also used because it does not require battery to establish communication, where the communication is established when a signal is sent by the transmitter to the tag [7], using a constant frequency. With this in mind, in backscatter circuits, the characterization necessity raises, to validate the performance to maintain communication. Consequently, the incident and reflected waves of these circuits should be read, to optimize them. The vector network analyzer (VNA) [25] can be used to characterize backscatter circuits, using traditional approaches. However, they operate using antennas, and for this reason, the backscatter circuits should be characterized to take into account antennas integration.

Thus, a characterization of high-order modulation backscatter systems is presented, where the optimum points of M-QAM modulation are extracted. By using this approach, the design of passive sensors with high-order backscatter is feasible, taking into account certain restrictions of error vector magnitude (EVM).

4.4.1 Characterization System

The main goal of backscatter modulation characterization is to measure the S_{11}. In this sense, two backscatter modulator circuits (the modulated high-bandwidth backscatter with WPT capabilities) were used working at 2.45 GHz. For this purpose, a characterization measurement system was developed in order to characterize the backscatter modulation circuits and to extract the optimum point for M-QAM modulations.

In order to measure the S_{11} in the RF port of backscatter modulation circuit, a DC sweep in the DC ports of the circuit should be done (to switch the transistors gate voltage). The first characterization setup is presented in Figure 4.23, where the calibration plane is before device under test (DUT) RF port. As can be seen,

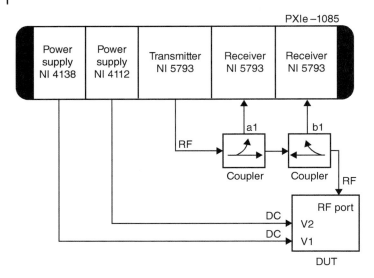

Figure 4.23 Block diagram of the implemented system. Source: Jordão et al. 2018 [26]. Reproduced with permission of IEEE.

the setup is composed by two power supplies, to generate the DC voltage in the backscatter circuits. In addition, the setup contains one transmitter (to generate the RF signal) and two receivers, which are connected to couplers in order to read the incident (a1) and reflected waves (b1). The first coupler is connected to the transmitter and the first receiver, to read the incident wave. The second coupler is connected to the first coupler and to the second receiver, to read the reflected wave. The laboratorial setup can be seen in Figure 4.24.

In the second characterization setup, a transmitting antenna and a receiving antenna were inserted in to the system in order to characterize the backscatter modulation in real scenario, as can be seen in Figure 4.25. In this second characterization approach, two calibration planes were tested in order to understand the importance of channel characterization in the backscatter modulation performance, as well as antennas. The laboratorial setup for this second characterization approach is shown in Figure 4.26.

It should be noted that, in both setups, a calibration procedure [27] was applied based on short open load (SOL) calibration and using a power meter to calibrate the magnitude.

To perform the measurement and characterization system, a LabVIEW application (Figure 4.27) was developed in order to control the setup and extract the optimum points for M-QAM modulation. First, in the LabVIEW application, the voltage values for each DC port are configured by selecting the start, the step, and

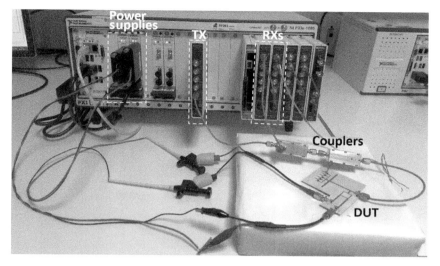

Figure 4.24 Backscatter modulator characterization system setup. Source: Jordäo et al. 2018 [26]. Reproduced with permission of IEEE.

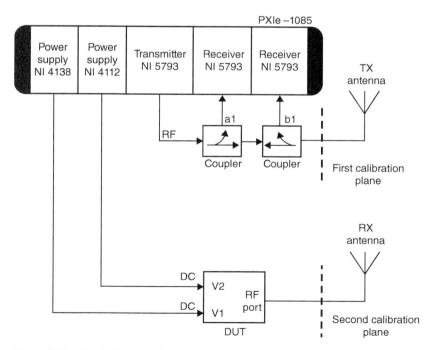

Figure 4.25 Block diagram of the implemented system with antennas.

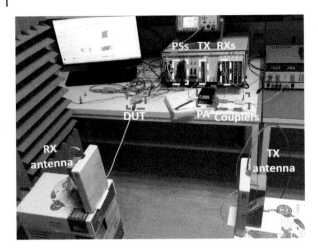

Figure 4.26 Backscatter modulator characterization system setup with antennas.

the stop values. Then, the short, open, load and power meter are connected to the RF port, and they are measured to apply the calibration. After calibration process is finished, the DUT (backscatter modulator) circuit is connected to the RF port, and the S_{11} is measured for each DC voltage combination. At this point, when all measurements are made, the optimum point to build the M-QAM constellations appears in the screen, as well as the respective voltage for this point. This procedure is summarized in Figure 4.28.

4.4.2 Measurements

The backscatter modulator circuit (the modulated high-bandwidth backscatter with WPT capabilities) was characterized, using the first setup approach. From this characterization process results the 16-QAM, 32-QAM, and the 64-QAM constellations, which can be seen in Figure 4.29. For these three constellations, EVM values of 1.098, 1.145, and 1.102 were obtained, respectively.

With the aim of study of the backscatter modulator performance in terms of power variation, several measurements were made. In Table 4.1 can be seen the EVM results deteriorate when the power decreases, as expected.

A second circuit working at the same frequency was characterized, using the second characterization setup (Figure 4.25), with the first and second calibration planes. By using the first calibration plane was concluded that it is not possible to demodulate because the constellation is very small. This fact happens because there is an impedance transformation of the antennas and channel interface. Thus, the S-parameters of the backscatter circuit are influenced.

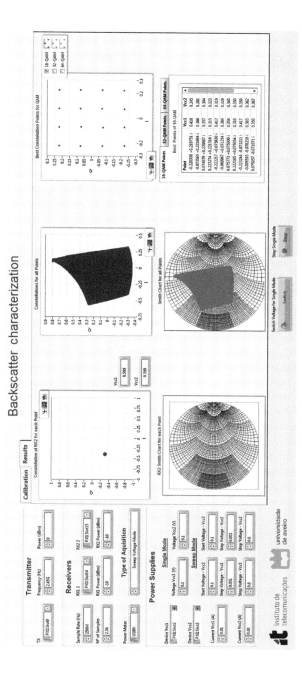

Figure 4.27 LabVIEW application. Source: Jordão et al. 2018 [26]. Reproduced with permission of IEEE.

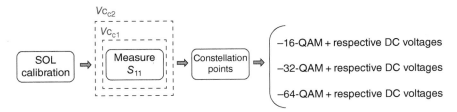

Figure 4.28 LabVIEW application flowchart. Source: Jordão et al. 2018 [26]. Reproduced with permission of IEEE.

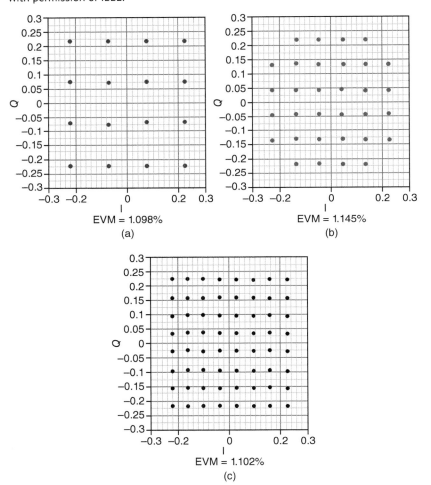

Figure 4.29 (a) 16-QAM, (b) 32-QAM, and (c) 64-QAM results and the respective EVM values. Source: Jordão et al. 2018 [26]. Reproduced with permission of IEEE.

Table 4.1 EVM results for different values of input power.

Power (dBm)	16-QAM	32-QAM	64-QAM
0	1.09	1.145	1.102
−10	1.89	1.97	2.12
−20	1.80	1.96	2.01
−30	1.93	2.70	2.30

Using first
calibration plane

Using second
calibration plane

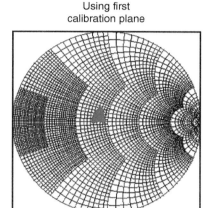 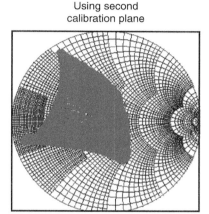

Figure 4.30 First calibration plane and second calibration plane.

Table 4.2 EVM results for different distances.

Distance (m)	16-QAM	32-QAM	64-QAM
0.5	2.45	3.70	3.89
1	5.42	5.52	5.82

However, in the second calibration plane, the antennas, as well as the channel, are accounted in the characterization process. The Smith Chart of these two different calibration planes position can be compared in Figure 4.30.

Using these second characterization planes, the same backscatter modulator circuit was placed at two different distances, to verify the calibration process at different distances. As can be seen in Table 4.2, the EVM results deteriorate when the distance is increasing, as might be expected.

References

1 Skolnik, M.I. (2001). *Introduction to Radar Systems*, 3e. New York: McGraw-Hill.

2 Sample, A. P., Braun, J., Parks, A., and Smith, J.R. (2011). Photovoltaic enhanced UHF RFID tag antennas for dual purpose energy harvesting. *2011 IEEE International Conference on RFID*, pp. 146–153.

3 Roundy, S., Wright, P.K., and Rabaey, J. (2003). A study of low level vibrations as a power source for wireless sensor nodes. *Computer Communications* 26 (11): 1131–1144.

4 Sample, A. and Smith, J.R. (2009). Experimental results with two wireless power transfer systems. In: *2009 IEEE Radio and Wireless Symposium*, 16–18. IEEE.

5 Thomas, S.J., Wheeler, E., Teizer, J., and Reynolds, M.S. (2012). Quadrature amplitude modulated backscatter in passive and semipassive UHF RFID systems. *IEEE Transactions on Microwave Theory and Techniques* 60 (4): 1175–1182.

6 Thomas, S. and Reynolds, M.S. (2010). QAM backscatter for passive UHF RFID tags. In: *2010 IEEE International Conference on RFID (IEEE RFID 2010)*, 210–214. IEEE.

7 Thomas, S.J. and Reynolds, M.S. (2012). A 96 Mbit/sec, 15.5 pJ/bit 16-QAM modulator for UHF backscatter communication. In: *2012 IEEE International Conference on RFID*, 185–190. IEEE.

8 Correia, R., Borges Carvalho, N., and Kawasaki, S. (2016). Continuously power delivering for passive backscatter wireless sensor networks. *IEEE Transactions on Microwave Theory and Techniques* 64 (11): 3723–3731.

9 Correia, R., Carvalho, N.B., and Kawasaki, S. (2016). Continuously power delivering for passive backscatter wireless sensor networks. *IEEE Transactions on Microwave Theory and Techniques*: 1–9.

10 Pereira, F., Correia, R., and Carvalho, N. (2017). Passive sensors for long duration internet of things networks. *Sensors* 17 (10): 2268.

11 Correia, R., Boaventura, A., and Borges Carvalho, N. (2017). Quadrature amplitude backscatter modulator for passive wireless sensors in IoT applications. *IEEE Transactions on Microwave Theory and Techniques* 65 (4): 1103–1110.

12 Correia, R. and Carvalho, N.B. (2017). Ultrafast backscatter modulator with low-power consumption and wireless power transmission capabilities. *IEEE Microwave and Wireless Components Letters* 27 (12): 1152–1154.

13 Silicon Labs (2013). The Evolution of Wireless Sensor Networks. White Papers.

14 Varshney, A., Harms, O., Pérez-Penichet, C. et al. (2017). LoRea: A Backscatter Architecture that Achieves a Long Communication Range.

15 Correia, R., de Carvalho, N.B., Fukuday, G. et al. (2015). Backscatter wireless sensor network with WPT Capabilities. *2015 IEEE MTT-S International Microwave Symposium*, pp. 1–4.

16 Belo, D., Correia, R., Pereira, F., and De Carvalho, N.B. (2017). Dual band wireless power and data transfer for space-based sensors. *2017 Topical Workshop on Internet of Space (TWIOS)*, pp. 1–4.

17 Texas Instruments (2012). MSP430F21x2 Mixed Signal Microcontroller.

18 Belo, D., Ribeiro, D.C., Pinho, P., and Carvalho, N.B. (2019). A selective, tracking, and power adaptive far-field wireless power transfer system. *IEEE Transactions on Microwave Theory and Techniques* 67 (9): 3856–3866.

19 Belo, D., Ribeiro, D.C., Pinho, P., and Carvalho, N.B.. (2018). A low complexity and accurate battery-less trackable device. *2018 IEEE MTT-S International Microwave Symposium (IMS)*, Philadelphia, PA (10–15 June 2018), pp. 1269–1271.

20 Belo, D., Correia, R., Pinho, P., and Carvalho, N.B. (2017). Enabling a constant and efficient flow of wireless energy for IoT sensors *2017 IEEE MTT-S International Microwave Symposium (IMS)*, Honolulu, HI (4–9 June 2017), pp. 1342–1344.

21 Balanis, C. (1997). *Antenna Theory, Analysis and Design*, 2e. New York: Wiley.

22 Boaventura, A., Santos, J., Oliveira, A., and Carvalho, N.B. (2016). Perfect isolation: dealing with self-jamming in passive RFID systems. *IEEE Microwave Magazine* 17 (11): 20–39.

23 Razavi, B. and Behzad, R. (1998). *RF Microelectronics*. Prentice Hall.

24 Nikitin, P.V. and Ra, K.V.S. (2007). Correction in the contribution to the AMTA Corner. Theory and measurement of backscattering from RFID tags. [Dec 06 212-218]. *IEEE Antennas and Propagation Magazine* 49 (2): 81–81.

25 Couraud, B., Deleruyelle, T., Kussener, E., and Vauché, R. (2018). Real-time impedance characterization method for rfid-type backscatter communication devices. *IEEE Transactions on Instrumentation and Measurement* 67 (2): 288–295.

26 Jordão, M., Correia, R., and Carvalho, N.B. (2018). High order modulation backscatter systems characterization. *2018 IEEE Topical Conference on Wireless Sensors and Sensor Networks (WiSNet)*. Anaheim, CA.: 44–46. https://doi.org/10.1109/WISNET.2018.8311560.

27 Heuermann, H. (2008). Calibration of network analyzer without a thru connection for nonlinear and multiport measurements. *IEEE Transactions on Microwave Theory and Techniques* 56: 2505–2510.

5

Ambient FM Backscattering Low-Cost and Low-Power Wireless RFID Applications

Spyridon N. Daskalakis[1], Ricardo Correia[2], John Kimionis[3], George Goussetis[1], Manos M. Tentzeris[3], Nuno B. Carvalho[2], and Apostolos Georgiadis[1]

[1] School of Engineering and Physical Sciences, Heriot-Watt University, EH144AS, Edinburgh, Scotland
[2] Departamento de Electrnica, Telecomunicaes e Informtica, Instituto de Telecomunicaes, Universidade de Aveiro, 3810-193, Aveiro, Portugal
[3] School of Electrical and Computer Engineering, Georgia Institute of Technology, Atlanta, GA, 30332-250, USA

5.1 Introduction

Internet-of-Things (IoT) has become the trend for networking our everyday devices to automate our life and make it easier. The most important challenge for IoT sector is the minimization of the cost and energy dissipation of the sensors nodes. Keeping the number of energy-constrained IoT sensors active with low-cost designs is a real challenge. Commercial radio components that used in IoT devices use to have power-hungry radio frequency (RF) chains including oscillators, mixers, and digital-to-analog converters (DACs) resulting in large power consumption and thus significant limitations of the battery life. One promising approach to alleviate these issues is backscatter communication technique [1] that allows sensor nodes/tags to transmit data by reflecting and modulating an incident RF signal [2]. Communication using backscatter principles has been widely deployed in the application of radio frequency identification (RFID) for passive tags. The RF front-end part of the tags consists of only one RF transistor or one RF switch. In commercial RFID applications, the tags are battery-free and can operate using only RF power transmitted from a RFID reader resulting in communication ranges up to several meters [3, 4]. The tags operate in monostatic architecture where the continuous wave (CW) emitter and the reader are in the same box. Alternatively, semi-passive tags (energy assisted) are built in bistatic architectures where the CW emitter and the reader are not co-located [5].

Wireless Power Transmission for Sustainable Electronics: COST WiPE - IC1301,
First Edition. Edited by Nuno Borges Carvalho and Apostolos Georgiadis.
© 2020 John Wiley & Sons, Inc. Published 2020 by John Wiley & Sons, Inc.

The tags are supplied by small batteries, and longer communication distances can be achieved. For example, in a recent work [6], an effective communication was observed over a tag-to-reader distance more than 250 m. In [7], a FPGA based tag can create up to 11 Mbps Wi-Fi and ZigBee compatible signals by backscattering Bluetooth transmissions. Binary amplitude shift keying (ASK) or phase shift keying (PSK) modulation is commonly used for the communication between the tag and reader, such that information is encoded using two states of the amplitude or the phase of the reflected CW. For example, in the WISP platform [4], the electronic product code (EPC) protocol employs 2-ASK modulation to encode the bits 1 and 0 with long and short gaps in RF power, respectively.

Recent works [8, 9] have shown that backscatter communication can be extended to include higher order modulation schemes, such as four-state quadrature amplitude modulation (4-QAM), 16-QAM, and four-pulse amplitude modulation (4-PAM). In [8], authors used a 16-to-1 Mux with SP4T RF switches to modulate the antenna impedance between 16 impedance states. In [9], a novel circuit including a Wilkinson power divider and two transistors was presented. Schemes of M-state quadrature amplitude modulation (M-QAM) or M-pulse amplitude modulation (M-PAM) can be effectively implemented as each transistor can be switched with different voltage levels to achieve different reflection coefficient values. In [10], 4-frequency-shift keying (4-FSK) was used to transmit data over the ambient frequency modulated (FM) signals.

Ambient backscattering is another idea based on the bistatic backscatter philosophy and could constitute a very promising novel approach for extremely low-power and low-cost communication schemes. It utilizes ambient signals for backscattering, and the communication scheme is simplified since it requires only a receiver eliminating the need for a dedicated CW emitter. Ambient backscattering devices, such as RFID tags, can communicate with a reader by backscattering ambient RF signals that are available from multiple sources, such as mobile communications, FM-AM radio [10], television [11], and Wi-Fi [12]. These types of signals are typically widely available in urban areas indoors and outdoors during day and night. In [11] is described a deployment where two battery free tags communicate via ambient backscatter TV signals. In [12], a Wi-Fi backscatter system was designed to connect battery-free devices with off-the-shelf Wi-Fi devices.

A full-duplex Wi-Fi ambient communication system was also introduced in [13]. An access point (AP) can cooperate with backscatter IoT sensors with high data rate. The use of ambient RF signals as the only source of both the CW carrier and the tag power is an extremely energy-efficient communication technique compared to the general backscattering technique. In ambient backscatter communication, there are issues with the signal detection; thus, we do not have a pure CW signal. The receivers adopt the differential encoding (FM0, Manchester, and

Miller encoding) to eliminate the necessity of channel estimation. In [12–15] is presented a fully developed theory on signal processing and performance analysis for ambient backscatter systems. A practical transmission model for an ambient backscatter system is presented in [16] where they assume that the tag sends some low-rate data to the reader using an ambient RF signal source. Reference [16] provides fundamental studies of noncoherent symbol detection when all channel state information of the system is unknown. In [16], the tag sends low bit rate messages to a reader using an ambient RF signal source. It provides all the fundamental studies of noncoherent symbol detection when all channel state information of the system is unknown.

In our conference work of [17], preliminary results for a wireless sensor node prototype for agricultural monitoring were presented. The sensor node measures the temperature difference between the leaf and the air to estimate the water stress of a plant [18]. The tag modulates and reflects a fraction of the ambient FM station signals back to the reader. The concept of [17] is described in Figure 5.1, and more information about the tag and the sensor board will be explained in the following. In this work, we propose an improved version of this system for generic environmental monitoring applications by designing two new sensor nodes/tags and two improved receiver algorithms. In addition to the receiver implementation, we provide additional details about the tag circuitry, a theoretical tag-receiver framework for the operation of the ambient backscatter system and a series of packet error rate (PER) and bit error rate (BER) measurements in a proof-of-concept indoor

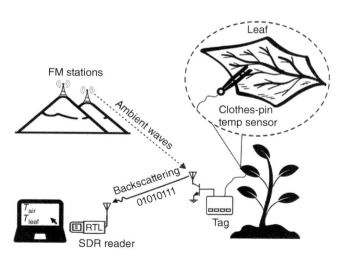

Figure 5.1 Deployment of ambient backscattering in smart agriculture applications. The differential temperature ($T_{leaf} - T_{air}$) is measured by the tag-sensor and is transmitted back to an SDR receiver.

environment. In our second tag design, we introduce for the first time 4-PAM for backscattering over ambient FM signals.

Each tag consists of a microcontroller (MCU) and an RF communication front-end. The tag reads the information from the sensors and generates pulses that control an RF switch or RF transistor. For the first tag, On-Off keying (OOK) with FM0 encoding [19] was selected for the binary modulation, similarly to conventional passive RFID tags. For the second design, the 4-PAM was selected mainly due to its hardware simplicity and low power consumption, as it can be implemented only with one RF transistor and an antenna. Furthermore, a real-time receiver was implemented using an ultra-low-cost software defined radio (SDR). The prototypes were demonstrated in the lab using an existing FM transmitter broadcasting 34 km away from the tags. Operation over a 5 m tag-to-reader distance was achieved by backscattering sensor data at 0.5, 1, and 2.5 kbps bit rates for the binary modulation. For the 4-PAM modulation, a communication range over 2 m was demonstrated for a bit rate 328 bps.

Our work is different from [11], which first proposed ambient backscattering. In [11], it used ambient digitally modulated television signals (DTV), whereas the system proposed in this work uses analog FM signals. Also, a moderately expensive software defined USRP-N210 radio used in [11] to receive and decode the signals, whereas in our work, a low-cost Realtek (RTL) SDR was used. In [10] is also proposed ambient backscattering using FM signals but only for 2-FSK and high order modulated 4-FSK signals. In addition, an arbitrary waveform generator was used in [10] to generate the ambient FM signals contrary to signals from existing broadcast FM stations in this paper. Our tag with high-order modulation is a different approach from [8, 9] because ambient analog FM signals are used as the carrier instead of a pure transmitted CW signal.

5.2 Ambient Backscattering

A typical backscatter system consists of three devices: a backscatter node (i.e. a tag), a reader, and a CW emitter. The tag receives a CW carrier signal with frequency F_c and scatters a fraction of it back to the reader as shown in Figure 5.2. It adds the sensor information on top of the carrier by appropriately changing the load connected to its antenna terminals according to [20]: $\Gamma_i = (Z_i - Z_a^*)/(Z_i + Z_a)$ with Z_i and Z_a denoting the load and the tag antenna impedance (50 Ω). For binary modulation, the reflected signal is modulated by switching the load between two discrete values Z_1 and Z_2 (Figure 5.3) effectively resulting in two reflection coefficient values, (Γ_1 and Γ_2) over time (Figure 5.4a). The 180° difference between the two load values (Figure 5.2, right) is necessary for maximization of backscatter performance. The 4-PAM scheme is based on a four distinct antenna-load values

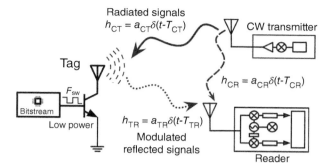

Figure 5.2 Bistatic backscatter principle. The emitter transmits a carrier signal, and the tag reflects a small amount of the approaching signal back to the reader.

Figure 5.3 Backscatter radio principle: an RF transistor alternates the termination loads Z_i of the antenna corresponding to different reflection coefficients Γ_i. Four reflection coefficients ($n = 4$) could create a four-pulse amplitude modulation (4-PAM), and two reflection coefficients ($n = 2$) could create a binary modulation (OOK).

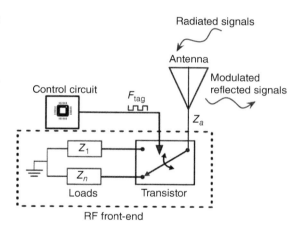

with reflection coefficients Γ_i with $i = 1, 2, 3, 4$ (Figure 5.4b). An RF transistor is used in this case to change the antenna termination load value. Each state can create a specific impedance for each transmitted symbol effectively introducing modulation through the change of each Γ_i over time. The performance of 4-PAM is maximized when Γ_i lie on the same line in Smith Chart with equal distances between different states.

The SDR reader captures the reflected signal at a frequency $f_c + \Delta f$ with an additional phase φ and then removes the high frequency components. The Δf is the carrier frequency offset (CFO) between the emitter and the reader. The received signal can be expressed in the following complex baseband form [5]:

$$y_r(t) = n(t) + \frac{A_c}{2}e^{-2\pi\Delta ft}[a_{CR}e^{-2j\varphi_{CR}} + sa_{CT}a_{TR}e^{-2j\varphi_{CTR}}\Gamma(t - \tau_{TR})] \tag{5.1}$$

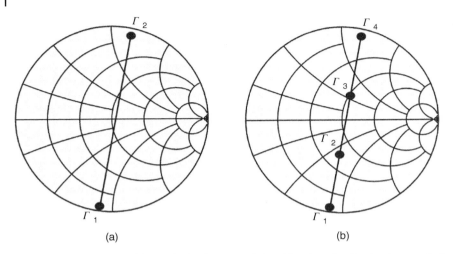

Figure 5.4 The tag modulates the backscattered signal by changing the load connected to its antenna terminals resulting in a Γ_i change between two or four values (states). (a) The binary modulation requires two-state antenna Γ_i parameters on a Smith chart. (b) Smith chart with required 4-PAM antenna Γ_i parameters.

where A_c is the amplitude of the carrier. The τ_{TR} is the time delay constant of the tag-reader channel. Term s is related to the tag scattering efficiency and tag antenna gain at a given direction. The term $a_{CR}e^{-2j\varphi_{CR}}$ defines the component which depends on the emitter-to-reader channel (h_{CR} in Figure 5.2). The tag signal is a direct function of Γ over time, and the term $sa_{CT}a_{TR}e^{-2j\varphi_{CTR}}$ scales and rotates the modulated part of the tag signal. This term depends on the transmitter-to-tag and tag-to-reader channel parameters (h_{CT} and h_{TR} in Figure 5.2). Finally, $n(t)$ is the complex thermal Gaussian noise at the receiver.

5.2.1 Ambient FM Backscattering

The FM broadcasting technique was first utilized in 1940 radio-audio transmissions, and today FM radio broadcasts take place between frequencies of 88–108 MHz with a bandwidth of 200 kHz per channel. Each FM station uses FM to transmit the audio signals and the information signals varying the frequency of a carrier wave accordingly. The FM output signal of a station is given by the following equation [21]:

$$x_{FM}(t) = A_c \cos\left(2\pi f_c t + 2\pi K_{VCO}\Delta f \int_0^t m(x)dx\right) \tag{5.2}$$

where $m(x)$ is the baseband message signal, and Δf is the frequency deviation which is equal to the maximum frequency shift from f_c while K_{VCO} is the gain of

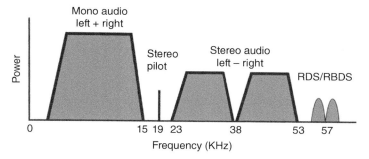

Figure 5.5 Baseband spectrum of an FM audio station. The signal contains left (L) and right (R) channel information for monophonic and stereo reception.

the transmitter voltage-controlled oscillator (VCO). It is difficult to analyze the properties of $x_{FM}(t)$ in theory due to its nonlinear dependence to the $m(x)$. The baseband message signal of a typical FM station is as shown in Figure 5.5. From the picture, we can observe that there are "stereo left" and "stereo right" channels as well as the RDS(t) signal, which is the signal of the radio data system (RDS) and radio broadcast data system (RBDS). The 0–15 kHz part of the signal consists of the left and right channel information [(Left) + (Right)] for monophonic sound. Stereophonic sound is the result of the amplitude modulation of the [(Left) − (Right)] message onto a suppressed 38 kHz subcarrier in the 23–53 kHz region of spectrum. Furthermore, there is a 19 kHz pilot tone to enable receivers to recognize the two stereo channels. Today, FM radio signals also include a 57 kHz subcarrier that carries RDS and RBDS data.

In the case of typical ambient FM backscatter systems, the incident "CW carrier" to the tag antenna is the signal $x_{FM}(t)$. The SDR receiver receives the superposition of this signal and the backscattered tag signal. Following the same procedure described in [5], but using a FM modulated carrier instead of a CW signal one, the following complex baseband signal is obtained at the receiver:

$$y_{amb}(t) = n(t) + \frac{A_c}{2}e^{-2\pi\Delta ft}[a_{CR}e^{-2j\varphi_{CR}} + sa_{CT}a_{TR}e^{-2j\varphi_{CTR}}e^{-jM(t-\tau_{TR})}\Gamma(t - \tau_{TR})]$$

(5.3)

$$M(t) = 2\pi K_{VCO}\Delta f \int_o^t m(x)dx$$

(5.4)

The received signal $y_{amb}(t)$ contains the desired information Γ but also the carrier, the FM modulation, and the frequency offset. In the next Sections 5.2.2 and 5.2.3, we show experimentally that it is possible to successfully decode the signal provided there is a sufficiently high signal-to-noise ratio (SNR).

5.2.2 Binary Modulation Tag

The main digital part of the first proposed tag is based on a 16-bit MCU development board MSPEXP430FR5969 [22] (Figure 5.6). The development board is powered from a 0.1 F supercapacitor. The tag also includes a real-time clock (RTC) to wake up the MCU from the "sleep" operation mode, where the current consumption of the board is 0.02 μA. The MCU generates 50% duty cycle pulses that control the RF switch, thus generating an OOK modulated backscattered signal. The OOK modulation is described in more detail in the following section. The MCU was programmed at 1 MHz clock speed using the internal local oscillator. The current consumption at 1 MHz was 126 μA at 2.3 V (290 μW). The MCU has a 16 channel, 12 bit analog-to-digital converter (ADC), which was used to read analog output signals from sensors. In this work, the tag is programmed to read four analog inputs and the voltage level of the super capacitor. When a tag wants to communicate with the reader, it sends a packet that contains the information of only one sensor each time.

In [17] only two ADC inputs for two high precision, analog temperature sensors were used. The sensor board consists of two high precision, (0.1 °C) analog temperature sensors LMT70A [23] (IDD: 10 μA at 2.3 V) in a "clothes-pin" prototype to be easily fixed on the leaf (Figure 5.6A). The first sensor on top measures the air temperature (T_{air}), and the second under the leaf surface measures the leaf temperature (T_{leaf}).

The backscatter communication of the tag is achieved with a separate RF front-end board. It consists of a 1.5 m wire dipole antenna to resonate within the

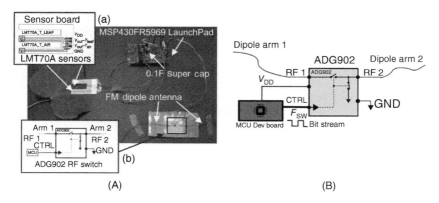

(A)   (B)

Figure 5.6 The first proposed tag prototype comprises of an MSP430 development board connected with RF front-end board. The RF front-end consists of the ADG902 RF switch and was fabricated using inkjet printing technology on a paper substrate. An MCU digital output pin was connected with the control signal of the RF switch (B). The operation power of RF front-end was supplied by the MCU development board, and the hole system was supplied by an embedded super capacitor for duty cycle operation.

FM band (95 MHz) and a single-pole, single-throw (SPST) RF switch ADG902 by analog devices. The circuit schematic of the front-end is provided in Figure 5.6B, while the fabricated prototype is shown in Figure 5.6A. The switch element varies the antenna load between two impedance values and is selected due to its low insertion loss and high off isolation. The RF switch is a reflected CMOS switch with high off-port voltage standing wave ratio (VSWR) and consumes less than 1 µA at 2.75 V according to the product manual [24]. It is driven by a digital output of the MCU as is shown in Figure 5.6. The front-end printed circuit board (PCB) was fabricated using inkjet printing technology on a paper substrate. The characteristics of the substrate were Er = 2.9, tan δ = 0.045, and substrate height 210 µm. The traces were printed with conductive silver nanoparticle (SNP) ink, and conductive epoxy deposition was used to attach the switch to the substrate. In order to minimize the average power consumption, a duty cycle operation was programmed where the tag was active only for a desired minimum period. The duty cycle operation was set using the RTC and the sleep mode of the MCU. A future challenge for the tag is to employ RF harvester in combination with a solar cell for powering as it is shown for example in [25, 26].

5.2.3 4-PAM Tag

The second sensor node uses 4-PAM high-order modulation for backscattering over ambient FM signals. The proof-of-concept sensor node/tag consists of a different MCU and a new RF front-end as depicted in the block diagram of Figure 5.7. The 8-bit PIC16LF1459 MCU from Microchip was selected with a power consumption of only 25 µA/MHz at 1.8 V [27]. The 31 kHz low-power internal oscillator was utilized as a clock source with an ultra-low-power consumption. The tag can collect data from sensors using the embedded 10-bit ADC, and it can drive the RF front-end through a 5-bit DAC. The external voltage regulator XC6504 was used to supply the tag with a stable reference voltage (V_{ref}) 1.8 V. The DAC switches the gate of the RF front-end transistor with 32 distinct voltage levels to change the antenna load impedance according to the chosen modulation. This tag design does not focus on sensing aspect but only on the novel telecommunication aspect of the system. The tag was powered by the flexible solar panel, SP3-37 provided by PowerFilm Inc. [28]. The solar panel charges a 220 µF tantalum capacitor instead of a battery through a low voltage-drop Schottky diode.

The RF front-end consists of an RF transistor ATF52189 [29] and a monopole antenna SRH788 for FM signal reception (Figure 5.8a). This part is responsible for the modulation over the backscattered FM signals with the main challenge being the appropriate change of the drain impedance by varying the voltage at the gate only between 0 and 0.6 V. The RF front-end circuit was optimized using the advanced design system (ADS) from Keysight to perform four different discrete

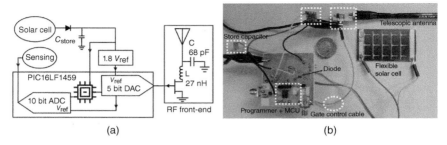

(a)   (b)

Figure 5.7 (a) The proof-of-concept 4-PAM tag consists of an MCU and an RF front-end (transistor and antenna). An ADC collects data from integrated sensors, and a DAC generates the appropriate gate voltages of the 4-PAM modulation. (b) The fabricated tag prototype with the RF front-end board. The tag is powered by a solar panel.

Γ_i values at 95.8 MHz. The matching network between the transistor and the antenna was composed by a capacitor and an inductor as depicted in Figure 5.7. The Large-Signal S-Parameter simulation was used to perform the backscatter modulation, and the components were optimized at 68 pF and 27 nH to maximize the distance between the consecutive Γ_i values. The RF front-end prototype was fabricated on Astra MT77 substrate with thickness 0.762 mm, Er = 3.0, and tan δ = 0.0017. The fabricated board was measured using a vector network analyzer (VNA) with $P_{in} = -20$ dBm over FM frequencies 87.5–108 MHz. The CW signal of VNA was used for the measurements, and the Γ_i that corresponds to the four levels for 4-PAM is presented in Figure 5.8b. The frequency sweep demonstrates the fact that each state becomes an "arc" on the Smith Chart, and a set of the four states (line) rotates clockwise as the frequency increases. By

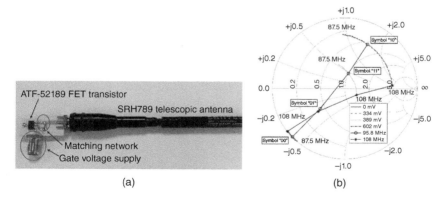

(a)   (b)

Figure 5.8 (a) Fabricated RF front-end prototype board. (b) Smith chart with measured reflection coefficient values for four different voltage levels at the gate of transistor. The P_{in} was fixed at −20 dBm for frequencies 87.5–108 MHz.

generating four voltages at the DAC (0, 334, 389, and 602 mV, respectively), the Γ_i symbols for the fixed frequency 95.8 and 108 MHz are depicted in Figure 5.8b.

5.2.4 Binary Telecommunication Protocol

The first tag uses ASK modulation to transmit its data via backscattering. More specifically, by changing the RF switch states between "on" and "off" and backscattering the ambient FM broadband signals, a binary ASK-modulated signal of OOK type can be created described by $y_{amb}(t)$ signal. Using OOK modulation, the information containing received tag signal of $y_{amb}(t)$ can be expressed as [5]:

$$\Gamma(t - \tau_{TR}) = \sum_{n=0}^{N-1} x_n \Pi[\tau - nT_{symbol} - \tau_{TR}] \tag{5.5}$$

where $x_n \in \{-1, 1\}$ are the N transmitted symbols, and $\Pi(t)$ is the pulse (symbol) with duration T_{symbol}. In addition to the OOK modulation, the low-power consuming FM0 technique is utilized to encode the sensor data. For binary OOK, x_n would be the bits, and for FM0-coded OOK, x_n are the binary symbols. In FM0 encoding, there is an inversion of the phase at every bit boundary (at the beginning and at the end of every bit), and additionally bit "0" has an additional phase inversion in the middle (Figure 5.9a). Each bit includes two symbols, as shown in Figure 5.9. The duration of a bit and of a symbol are denoted as T_{bit} and T_{symbol}, respectively. The data bit rate is $1/T_{bit}$ bits per second (bps). The FM0 encoding always ends with a dummy "1" bit to detect easily the end of the bitstream. In the case that the received backscatter waveform finishes with a "LOW", it would be indistinguishable from receiving the reader's CW only (i.e. no packet transmission). The tag is programmed to send the data in packets to the reader, and the reader tries to receive and decode them. The length of each packet is fixed. Figure 5.10a shows a typical packet format. The packet has the length of 26 bits and begins with the preamble bits. After that follow the "Tag ID" bits, the "Sensor ID" bits, and finally the "Sensor Data" bits. The preamble is useful for bit-level synchronization at the

Figure 5.9 (a) FM0 encoding, the boundaries of the bits must always be different. (b) FM0 decoding technique, after shifting by T_{symbol}, receiver need to detect only two possible pulse shapes (line square or dash line square).

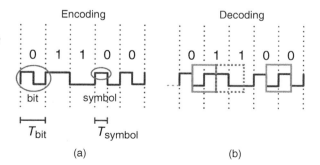

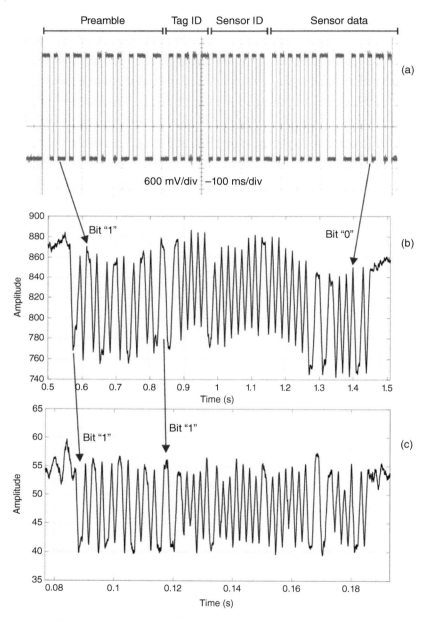

Figure 5.10 Time domain backscatter packet signals, (a) oscilloscope measurement of transmitted rectangular pulses, (b) received packet pulses at $T_{symbol} = 10\,ms$, and (c) received packet pulses at $T_{symbol} = 1\,ms$.

receiver and was fixed to be 1010101111 (10 bits) in our proof-of-concept tests. The "Tag ID" (2 bits) is utilized in the case of simultaneous multiple tag utilization. As mentioned previously, the tag can support up to four sensors, and therefore, the "Sensor ID" (2 bits) is used to identify the sensor that the data are coming from.

5.2.5   4-PAM Telecommunication Protocol

In the 4-PAM, there are four symbols, and each symbol corresponds to a pair of 2 bits. Each bit duration is denoted as T_{bit}, and the data bit rate is $2/T_{bit}$ bps. According to 4-PAM, it is possible to transmit 2 bits with each symbol/pulse, for example, by associating the amplitudes of $-3, -1, +1, +3$, with 4-bit choices 00, 01, 11, and 10. The symbols $\pm1, \pm3$ are shown in Figure 5.11, and the bit representation of the symbols is Gray coded [30]. To transmit a digital stream, it must be converted into an analog signal. After conversion of the bits into symbols, the analog form of a 4-PAM modulation signal can be expressed with the $\Gamma(t - \tau_{TR})$ signal from binary modulation, but in this case, we have $x_n \, \varepsilon \, \{-3, -1, +1, +3\}$. The four distinct pulse amplitudes ("symbols") with period T_{symbol} are used to convey the information. In this case the data rate is doubled, and 4-PAM is twice bandwidth-efficient as conventional binary OOK modulation. This scheme requires a high SNR because the extra two amplitude levels reduce the level spacing by a factor of 3, therefore making the modulation more susceptible to noise than the binary modulation.

5.2.6   Receiver

In this work, the low cost (US $22) RTL SDR (Nooelec NESDR SMArt) was used as receiver (Figure 5.12). It is an improved version of RTL SDR dongle that was

Figure 5.11   The 4-PAM symbols. Three thresholds are calculated for the decision.

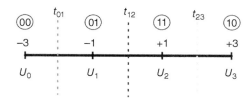

Figure 5.12   USB software defined radio receiver and telescopic monopole antenna for FM signals reception.

used in [6, 17], and it is based on the same RTL2832U Demodulator/USB interface IC and R820T2 tuner. The new version provides a better oscillator, temperature stability, and antenna improvements compared to the old one. It comes with an ultra-low phase noise 0.5 PPM temperature compensated crystal oscillator. Power consumption has been reduced by an average of 10 mA according to manufacturer [31]. A custom heatsink is affixed to the primary PCB for temperature improvement, and it comes with a low-loss RG58 feed cable and SMA antenna connector for better signal reception. In general, the RTL SDR has a tuning frequency range from 24 to 1850 MHz, and it can support sampling rates up to 2.8 MS/s, and its noise figure is specified to be about 3.5 dB. The SDR downconverts the received RF signal to baseband and sends in-phase (*I*) and quadrature (*Q*) samples to the PC through the USB interface, while it is connected to an improved telescopic monopole antenna to receive the FM signals.

5.2.7 Software Binary Receiver

A real-time receiver with digital signal processing was implemented to read the backscattered information sent from the binary modulation tag. The steps of the algorithm are briefly shown in Figure 5.13 (A), and the software that was used was MATLAB and GNU radio framework. The GNU radio provides the *I* and *Q* samples to MATLAB through a First in, first out (FIFO) file, and the samples are interleaved for further processing. The received digitized signal after sampling with a sampling period T_s can be written as:

$$y_r[k] = y_{amb}(kT_s + \tau_{TR}) \tag{5.6}$$

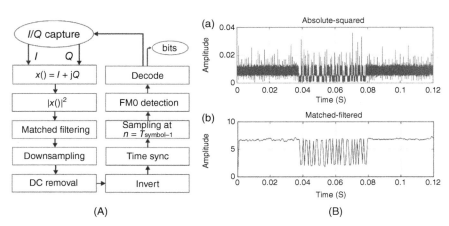

(A)   (B)

Figure 5.13 (A) Flow chart of the real-time receiver algorithm for binary modulation. (B) Received signal including a data packet. (a) Squared absolute value signal. (b) Received signal after matched filtering for a symbol period, $T_{symbol} = 1$ ms.

The term $y_r[k]$ includes the noise signal and a signal without noise that consists of a DC component, a modulated component, and the ambient FM signal utilized for the backscattering. The algorithm collects and processes the data in a window with duration: 3*packet duration. The first step of the signal processing algorithm is the CFO correction. In our case, CFO is the frequency difference between the FM transmitter and SDR reader, and if not properly removed, it causes a performance loss at the receiver. To eliminate this term without using a priori CFO estimation and correction algorithm, the absolute value $|y_r(t)|^2$ was taken [30].

A matched filter was then applied to the samples to filter out noise and interference terms and maximize the SNR, consisting of a square pulse with duration T_{symbol}. Figure 5.13 (B, b) depicts the received packet of Figure 5.10a) after absolute square operation. The same packet after matched filtering is shown in Figure 5.13 (B, b). Also Figure 5.10b,c show the received packet after the matching filtering for $T_{symbol} = 10$ ms and $T_{symbol} = 1$ ms, respectively. By comparing Figure 5.10a–c, one can see that the reader can correctly receive the transmitted packet at the correct sampling instants. In FM0 decoding, the reader distinguishes the "on" and "off" stages by observing voltage changes in the envelope magnitude. Measuring the duration between two changes, bit decoding can be achieved as it is explained in the following in more detail. Matched filtering was followed by downsampling by a factor of 10 to reduce the computational cost of the subsequent operations without compromising the detection quality.

The DC offset of the received window was estimated by averaging some samples when the tag is not transmitting data. The DC offset was removed by subtracting the aforementioned estimate from all the values within the receive window. The outcome of this step can be an upright or an inverted waveform. Upright or inverted waveforms may result due to the channel propagation characteristics. If an inverted waveform is detected after the DC offset removal, it is flipped so that only upright waveforms are forwarded to the synchronization block. The received signal must be symbol-synchronized to determine when the packet starts. To find the starting sample (I_{start}) of the packet, cross-correlation with the known preamble symbol sequence (11010010110100110011) was used. The starting point of the packet is defined as the point that the maximization for the cross-correlation happens.

The OOK modulation is known as biphase-space modulation [32], and in FM0 encoding, four possible waveforms are transmitted corresponding to a two-dimensional biorthogonal constellation. However, if one observes the FM0 signal shifted by half a symbol, only two possible waveforms exist, which are the ones of symbol "0". These two waveforms are depicted in Figure 5.9b with a solid line square and a dash line square and correspond to a one-dimensional antipodal constellation, which is easier to decode. The received bits can be determined by comparing two neighboring symbols. To begin decoding, the signal is shifted to sample $I_{start} + P + T_{symbol}$, where P is the length of preamble. The two possible orthogonal pulse waveforms can be received, as shown in [32] and used in [33]. With this observation, the algorithm must easily decode two adjacent received symbols to detect a whole bit. This method gives a gain of 3 dB compared to maximum likelihood symbol-by-symbol detection [34]. The two orthogonal waveforms are as follows:

$$D_1[k] = \begin{cases} +1 & \text{if } a \\ -1 & \text{if } b \end{cases} \quad \text{and} \quad D_2[k] = -D_1[k] \tag{5.7}$$

with $a: \frac{T_{symbol}/T_s}{2} < k < \frac{T_{symbol}}{T_s}$ and $b: 0 < k < \frac{T_{symbol}/T_s}{2}$

The shifted signal is correlated with $D_1[k]$ and $D_2[k]$, and it is possible to determine which bit has been sent according to [35]:

$$S_k = \begin{cases} 1 & \text{if } \sum_{i=1}^{N_s} y_{sh}[i]D_1[k] > \sum_{i=1}^{N_s} y_{sh}[i]D_2[k] \\ 0 & \text{elsewhere} \end{cases} \tag{5.8}$$

and $y_{sh}[i]$ is the shifted version of waveform $y_r[k]$. The results from the aforementioned calculation were stored in a vector L, and the estimated bit $a_k + 1$ that was sent is determined by the following equation:

$$S_k = \begin{cases} 0 & \text{if } L_k = L_{k+1} \\ 1 & \text{elsewhere} \end{cases} \tag{5.9}$$

It is noticed that the first waveform derived by this decoding procedure is from the last preamble symbol. The following waveforms will be either D_1 or D_2. This means that if the first waveform is D_1 and the second is D_2 and vice versa, the bit "1" was sent, otherwise the bit "0" was transmitted.

5.2.8 Software 4-PAM Receiver

A modified version of previous algorithm was used for our high-order modulation, real-time receiver algorithm. The $y_r[k]$ signal includes the modulated useful information and a component based on the FM message. The absolute squared value of $y_r[k]$ was taken, and a matched filtering was utilized to maximize the SNR. The $|y_r(t)|^2$ signal is CFO corrected as it is explained in the previous Section 5.2.7. The matched filter is a square pulse signal with T_{symbol} duration, and Figure 5.14a shows an example of a received packet in time domain after the absolute square operation. The packet after the low-pass filtering is depicted in Figure 5.14b.

5.2.9 Experimental and Measurement Results

The proposed system with the two tags was tested indoors at the Heriot-Watt electromagnetics lab, choosing the most powerful FM station as the ambient RF source

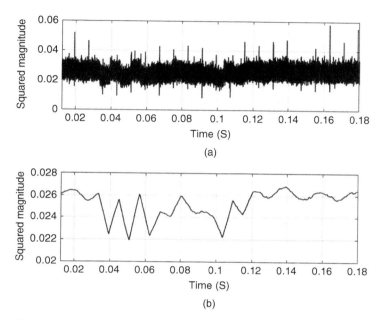

Figure 5.14 Received packet signal. (a) Signal after squared absolute operation. (b) Signal after matched filtering for $T_{symbol} = 5.4$ ms.

to use in backscattering. Our receiver was tuned to BBC 95.8 MHz station with 1 MS/s sampling rate, and the station is located 34.5 km away at the "Black Hill" location between the town of Edinburgh and Glasgow as depicted in the map of Figure 5.15a. The transmission power of the station is 250 kW, and the power of

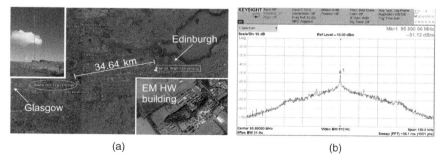

Figure 5.15 (a) FM radio deployment near Edinburgh (UK). The BBC 95.8 MHz station was selected for experimentation. The FM transmitter was 34.5 km away from our lab, and its transmission power was 250 kW. (b) The power of the FM station carrier signal was measured next to the tag antenna in the lab at −51 dBm.

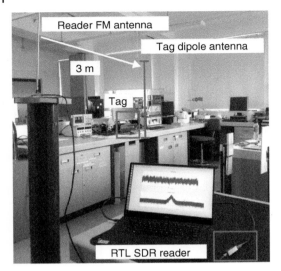

Figure 5.16 Indoor experimental setup. The tag with the FM antenna was set in a vertical position, and the receiver was tuned at the most powerful FM station. For communication experiments, the receiver antenna was placed at a maximum of 5 m away.

the FM station carrier signal was measured next to the tag antenna in the lab at −51 dBm (Figure 5.15b). The reader was placed close to the tag at different reader-to-tag distances with a maximum range of 5 m (Figure 5.16). The antenna of the reader was placed on top of a plastic stick with height 1.5 m for better reception. The receiver uses a commercial passive FM antenna with gain 2.5 dBi.

The tag with binary modulation had a wire dipole as it is described previously, and it was programmed to send packets with the fixed information bits for the following different bit rates: 50, 100, 500, 1000, 1250, and 2000 bps. The oscilloscope measurement of the packet transmitted at 500 bps is presented in Figure 5.10a. This transmitted packet corresponds to a measurement of 965 mV from T_{air} sensor as described in [17]. The binary representation of 965 is 0000001111000101 as depicted in the sensor-data section of the packet. Using the third-order transfer function of LMT70A sensor [23], the output temperature was calculated to be 25.81 °C.

The received packets for 50 and 500 bps after the matched filtering step are illustrated in Figures 5.10c,b respectively. One can see that the packets are inverted due to the channel conditions i.e. random, unknown channel phase. There is trade-off between bit rate and efficient filtering. In case that a high bit rate is employed (Figure 5.10b), there is less channel fluctuation, and the matched filtering operation is not able to remove the high-frequency components of the ambient FM signal, due to the wider bandwidth of the matched filter. In the case of low bit rate transmission (Figure 5.10c), the filtering operation is more effective, corresponding to a higher SNR, but a channel fluctuation effect

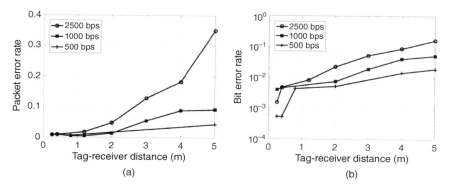

Figure 5.17 (a) Measured packet error rate (PER) versus the tag-receiver distance for 0.5, 1, and 2.5 kBps. (b) Measured bit error rate (BER) versus the tag-receiver distance for 0.5, 1, and 2.5 kbps.

is visible. When channel fluctuation is present, it is more difficult to decode the packet due to the fast-varying signal level.

To validate the effectiveness of our digital backscatter communication system, numerous range measurements were performed indoors with the setup described previously. Figure 5.17 displays the BER and PER performance as a function of the tag-to-reader distance for the three different data rates. The minimum PER and BER value at 5 m was measured to be 0.043 and 0.0019, respectively. As the tag-to receiver distance decreases, the reader can decode successfully more the bit packets. It is also seen that for a given distance value, reducing the bit rate improves the PER and BER performance.

However, transmitting packets at lower bit rates results in increased transmission time and energy per packet while the MCU and the front-end staying in "on" state for longer time. There is a direct and inversely proportional relationship between the bit rate and the energy that a tag consumes sending a packet as shown in Table 5.1, where the energy per packet for 6 bit rates is presented. The

Table 5.1 Tag power characteristics.

Bit rate (bps)	Power (mW)	Energy/packet (µJ)
2000	2.838	36.9
1250	2.087	43.4
1000	1.785	46.4
500	1.283	66.73
100	0.751	195.45
50	0.677	352.15

table also provides the tag power consumption for each bit rate. A higher power consumption of the MCU electronics is observed when operating at a higher bit rate. To compile the measurements shown in Table 5.1, the tag was programmed to wake up every three seconds, transmit a packet, and go to sleep mode, while being powered from the supercapacitor.

The power consumption of the binary modulation tag can be reduced by replacing the MCU with a new one. It is possible to use the more energy efficient MCU PIC16LF1459 [27] that it is used in high-order modulation tag. Similarly, one can select sensing elements with minimum power dissipation or even employ some passive sensing technique such as for example [36, 37]. Second, the RF front-end can be modified to use instead of an off-the-shelf switch, a single transistor-based switch such as the ones in [9, 38] with pJ/bit energy consumption. The tag with high-order modulation is the improved version of binary tag using all the aforementioned suggestions.

In addition to reducing the circuit consumption, battery-less operation can be achieved by exploring energy harvesting techniques. There are several studies related to the availability of ambient RF energy [39–42] as well as demonstrations of sensors powered by harvesting ambient RF energy from TV [26], WiFi [43], or even microwave oven signals [44], which could be used for smart house-targeted sensors. In addition, multiple technology of energy harvesters such as solar and electromagnetic energy harvesters can be employed to combine the different forms of ambient energy availability [25, 45].

Further improvement to the range could be achieved by adding encoding at our packet information. The last three parts of the packet could be encoded into a 19-bit word using an (13,19) Hamming code, which provides single error correction and double error detection (SECDED), providing a more reliable transmission through the noisy FM ambient signals. After the Hamming encoding, the hole packet is encoded again with FM0 encoding.

The high-order modulation tag was tested in the lab using the same setup shown in the Figure 5.16 previous section. The proof-of-consent prototype (Figure 5.7b) was programmed to send a fixed packet bit-stream (1010101111-01-01-0011010001) to the reader. The symbol representation of bit-sequence is depicted in Figure 5.18a with the oscilloscope measurement showing the four levels of the transmitted symbols that are used to drive the transistor. Due to the nonlinear relationship between the transistor gate voltage and the Γ_i, the small variation between the gate voltages corresponding to the states 01 and 11 (Figure 5.18a) leads to the maximum distance between the respective Γ_i, (Figure 5.18b). To test the performance of the backscatter communication link, the RF front-end antenna was placed 2 m away from the receiver antenna, while the FM station antenna was 34 km away. The received packet waveform, in time domain, after the low-pass filtering is shown in Figure 5.18b. The maximization of received

Figure 5.18 (a) Oscilloscope measurement of the time-domain backscatter packet at the gate of the transistor. Voltage levels correspond to the 4-PAM symbols. (b) Received packet after matched filtering. A good agreement with left picture is observed and the respective symbols can be detected using nearest neighbor method.

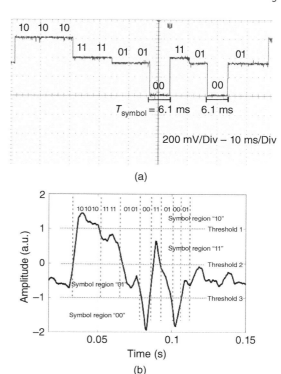

(a)

(b)

packet energy could be used in the future to find the start of the transmitted packet and synchronize it at the correct sampling instants. After that, maximum likelihood method is required to detect the transmitted symbols of Figure 5.18a. The signal could be quantized to the nearest element of the 4-PAM alphabet using the nearest neighbor method. Using three amplitude thresholds, as depicted in Figure 5.11, the specific symbol region corresponding to a received sample for a given T_{symbol} can be easily determined. Finally, the transmitted bits could be easily extracted through the detected symbols.

For comparison and validation purposes, OOK with FM0 encoding was implemented using the hardware of the high-order modulation tag; thus, this modulation requires only a digital output pin from MCU [6]. The minimum T_{symbol} achieved was 3.4 ms for a bit rate of 147 bps. Current consumption results for OOK and 4-PAM are presented in Table 5.2. In OOK modulation, the dissipated current was measured 3.6 µA when the ADC was off and 220 µA when the ADC was activated. In this work, the transistor switches between four states and the minimum T_{symbol} achieved was 6.1 ms. Every symbol contains 2 bits, thus leading to a calculated maximum bit rate of 328 bps ($2T_{symbol}$). The current consumption on

Table 5.2 Tag current consumption.

Tag operation mode @VDD = 1.8 V	μA	Bit rate (bps)
Sleep: (no DAC, no ADC)	0.6	0
Active: OOK (no DAC, no ADC)	3.6	147
Active: OOK (no DAC, ADC)	220	147
Active: 4PAM (DAC, no ADC)	16	328
Active: 4PAM (DAC, ADC)	232	328

4-PAM was measured at 16 μA with ADC disabled and 232 μA with ADC enabled. Table 5.2 shows that there is a clear trade-off between the bit rate and the power consumption between the two modulation schemes.

Finally, in this work, the interference of the tag to the commercial FM receivers was also studied. In USA, according to the Federal Communications Commission (FCC), it is illegal to broadcast signals on FM band (88–108 MHz) from high-power transmitters [46]. However, devices that communicate with backscatter signals (e.g. RFID tags) have not been reviewed by FCC. The reason is that the RF front ends of backscatter tags are not active circuits (they have no amplifiers), and they only modulate the reflections of the incoming signals. Also, the power emission of the reflected signal has low level. The Ambient backscatter operation and our system belong to the category of RFID tags, so it is legal under current rules. The reflected signals of existing FM signals could be synchronized with FM transmissions to interference the commercial FM receivers. The transitions do not affect the FM receivers because the backscattered signals are very weak, and they are also modulated with a different technique, amplitude modulation. The commercial FM receivers detect the FM signals, and we use ASK (OOK or 4-PAM) to modulate our information. We used a smartphone FM Receiver (Xiaomi Redmi note 4) to test the interference of the system. For the testing, we programmed the tag in the worst-case scenario where always backscatters random data with OOK modulation. The transmit antenna of the tag is placed parallel next to the FM receiver antenna (headphones). When we turned on the FM Receiver, we did not observe any glitches in the sound.

5.3 Conclusions

In this work, we present two novel FM backscatter tags and a receiver system. The tags communicate with a low-cost SDR reader by backscattering the ambient FM signals. Data acquisition from sensors with low-power operation and communication ranges up to 5 m has been demonstrated experimentally. The communication was implemented with OOK and 4-PAM modulation over the modulated carrier of the most powerful FM station. This concept can be the next novel way for low power and low cost long-range communication. Also, the high-order modulation approach is the first demonstration of backscatter 4-PAM modulation on ambient FM signals and paves the way toward enabling practical deployment for short-range, ultra-low-power RFID sensors such as wearable body area-network ID-enabled sensors.

Acknowledgments

This work was supported by EU COST Action IC1301 WiPE, Lloyd's Register Foundation (LRF) and the International Consortium in Nanotechnology (ICON). This work of R. Correia and N. B. Carvalho was supported by the European Regional Development Fund (FEDER), through the Competitiveness and Internationalization Operational Programme (COMPETE 2020) of the Portugal 2020 framework, Project, MOBIWISE, POCI-01-0145-FEDER-016426 and is funded by FCT/MEC through national funds and when applicable co-funded by FEDER PT2020 partnership agreement under the project UID/EEA/50008/2013. The work of A. Georgiadis was supported by EU H2020 Marie Sklodowska-Curie Grant Agreement 661621 and by COST Action IC1301 Wireless Power Transmission for Sustainable Electronics. The work of J. Kimionis and M. M. Tentzeris was supported by the National Science Foundation (NSF) and the Defense Threat Reduction Agency (DTRA). The authors would like to thank all members of Agile Technologies for High-frequency Electromagnetic Applications (ATHENA) Group, Georgia Institute of Technology, Atlanta, GA and all members of Antenna Engineering Research Group, Heriot-Watt University, Edinburgh, Scotland for their help in various steps throughout this work.

References

1 Liu, W., Huang, K., Zhou, X., and Durrani, S. (2017). Backscatter communications for internet-of-things: theory and applications. *Computing Research Repository (CoRR)* abs/1701.07588 https://pdfs.semanticscholar.org/e15b/cfc0229886d9fdab12469dc7b7665925992b.pdf.

2 Stockman, H. (1948). Communication by means of reflected power. *Proceedings of the IRE* 36 (10): 1196–1204.

3 Sample, A.P., Yeager, D.J., Powledge, P.S. et al. (2008). Design of an RFID-based battery-free programmable sensing platform. *IEEE Transactions on Instrumentation and Measurement* 57 (11): 2608–2615.

4 Naderiparizi, S., Parks, A.N., Kapetanovic, Z. et al. (2015). Wispcam: a battery-free RFID camera. *Proceedings of the IEEE International Conference on RFID*, San Diego, CA, USA (15–17 April 2015).

5 Kimionis, J., Bletsas, A., and Sahalos, J.N. (2014). Increased range bistatic scatter radio. *IEEE Transactions on Communications* 62 (3): 1091–1104.

6 Daskalakis, S.N., Assimonis, S.D., Kampianakis, E., and Bletsas, A. (2016). Soil moisture scatter radio networking with low power. *IEEE Transactions on Microwave Theory and Techniques* 64 (7): 2338–2346.

7 Iyer, V., Talla, V., Kellogg, B. et al. (2016). Intertechnology backscatter: towards internet connectivity for implanted devices. *Proceedings of the ACM Special Interest Group on Data Communication Conference (SIGCOMM)*, Florianopolis, Brazil (22–26 August 2016), pp. 356–369.

8 Thomas, S.J. and Reynolds, M.S. (2012). A 96 Mbit/sec, 15.5 pJ/bit 16-QAM modulator for UHF backscatter communication. *Proceedings of the IEEE International Conference on RFID*, Orlando, FL, USA (3–5 April 2012), pp. 185–190.

9 Correia, R., Boaventura, A., and Carvalho, N.B. (2017). Quadrature amplitude backscatter modulator for passive wireless sensors in IoT applications. *IEEE Transactions on Microwave Theory and Techniques* 65 (4): 1103–1110.

10 Wang, A., Iyer, V., Talla, V. et al. (2017). FM backscatter: enabling connected cities and smart fabrics. *Proceedings of the USENIX Symposium on Networked Systems Design and Implementation (NSDI)*, Boston, MA (27–29 March 2017), pp. 243–258.

11 Liu, V., Parks, A., Talla, V. et al. (2013). Ambient backscatter: wireless communication out of thin air. *ACM SIGCOMM Computer Communication Review* 43 (4): 39–50.

12 Kellogg, B., Parks, A., Gollakota, S. et al. (2014). Wi-Fi backscatter: Internet connectivity for RF-powered devices. *Proceedings of the ACM Special Interest Group on Data Communication Conference (SIGCOMM)*, Chicago, IL, USA. 44 (4): 607–618.

13 Bharadia, D., Joshi, K.R., Kotaru, M., and Katti, S. (2015). BackFi: high throughput WiFi backscatter. *Proceedings of the ACM Special Interest Group on Data Communication Conference (SIGCOMM)*, London, UK. 45 (4): 283–296.

14 Wang, G., Gao, F., Fan, R., and Tellambura, C. (2016). Ambient backscatter communication systems: detection and performance analysis. *IEEE Transactions on Communications* 64 (11): 4836–4846.

15 Darsena, D., Gelli, G., and Verde, F. (2016), Performance analysis of ambient backscattering for green Internet of Things. *Proceedings of the IEEE International Symposium on Personal, Indoor, Mobile Radio Communications (PIMRC)*, Valencia, Spain (4–7 September 2016), pp. 1–6.

16 Qian, J., Gao, F., Wang, G. et al. (2017). Noncoherent detections for ambient backscatter system. *IEEE Transactions on Wireless Communications* 16 (3): 1412–1422.

17 Daskalakis, S.N., Kimionis, J., Collado, A. et al. (2017). Ambient FM backscattering for smart agricultural monitoring. *Proceedings of the IEEE MTT-S International Microwave Symposium (IMS)*, Honolulu, HI, USA (4–9 June 2017).

18 Palazzari, V., Mezzanotte, P., Alimenti, F. et al. (2017). Leaf compatible eco-friendly temperature sensor clip for high density monitoring wireless networks. *Wireless Power Transfer* 4 (1): 55–60.

19 Dobkin, D.M. (2008). *The RF in RFID Passive UHF in Practice*. Amsterdam, USA: Newness.

20 Kurokawa, K. (1965). Power waves and the scattering matrix. *IEEE Transactions on Microwave Theory and Techniques* 13 (2): 194–202.

21 Der, L. (2008). *Frequency Modulation FM Tutorial*. Silicon Laboratories Inc.

22 Texas Instruments (2015). MSP430FR5969 launchpad development kit, product manual. http://www.ti.com/lit/ug/slau535b/slau535b.pdf (accessed 06 September 2019).

23 Texas Instruments (2015). LMT70A high precision analog temperature sensor, product manual, Online available: http://www.ti.com/lit/ds/symlink/lmt70a.pdf.

24 Analog Devices (2005). ADG902 RF switch, product manual. . http://www.analog.com/media/en/technical-documentation/data-sheets/ADG901902.pdf (accessed 06 September 2019).

25 Niotaki, K., Collado, A., Georgiadis, A. et al. (2014). Solar/electromagnetic energy harvesting and wireless power transmission. *Proceedings of the IEEE* 102 (11): 1712–1722.

26 Kim, S., Vyas, R., Bito, J. et al. (Nov. 2014). Ambient RF energy-harvesting technologies for selfsustainable standalone wireless sensor platforms. *Proceedings of the IEEE* 102 (11): 1649–1666.

27 Microchip Technology Inc. (2014). PIC16LF1459, USB microcontroller with extreme low-power technology, product manual. http://www.microchip.com/downloads/en/DeviceDoc/40001639B.pdf (accessed 06 September 2019).

28 PowerFilm (2009). SP3-37 flexible solar panel 3V@22mA, product manual. https://goo.gl/q5ECXh (accessed 06 September 2019).

29 Broadcom (2006). ATF-52189 High Linearity Mode Enhancement Pseudomorphic HEMT FET Transistor, product manual. https://www.broadcom.com/products/wireless/transistors/fet/atf-52189# (accessed 06 September 2019).

30 Proakis, J.G. (2001). *Digital Communications*, 4e. New York, NY: McGraw-Hill Companies, Inc.

31 NooElec Inc. (2017). NESDR SMArt bundle-premium RTL-SDR, product manual. http://www.nooelec.com/store/nesdr-smart.html (accessed 06 September 2019).

32 Simon, M. and Divsalar, D. (2006). Some interesting observations for certain line codes with application to RFID. *IEEE Transactions on Communications* 54 (4): 583–586.

33 Bletsas, A., Kimionis, J., Dimitriou, A.G., and Karystinos, G.N. (2012). Single antenna coherent detection of collided FM0 RFID signals. *IEEE Transactions on Communications* 60 (3): 756–766.

34 Kargas, N., Mavromatis, F., and Bletsas, A. (2015). Fully-coherent reader with commodity SDR for Gen2 FM0 and computational RFID. *IEEE Wireless Communications Letters* 4 (6): 617–620.

35 Bamiedakis-Pananos, M. (2015). Synchronization and detection for Gen2 RFID signals. Master's thesis. School of Electrical and Computer Engineering, Technical University of Crete, Greece, online available: https://dias.library.tuc.gr/view/manf/23965

36 Kim, S., Kawahara, Y., Georgiadis, A. et al. (2013). Low-cost inkjet-printed fully passive RFID tags using metamaterial inspired antennas for capacitive sensing applications. *Proceedings of the IEEE MTT-S International Microwave Symposium (IMS)*, Seattle, WA, USA (Jun 2013), pp. 1–4.

37 Bhattacharyya, R., Floerkemeier, C., and Sarma, S. (2010). Low-cost, ubiquitous RFID-tag-antenna-based sensing. *Proceedings of the IEEE* 98 (9): 1593–1600.

38 Kimionis, J. and Tentzeris, M.M. (2016). Pulse shaping: the missing piece of backscatter radio and RFID. *IEEE Transactions on Microwave Theory and Techniques* 64 (12): 4774–4788.

39 Visser, H.J., Reniers, A.C., and Theeuwes, J.A. (2008). Ambient RF energy scavenging: GSM and WLAN power density measurements. *Proceedings of the IEEE European Microwave Conference (EuMC)*, Amsterdam, Netherlands (28–30 June 2008), pp. 721–724.

40 Guenda, L., Santana, E., Collado, A. et al. (2014). Electromagnetic energy harvesting global information database. *Transactions on Emerging Telecommunications Technologies* 25 (1): 56–63.

41 Piñuela, M., Mitcheson, P.D., and Lucyszyn, S. (2013). Ambient RF energy harvesting in urban and semi-urban environments. *IEEE Transactions on Microwave Theory and Techniques* 61 (7): 2715–2726.

42 Mimis, K., Gibbins, D., Dumanli, S., and Watkins, G.T. (2015). Ambient RF energy harvesting trial in domestic settings. *IET Microwaves, Antennas and Propagation* 9 (5): 454–462.

43 Gudan, K., Chemishkian, S., Hull, J.J. et al. (2014). A 2.4 GHz ambient RF energy harvesting system with −20 dbm minimum input power and NiMH battery storage. *Proceedings of the IEEE Conference on RFID-Technologies and Applications (RFID-TA)*, Tampere, Finland (8–9 September 2014), pp. 7–12.

44 Kawahara, Y., Bian, X., Shigeta, R. et al. (2013). Power harvesting from microwave oven electromagnetic leakage. *Proceedings of the ACM International Joint Conference on Pervasive and Ubiquitous Computing*, Zurich, Switzerland (8–12 September 2013), pp. 373–382.

45 Bito, J., Bahr, R., Hester, J.G. et al. (2017). A novel solar and electromagnetic energy harvesting system with a 3-D printed package for energy efficient internet-of- things wireless sensors. *IEEE Transactions on Microwave Theory and Techniques* 65 (5): 1831–1842.

46 Federal Communications Commission (FCC). Permitted forms of low power broadcast operation, Public Notice 14089. https://apps.fcc.gov/edocs_public/ attachmatch/DOC-297510A1.pdf (accessed 06 September 2019).

6

Backscatter RFID Sensor System for Remote Health Monitoring

Jasmin Grosinger

Institute of Microwave and Photonic Engineering, Graz University of Technology, 8010 Graz, Austria

6.1 Introduction

Currently, state-of-the-art passive radio frequency identification (RFID) technologies are used in logistics and maintenance [1, 2]. Since 2010, researchers have been working consistently toward the goal of reaching beyond the ID in RFID by integrating sensing capabilities in backscatter RFID tag to additionally monitor the tag environment (e.g. temperature, curvature, liquid level, etc.) [3–5]. The existing literature shows that backscatter RFID sensor tags can be realized by the use of additional sensor circuitry, as presented in [6–10], or without the use of additional sensor circuitry, as outlined in the following. Without the use of additional sensor circuitry, sensor tags consume less power, which leads to a longer read range and thus to a reduced number of tags. This is attractive for applications, which have tight power and weight constraints, as for example remote health monitoring applications.

In general, the RFID tag consists of an antenna and a microchip that are characterized by an antenna impedance and two chip impedances. The backscatter modulation of the reflected reader signal is realized by switching between the two chip impedances, i.e. the chip impedance in the absorbing mode and the chip impedance in the reflecting mode (representing a logical "0" and "1"). Sensor tags without the use of additional sensor circuitry are realized by integrating passive transducers into tags that change their impedance with a change of tag environment. The reader then detects this change in transducer impedance, which additionally modulates the backscattered signal in amplitude and phase.

An obvious approach to integrate a transducer into a backscatter radio frequency (RF) tag is to use the tag antenna as the sensing device. By its very nature, the antenna and its impedance are sensitive to its closest environment [11].

Wireless Power Transmission for Sustainable Electronics: COST WiPE - IC1301,
First Edition. Edited by Nuno Borges Carvalho and Apostolos Georgiadis.
© 2020 John Wiley & Sons, Inc. Published 2020 by John Wiley & Sons, Inc.

A major benefit of this sensing approach is that the sensor tag can still rely on a state-of-the-art ultra-high frequency (UHF) RFID tag chip. Several sensor tags have already been prototyped for antenna based sensing, such as tags to sense temperature [3, 12–17], displacement and orientation [3, 14, 18–21], filling level [3, 14, 22–24], strain and crack [14, 25–27], humidity [14, 28–32], gas [33–35], corrosion [36], touch [37], pH-value [38], or epidermal properties [39].

Remote health monitoring applications are often based on a wireless body area network (WBAN) and support medical experts, patients, and humans with physiological data to improve the quality of the clinical environment. In addition, such applications favor therapy support at home and improve the well-being of people. WBAN connect sensor nodes situated in clothes, on the body, or under the skin of a person through a wireless communication channel [40, 41]. Backscatter RFID at UHF is a promising communication technology for WBAN. RFID sensor tag, as previously listed, can be used to monitor the physiological parameters of persons (e.g. blood pressure, temperature, heartbeat, body motion). As an exemplary application, sensor tags that monitor epidermal properties are used for remote temperature monitoring of patients [42, 43].

In the following, this work investigates the performance of on-body backscatter RFID sensor systems for remote health monitoring applications at 900 MHz and 2.45 GHz. Section 6.2 describes the arrangement of the on-body RFID reader and sensor tag as well as the design of their antennas. Section 6.3 deals with on-body radio channel measurements, are used in Section 6.4 to evaluate the system performance in the forward and backward links of the system.

6.2 On-Body System

Figure 6.1 shows the arrangement of the investigated RFID sensor system situated on the body of an adult female. In particular, the positions of the RFID reader and tag are highlighted. The RFID reader is situated on the stomach of the female, while four RFID tags are placed at various positions on the female's body: on the right chest, on the middle of the back, on the left side of the head, and on the right wrist. These links represent two trunk-to-trunk, a trunk-to-head, and a trunk-to-limb link following a classification of on-body links introduced in [44].

6.2.1 Body Model

The antennas of the RFID reader and tag are influenced by the close proximity to the human body. Proximity effects, which are experienced by an antenna attached to the body, include a shift in the antenna resonant frequency, a distortion of its

Figure 6.1 On-body RFID system: the sensor tag on the female's back is represented by the orange circle.

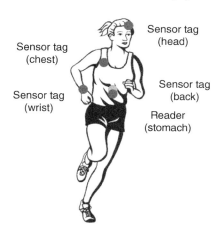

Sensor tag (chest)

Sensor tag (wrist)

Sensor tag (head)

Sensor tag (back)

Reader (stomach)

radiation pattern, and a reduction in its radiation efficiency [45]. The strength of these effects depends on the antenna type, the antenna placement on the body, and the antenna-to-body separation distance.

Previous investigations and measurements of human tissues have shown that the electromagnetic properties vary significantly with tissue type and frequency. These properties – the relative permittivity, ε_r, and the electric conductivity, σ – are plotted versus frequency in Figures 6.2 and 6.3 for muscle, skin, and fat tissues. The

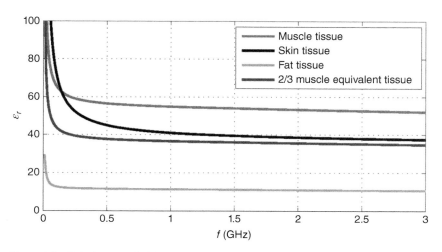

Figure 6.2 Relative permittivity, ε_r, of muscle, skin, fat, and two-third muscle equivalent tissues versus frequency.

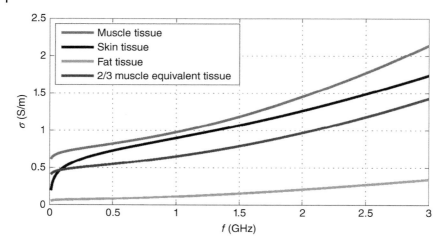

Figure 6.3 Electric conductivity, σ, of muscle, skin, fat, and two-third muscle equivalent tissues versus frequency.

curves are based on a parametric model for the complex relative permittivity $\tilde{\varepsilon}_r$ of human tissues presented in [46]:

$$\tilde{\varepsilon}_r(\omega) = \varepsilon_\infty + \sum_{m=1}^{4} \frac{\Delta\varepsilon_m}{1 + (j\omega\tau_m)^{1-\alpha_m}} + \frac{\sigma_i}{j\omega\varepsilon_0} \tag{6.1}$$

where $\omega = 2\pi f$ is the angular frequency, ε_∞ is the material permittivity at terahertz frequencies, $\varepsilon_0 = 8.854\,\text{pF/m}$ is the free space permittivity, σ_i is the ionic conductivity, and $\Delta\varepsilon_m$, τ_m, and α_m are material parameters for each dispersion region. The relative permittivity $\varepsilon_r(\omega)$ is the real part of the complex relative permittivity:

$$\varepsilon_r(\omega) = \Re\{\tilde{\varepsilon}_r(\omega)\} \tag{6.2}$$

The electric conductivity $\sigma(\omega)$ is defined by [47],

$$\sigma(\omega) = -\omega\varepsilon_0\varepsilon_r''(\omega) \tag{6.3}$$

where $\varepsilon_r''(\omega) = I\{\tilde{\varepsilon}_r(\omega)\}$ is the imaginary part of the complex relative permittivity. Table 6.1 lists the individual parameters for different body tissues, which are used to calculate $\tilde{\varepsilon}_r(\omega)$ (see Eq. (6.1)). Figures 6.2 and 6.3 show that there is only a minor change in ε_r at UHF, while the losses increase with frequency.

Due to these adverse electromagnetic properties of human tissues, the performance of on-body antennas is rather difficult to predict. For an adequate antenna design and a computation of the on-body channel, a human body model has been created using Ansys HFSS [48]. Following a recommendation given in [47], the body model is composed of two-third muscle equivalent tissue, which varies with

Table 6.1 Material parameters for muscle, skin (dry), and fat (average infiltrated) tissues [44].

Parameter	Muscle	Skin	Fat
ε_∞	4	4	2.5
$\Delta\varepsilon_1$	50	32	9
τ_1 (ps)	7.234	7.234	7.958
α_1	0.1	0	0.1
$\Delta\varepsilon_2$	7000	1100	7000
τ_2 (ns)	353.678	32.481	353.678
α_2	01	0.2	0.1
$\Delta\varepsilon_3$	1.2e6	0	3.3e4
τ_3 (μs)	318.310	159.155	159.155
α_3	0.1	0.2	0.05
$\Delta\varepsilon_4$	2.5e7	0	1e7
τ_4 (ms)	2.274	15.915	15.915
α_4	0	0.2	0.01
σ_i	0.2	0.0002	0.035

frequency. ε_r and σ of this tissue type are also plotted in Figures 6.2 and 6.3. The body model is realized as a simple rectangular torso to efficiently solve the electromagnetic problem in a reasonable time. Figure 6.4 shows the HFSS model for the computation of the stomach-chest link using monopole antennas at 900 MHz. The spacing of the monopoles is 30 cm, which is approximately the length of the line of sight path on the female's body.

6.2.2 Antennas

The primary requirement for on-body antennas is that the mutual influence between the antenna and the human body is low [49]. This decoupling can be achieved by metallic shields, which are integrated in the antennas as groundplanes [50]. Suitable antenna types are for example monopoles or patch antennas, their groundplanes mounted parallel to the body surface.

6.2.2.1 Monopole Antennas

Practically, a monopole antenna is not suitable for WBAN applications because it is not low profile. However, monopoles show the best performance in on-body systems [45, 49]. A monopole shows an omnidirectional radiation pattern on the body, i.e. a maximum radiation along the body surface, and a vertical polarization. Such a radiation is favorable for on-body links, where the main mechanism for

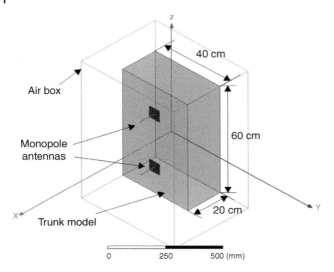

Figure 6.4 HFSS model for the channel computation of the stomach-chest link: monopole antennas operating at 900 MHz are attached to the human trunk. The length of the line of sight path is 30 cm.

propagation around the body is via a surface wave [51]. Thus, monopoles are used as a best-case reference in this work and help to define an upper bound for the performance of practical system implementations.

In the following, monopole antennas resonant at 900 MHz and 2.45 GHz are designed by means of the human body model presented before and realized on FR-4 substrate with a thickness of 1.6 mm [45]. The design is done for an antenna-to-body separation distance of 5 mm. Pictures of the realized monopoles and their dimensions are depicted in Figure 6.5.

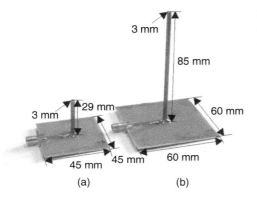

Figure 6.5 On-body monopole antennas resonant at 900 MHz (b) and 2.45 GHz (a).

The monopoles are matched to a 50 Ω feed and show an overall matching of some −10 dB (see Figures 6.11 and 6.12), which is sufficient for real-world systems. The planar feed geometry is essential for wearable applications because probe-fed antennas cannot be mounted close to the body surface. Furthermore, the addition of the microstrip line provides an extra degree of freedom for antenna matching [45].

The simulated radiation efficiency of the monopole antenna operating at 900 MHz is 77% in comparison to the monopole in vacuum which shows an efficiency of 97%. The maximum antenna gain compared to an isotropic radiator is $G_{\text{Max}} = 1.3$ dBi on-body and 2.1 dBi in vacuum. The monopole at 2.45 GHz shows an on-body radiation efficiency of 71% in comparison to an efficiency of 97% in vacuum. The maximum gain is 1.6 dBi on-body and 2.2 dBi in vacuum. The smaller efficiency at 2.45 GHz, although the groundplane is bigger than the radiator length in comparison to the 900 MHz monopole, is due to higher losses in the human tissue at higher frequencies (see Figure 6.3).

6.2.2.2 Patch Antennas

Patch antennas are low profile and show a broadside radiation pattern, i.e. a maximum radiation away from the body, when they are excited at their fundamental mode [52]. This radiation characteristic is suitable for on-body links where the propagation path is a free space path or a shadowed free space path with diffraction around the body (e.g. the stomach-wrist link) [49]. In addition, there is the possibility to realize patch antennas, which operate at a higher mode with a maximum radiation along the body surface [45]. This radiation pattern is suitable for on-body links, where the propagation is predominantly due to a surface wave (e.g. the stomach-chest link). Hence, patch antennas are especially suitable for on-body applications. In this work, the less efficient patch antennas – in comparison to the monopole antennas – represent typical tag antennas and provide an insight in the performance of practical system implementations.

An initial set of patch antennas is designed and realized on FR-4 substrate with a thickness of 1.6 mm for the 900 MHz and 2.45 GHz regime. Again, the antennas are designed for an antenna-to-body separation of 5 mm. Photographs and the dimensions of the patch antennas can be seen in Figure 6.6.

The antennas are fed by a microstrip line and matched to 50 Ω. The matching is about −10 dB in the operating frequency range (see Figures 6.13 and 6.14).

The patch antennas operate at their fundamental mode and thus show a broadside radiation. The patch at 900 MHz shows a simulated radiation efficiency of 22% and a maximum gain of −1.2 dBi. In comparison, the patch antenna in vacuum shows an efficiency of 24% and a gain of −0.7 dBi. While the patch antenna for the 2.45 GHz regime shows a radiation efficiency of 42% on the body and 47% in vacuum. The maximum gain is 2.7 dBi on the body and 3.7 dBi in vacuum. The

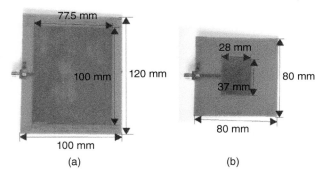

Figure 6.6 On-body patch antennas resonant at 900 MHz (b) and 2.45 GHz (a).

higher radiation efficiency in comparison to the patch antenna at 900 MHz is due to the larger size of the groundplane relative to the patch size.

6.3 Radio Channel

In a backscatter RFID system, a bidirectional radio link is established between the reader and tag – the reader-tag-reader link – which can be subclassified into the forward link and backward link [53]. The link budget of the backscatter radio channel is outlined in Figure 6.7.

In the forward link, the reader transmits RF power, $P_{TX,Reader}$, and data to the tag. The power absorbed by the tag's chip, P_{Chip}, is defined by [54]

$$P_{Chip} = \tau P_{Tag} = \tau |S_{21}|^2 P_{TX,Reader} \tag{6.4}$$

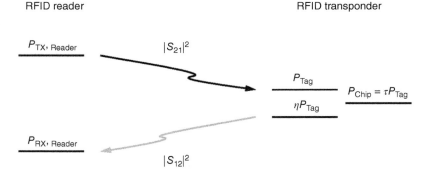

Figure 6.7 Link budget of the backscatter radio channel.

where P_{Tag} is the chip's input power and S_{21} is the channel transfer function of the forward link. P_{Chip} should be higher than the chip's sensitivity, T_{Chip}, which is defined as the minimum input power at the tag chip to turn on its circuitry. If P_{Chip} is smaller than T_{Chip}, the backscatter communication is limited in its forward link [55].

In the backward link, the tag responds to the reader by modulating the backscattered signal. The power of the tag's signal at the RX of the reader, $P_{\text{RX, Reader}}$, can be written as [54]

$$P_{\text{RX,Reader}} = |S_{12}|^2 \eta P_{\text{Tag}} = |S_{12}|^2 \eta |S_{21}|^2 P_{\text{TX,Reader}} \qquad (6.5)$$

where S_{12} is the channel transfer function of the backward link. $P_{\text{RX, Reader}}$ should be higher than the RX's sensitivity, $T_{\text{RX, Reader}}$, which is defined as the minimum input power at the reader to assure a successful reception of the tag's data. If $P_{\text{RX, Reader}}$ is smaller than $T_{\text{RX, Reader}}$, the communication system is limited in its backward link [55].

The channel transfer functions, S_{21} and S_{12}, depend on the antenna characteristics of the reader and tag (e.g. antenna gain, polarization) and the properties of the propagation channel (e.g. path loss, fading) [56].

In the following, the radio links of the on-body RFID sensor system, which is composed of monopole or patch antennas, are investigated. The realized antennas operate as both reader and tag antennas. The investigation of the on-body system is done by means of channel measurements at 900 MHz and 2.45 GHz. The channel transfer functions of the forward link S_{21} and the backward link S_{12} between the reader antenna on the stomach and the four different tag antennas are captured by means of a vector network analyzer (VNA) [56].

6.3.1 Measurement Setup

The channel measurement setup is depicted in Figure 6.8. The custom-built on-body antennas are connected via two coaxial cables to the VNA. The calibrated VNA measures the scattering parameters – reflection coefficients S_{11} and S_{22} and transmission coefficients S_{12} and S_{21} – at the antenna inputs and communicates via a LAN with a PC. The PC controls the measurement and defines parameters like measurement duration and rate.

With Rohde&Schwarz ZVA8 VNA, a good trade-off between accuracy and temporal resolution of the measurement is found. This trade-off manifests in a noise floor of about −100 dB for the transmission coefficients.

The coaxial cables are equipped with ferrite beads, which are used to reduce sheath currents and to mitigate influences in the measurement due to the cables [57]. This is adequate for frequencies up to 1 GHz. Additionally, the

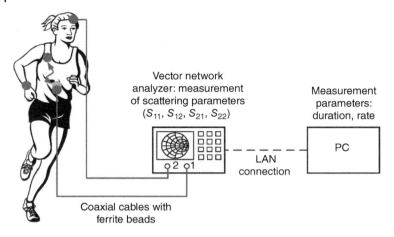

Figure 6.8 Measurement setup: the reflection and transmission properties of the on-body antennas are analyzed by means of a VNA.

positioning of the cables is arranged to minimize their influence. This is done by guiding the cables along the body. In this way, the near field of the on-body antenna is predominately influenced by the high complex permittivity of the human body and not by the measurement cables. The variations of the cable loss are less than 0.1 dB for movements similar to those performed during the measurement (see Table 6.2).

The scattering parameters are measured for 18 different stationary and moving body postures versus time [44] in an indoor scenario. The postures are listed in Table 6.2. Snapshots of these postures during the measurement of the stomach-head and stomach-back links can be found in [58]. Each posture is held 20 seconds. The repetition rate of the measurement is 5 Hz.

6.3.2 Comparison of Simulations and Measurements

In this section, results obtained by the simulation are compared with the measurement. The comparison is done for the on-body system composed of monopole antennas operating at 900 MHz.

Figures 6.9 and 6.10 compare the simulated and measured magnitudes of the scattering parameters in decibel versus a frequency range of 100 MHz – 3 GHz. The squared magnitudes of the reflection coefficients $|S_{11}|^2$ and $|S_{22}|^2$ define the reader and tag antenna matching at their respective feed points. It can be seen that a good matching occurs at about 900 MHz and 2.6 GHz. The squared magnitudes

Table 6.2 List of body postures performed during the channel measurement [44].

Number	Time (s)	Description
1	0–20	Standing, upright
2	20–40	Standing, body turned left
3	40–60	Standing, body turned right
4	60–80	Standing, body leaning forward
5	80–100	Standing, head leaning forward
6	100–120	Standing, head turned left
7	120–140	Standing, head turned right
8	140–160	Standing, arms stretched out to sides
9	160–180	Standing, arms above head
10	180–200	Standing, arms reaching forward
11	200–220	Standing, forearms forward
12	220–240	Sitting, arms hanging body
13	240–260	Sitting, hands in lap
14	260–280	Standing, upright
15	280–300	Standing, moving arms, head, and body randomly
16	300–320	Sitting, moving arms, head, and body randomly
17	320–340	Walking, back and forth
18	340–360	Walking, moving arms, head, and body randomly

of the channel transfer functions $|S_{12}|^2$ and $|S_{21}|^2$ define the channel gain. $|S_{21}|^2$ is defined as the ratio of the power received at the tag antenna output (port 2 of the VNA) to the power available at the reader antenna input (port 1 of the VNA), while $|S_{12}|^2$ is defined as the ratio of the power received at the reader antenna output (port 1 of the VNA) to the power available at the tag antenna input (port 2 of the VNA). It can be seen that the channel gain is maximum at about 900 MHz and 2.6 GHz.

Figure 6.9 shows a comparison of simulation and measurement for the stomach-back link, while Figure 6.10 shows results of the stomach-chest link. The measurement curves picture a snapshot in time at $t = 15$ s in the standing, upright position. It can be seen that the simulation results are well confirmed by the measurements, although the simulation does not model body movements due to respiration, the indoor multipath environment, and the exact shape of the body. Consequently, the simplified human body model is an appropriate tool to design on-body antennas and roughly evaluate on-body links.

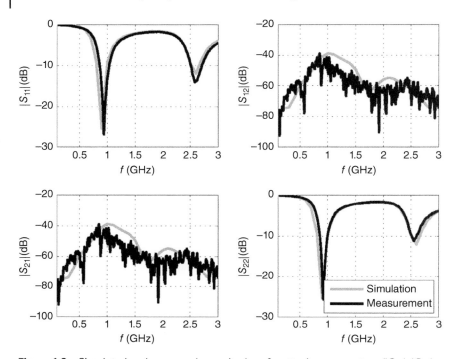

Figure 6.9 Simulated and measured magnitudes of scattering parameters ($|S_{11}|$, $|S_{12}|$, $|S_{21}|$, and $|S_{22}|$) in decibel versus frequency of the stomach-back link in a standing, upright position (measurement at $t = 15$ s).

6.3.3 Measurement Results

As previously mentioned, the scattering parameters S_{11}, S_{22}, S_{12}, and S_{21} are measured at 900 MHz and 2.45 GHz in a realistic, multipath environment versus stationary and moving body postures (see Table 6.2).

6.3.3.1 Antenna Matching

Figure 6.11 illustrates the matching of the reader monopole antenna $|S_{11}|^2$ and the matching of the tag antennas $|S_{22}|^2$ at 900 MHz versus stationary and moving body postures over time. Figure 6.12 illustrates the matching of the monopole antennas at 2.45 GHz, while Figures 6.13 and 6.14 depicts the matching of the patch antennas at 900 MHz and 2.45 GHz.

It can be seen that the antenna matching depends on the body posture, which influences the antenna-to-body separation distance, and on the antenna position on the body, which influences the reactive near field of the antenna due to a varying effective permittivity. Thus, the tag antennas show a wider variation in matching in comparison to the reader antennas, which are mounted on the female's

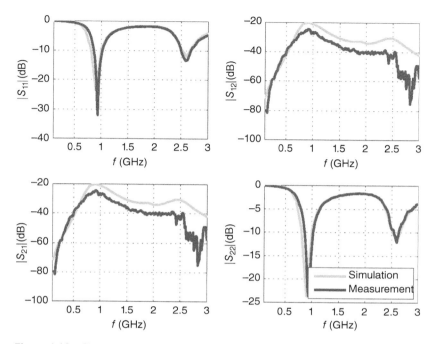

Figure 6.10 Simulated and measured magnitudes of scattering parameters ($|S_{11}|$, $|S_{12}|$, $|S_{21}|$, and $|S_{22}|$) in decibel versus frequency of the stomach-chest link in a standing, upright position (measurement at $t = 15$ s).

stomach. This behavior can be observed in Table 6.3, which lists the temporal mean of the antenna matching for all four on-body links. To account for varying proximity effects, the use of broadband antennas is advisable [59]. In practice, the antenna matching is better than -10 dB, which is deemed to be sufficient for real-world systems.

6.3.3.2 Channel Gain

Exemplarily, Figure 6.15 plots the measured channel gain in the forward link $|S_{21}|^2$ of the on-body system versus stationary and moving body postures. The reader and tag antennas are monopoles operating at 900 MHz. Another illustration of these measurement results is shown in Figure 6.16. The figure depicts the cumulative distribution function (CDF) of the channel gain $|S_{21}|^2$ for all four on-body links. The CDF describes the probability that the channel gain is less or equal to values plotted on the x-axis of Figure 6.16 [56]. The figures show that on-body links with shorter path lengths (e.g. the stomach-chest link) have a higher channel gain in comparison to links with longer distances (e.g. the stomach-back link). In addition, it can be observed that depending on the

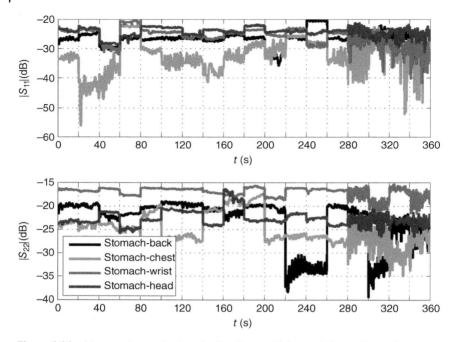

Figure 6.11 Measured magnitudes of reflection coefficients of the reader and tag monopole antennas $|S_{11}|$ and $|S_{22}|$ in decibel versus stationary (0–280 seconds) and moving (280–360 seconds) body postures over time at 900 MHz: each posture is held 20 seconds.

Table 6.3 Temporal mean of antenna matching for all four on-body links (stomach-back, stomach-chest, stomach-wrist, and stomach-head links).

Reader antennas	Back (dB)	Chest (dB)	Wrist (dB)	Head (dB)
Monopole @ 900 MHz	−26	−32	−26	−24
Monopole @ 2.45 GHz	−16	−19	−17	−17
Patch @ 900 MHz	−36	−36	−34	−33
Patch @ 2.45 GHz	−12	−12	−12	−12
Tag antennas	**Back (dB)**	**Chest (dB)**	**Wrist (dB)**	**Head (dB)**
Monopole @ 900 MHz	−23	−25	−17	−23
Monopole @ 2.45 GHz	−15	−17	−19	−17
Patch @ 900 MHz	−35	−32	−40	−33
Patch @ 2.45 GHz	−14	−13	−14	−14

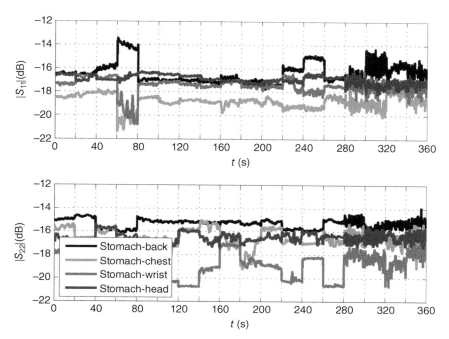

Figure 6.12 Measured magnitudes of reflection coefficients of the reader and tag monopole antennas $|S_{11}|$ and $|S_{22}|$ in decibel versus stationary (0–280 seconds) and moving (280–360 seconds) body postures over time at 2.45 GHz: Each posture is held 20 seconds.

on-body link, the link geometry and thus the channel gain is influenced by the body movements. For example, the stomach-chest link shows a smaller variation in channel gain in comparison to the stomach-wrist link, which is more affected by movements. In general, trunk-to-trunk links are less influenced by body movements in contrast to trunk-to-wrist and trunk-to-head links [44].

Any real propagation environment is symmetrical [60], i.e. the channel gains of the forward and backward links between the two communicating antennas are equal: $|S_{21}|^2 = |S_{12}|^2$. This is also found within these channel measurements. $|S_{21}|^2$ and $|S_{12}|^2$ are almost captured simultaneously during the measurement and are found to be virtually the same (see Figures 6.9 and 6.10).

6.4 System Performance

The evaluation of the performance of the on-body RFID system is done by means of the measured channel transfer functions S_{21} and S_{12}. As stated earlier, $|S_{21}|^2$

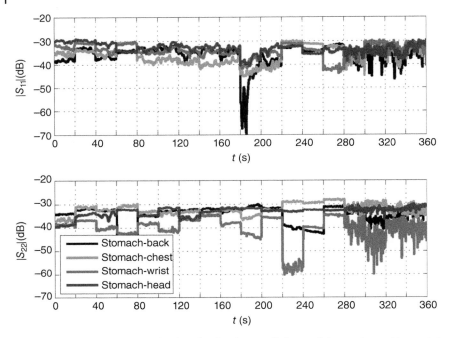

Figure 6.13 Measured magnitudes of reflection coefficients of the reader and tag patch antennas $|S_{11}|$ and $|S_{22}|$ in decibel versus stationary (0–280 seconds) and moving (280–360 seconds) body postures over time at 900 MHz: Each posture is held 20 seconds.

defines the channel gain in the forward link of the RFID system, while $|S_{12}|^2$ defines the channel gain in the backward link.

The CDF of $|S_{21}|^2$ directly relates to an outage probability in the system forward link [61], more precisely to the probability that the backscatter system operates at its limit. This probability is

$$P_F = P\{|S_{21}|^2 \le F_{Th}\} \tag{6.6}$$

The threshold F_{Th} is defined as the channel gain, which is necessary to realize $P_{Chip} = T_{Chip}$, i.e. the power absorbed by the chip is equal to the chip sensitivity (see Eq. (6.4)):

$$F_{Th} = \frac{T_{Chip}}{\tau P_{TX,Reader}} \tag{6.7}$$

The CDF of the product of the channel gain in the forward link and backward link $|S_{21}|^2|S_{12}|^2$ relates to an outage probability in the system backward link:

$$P_B = P\{|S_{21}|^2|S_{12}|^2 \le B_{Th}\} \tag{6.8}$$

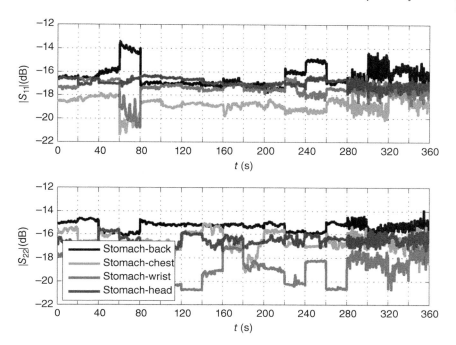

Figure 6.14 Measured magnitudes of reflection coefficients of the reader and tag patch antennas $|S_{11}|$ and $|S_{22}|$ in decibel versus stationary (0–280 seconds) and moving (280–360 seconds) body postures over time at 2.45 GHz: Each posture is held 20 seconds.

The threshold B_{Th} is the total channel gain of the forward and backward links that is necessary to realize a RX power at the reader equal to the RX sensitivity $P_{\text{RX, Reader}} = T_{\text{RX, Reader}}$ (see Eq. (6.5)):

$$B_{\text{Th}} = \frac{T_{\text{RX,Reader}}}{\eta P_{\text{TX,Reader}}} \tag{6.9}$$

With the probabilities defined in Eqs. (6.6) and (6.8), the backscatter system performance can be analyzed. Such an analysis allows to explore each system parameter, i.e. $P_{\text{TX,Reader}}$, τ, T_{Chip}, η, and $T_{\text{RX, Reader}}$, individually (see Eqs. (6.7) and (6.9)).

Usually, the outage probabilities which are allowed in an on-body system are governed by the application. In the case of a system which monitors life parameters of patients in clinical care, the outage probabilities should be close to zero. While systems used in sports analysis can deal with higher outage probabilities of e.g. 10%.

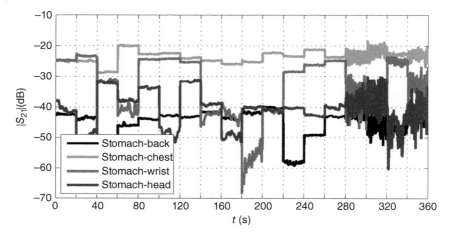

Figure 6.15 Measured magnitudes of the channel transfer function $|S_{21}|$ in decibel for all four on-body links using the 900 MHz monopoles versus stationary (0–280 seconds) and moving (280–360 seconds) body postures over time: Each posture is held 20 s.

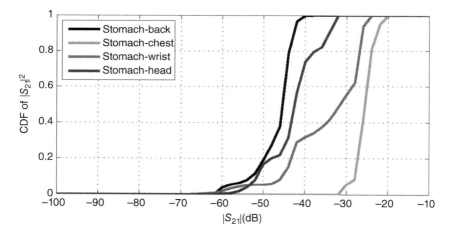

Figure 6.16 CDF of the channel gain $|S_{21}|^2$ for all four on-body links using the 900 MHz monopoles.

6.4.1 Forward Link

Figures 6.17–6.20 plot the outage probabilities in the forward link P_F versus the gain threshold F_{Th} for the respective links of the on-body system using all four antenna configurations, i.e. monopoles or patch antennas operating at 900 MHz or 2.45 GHz.

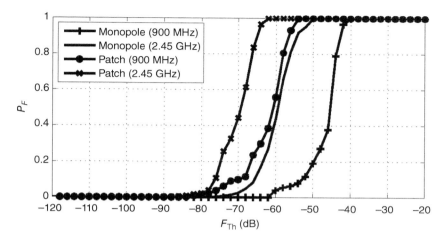

Figure 6.17 Outage probability P_F versus gain threshold F_{Th} of the stomach-back forward link.

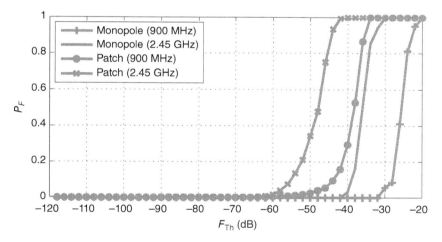

Figure 6.18 Outage probability P_F versus gain threshold F_{Th} of the stomach-chest forward link.

As observed in Section 6.3.3, the curves show that links with longer path lengths show higher outage probabilities in comparison to links with shorter distances. In addition, depending on the on-body link, the link geometry and thus the channel gain is influenced by body movements. Thus, on-body links with higher mobility, i.e. trunk-to-limb links, experience a wider range of outage probabilities than trunk-to-trunk links with lower mobility. These phenomena can be observed for both antenna types in both frequency ranges.

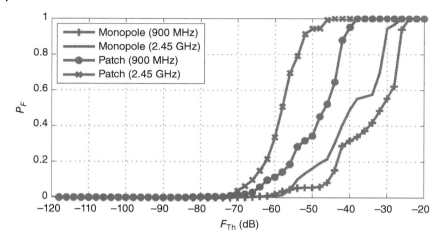

Figure 6.19 Outage probability P_F versus gain threshold F_{Th} of the stomach-wrist forward link.

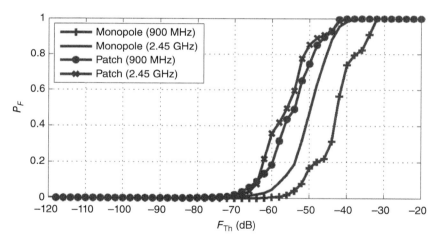

Figure 6.20 Outage probability P_F versus gain threshold F_{Th} of the stomach-head forward link.

As expected from theory, the outage probabilities at 900 MHz are lower than the probabilities at 2.45 GHz. This is due to an increased energy absorption in human tissues at higher frequencies (see Figure 6.3). This behavior can be observed for both antenna types, although there are quite some differences in the patch antennas radiation efficiencies (see Section 6.2.2).

In addition, the probability curves show that the monopoles are indeed a best-case reference for on-body systems.

6.4.1.1 System Example

Subsequently, the performance of an on-body RFID system is evaluated using state-of-the-art reader and tag chips.

A modern RFID system provides for example a gain threshold of $F_{\text{Th}} = -47\,\text{dB}$ (see Eq. (6.7), with $T_{\text{Chip}} = -17.8\,\text{dBm}$ [62], $P_{\text{TX,Reader}} = 30\,\text{dBm}$ [63], and $\tau = 100\%$). In Table 6.4, the corresponding outage probabilities of each on-body link and for each antenna configuration are listed. The table shows that the investigated RFID system is mostly limited in its forward link.

There are different strategies to overcome this limitation. An increase in the TX power at the on-body reader is not an option, because of the safety regulations and power constraints in on-body systems [64, 65]. A promising approach is to use semi-passive backscatter tag with chip sensitivities down to $-40\,\text{dBm}$ [66]. Such a sensitivity corresponds to a gain threshold of $F_{\text{Th}} = -70\,\text{dB}$. Table 6.5 shows that semi-passive tags lead to a good performance in the system forward links.

However, there is still a rather large limitation in the stomach-back link of the patch antenna systems. This constraint can be resolved by the use of more efficient patch antennas realized on a low-loss substrate or by the use of higher mode patches that benefit surface waves on the body. Another solution would be the use of a second RFID reader on the female's back to reduce the distance of the propagation path.

Table 6.4 Outage probabilities for a gain threshold of $F_{\text{Th}} = -47\,\text{dB}$.

Monopoles	Back (%)	Chest (%)	Wrist (%)	Head (%)
900 MHz	30	0	6	21
2.45 GHz	100	0	20	64
Patch antennas	**Back (%)**	**Chest (%)**	**Wrist (%)**	**Head (%)**
900 MHz	100	40	46	86
2.45 GHz	100	50	95	90

Table 6.5 Outage probabilities for a gain threshold of $F_{\text{Th}} = -70\,\text{dB}$.

Monopoles	Back (%)	Chest (%)	Wrist (%)	Head (%)
900 MHz	0	0	0	0
2.45 GHz	2	0	0	0
Patch antennas	**Back (%)**	**Chest (%)**	**Wrist (%)**	**Head (%)**
900 MHz	10	0	2	2
2.45 GHz	44	0	4	1

6.4.2 Backward Link

Figures 6.21–6.24 plot the outage probabilities of the backward links P_B versus the gain threshold B_{Th} for all antenna configurations. The same characteristic behavior of on-body links as described in Section 6.4.1 can be observed.

6.4.2.1 System Example

A state-of-the-art RFID system provides for example a gain threshold of $B_{Th} = -118$ dB (see Eq. (6.9), with $T_{RX, Reader} = -95$ dBm [63], $P_{TX,Reader} = 30$ dBm, and $\eta = 20\%$). In Table 6.6, the outage probabilities of each on-body link and for each antenna configuration are listed. Some limitations in the backward links of the system can be observed.

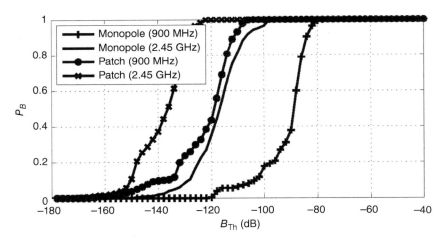

Figure 6.21 Outage probability P_B versus gain threshold B_{Th} of the stomach-back backward link.

Table 6.6 Outage probabilities for a gain threshold of $B_{Th} = -118$ dB.

Monopoles	Back (%)	Chest (%)	Wrist (%)	Head (%)
900 MHz	4	0	2	1
2.45 GHz	44	0	1	5
Patch antennas	**Back (%)**	**Chest (%)**	**Wrist (%)**	**Head (%)**
900 MHz	56	1	12	18
2.45 GHz	100	2	34	36

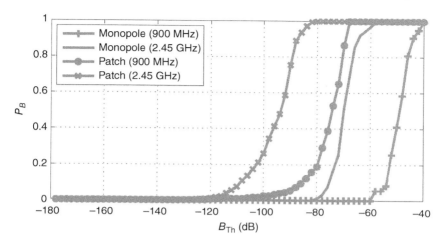

Figure 6.22 Outage probability P_B versus gain threshold B_{Th} of the stomach-chest backward link.

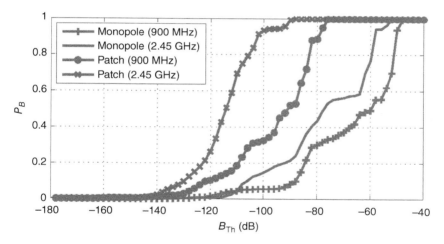

Figure 6.23 Outage probability P_B versus gain threshold B_{Th} of the stomach-wrist backward link.

To overcome these, a phase-modulated backscatter signal could be used. This modulation scheme provides a maximum modulation efficiency of $\eta = 81\%$ [67], which corresponds to gain threshold of about -124 dB. Table 6.7 lists the corresponding outage probabilities. It can be seen that the 6 dB difference in the threshold sufficiently improves the performance in the stomach-chest, stomach-wrist, and stomach-head links. If a phase-modulated backscatter signal is applied, there should be no substantial limitations in the system forward link

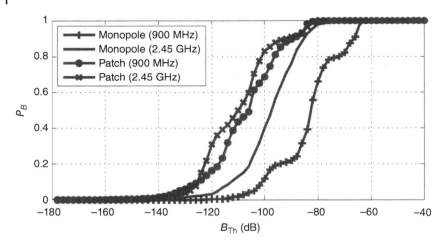

Figure 6.24 Outage probability P_B versus gain threshold B_{Th} of the stomach-head backward link.

Table 6.7 Outage probabilities for a gain threshold of $B_{Th} = -124$ dB.

Monopoles	Back (%)	Chest (%)	Wrist (%)	Head (%)
900 MHz	0	0	1	0
2.45 GHz	23	0	1	2
Patch antennas	**Back (%)**	**Chest (%)**	**Wrist (%)**	**Head (%)**
900 MHz	33	0	8	11
2.45 GHz	98	0	18	15

because of a reduced power transmission coefficient. Other strategies to overcome backward link limitations are, e.g. the use of a second reader unit, the realization of more sophisticated antennas, or the use of a reader RX with a lower sensitivity.

6.5 Conclusions

Backscatter RFID in the UHF and microwave frequency ranges is a promising communication technology for remote health monitoring applications. In particular, their low-power consumption makes backscatter tag appropriate for wireless sensing applications that require small, light-weight, and low-maintenance sensor nodes [68, 69]. Such sensor tags can be used, for example, to monitor the physiological parameters of a person, i.e. blood pressure, temperature, heartbeat, body motion. The backscatter bend sensor proposed in [70], which senses an object curvature, can be used for example to monitor the motion sequence of people for sports analysis or for human–computer interaction purposes, when a patient is undergoing physical therapy.

In backscatter RFID systems, it is vital to ensure a reliable wireless power transfer to the backscatter tag and to realize a robust wireless communication between the reader and tag. For example, if the power at the tag chip is smaller than the chip sensitivity, the backscatter communication system is limited in its forward link. If the power at the reader RX is smaller than the RX sensitivity, a limitation in the backward link occurs. Thus, the proper design of backscatter RFID devices, which are included in sensor tag, and the investigation of backscatter radio channels are key study areas.

This work investigates an on-body backscatter RFID system for remote health monitoring applications at 900 MHz and 2.45 GHz. The investigation is done for two different on-body antenna types. Monopole antennas act as a best-case reference, while less efficient patch antennas are used to give insight into practical RFID system implementations. The antennas are designed by means of a human body model to account for proximity effects that are caused by human tissues. The body model reflects the behavior of the electromagnetic properties of human tissues and is found to be an appropriate tool for the design of on-body antennas.

The system performance is evaluated by means of on-body channel measurements in a realistic indoor scenario. In particular, the channel transfer functions of the system are examined versus different body postures and lead to outage probabilities of the system forward and backward links. These probabilities help to identify limitations in the backscatter system and to evaluate strategies to overcome these barriers for the realization of a reliable on-body RFID system. An analysis of a state-of-the-art system example shows that the use of semi-passive chips leads to a reliable performance in the system forward link. A strategy to overcome limitations in the system backward link is to use a phase-modulated backscatter signal.

It is worth pointing out that the presented analysis can be performed for any kind of backscatter RFID system. The analysis gives an initial overview of a backscatter system and ultimately allows to realize a robust wireless power transfer and wireless communication.

Acknowledgments

This work was performed as part of the project Dependable, Secure, and, Time-Aware Sensor Networks that is supported by the Austrian Research Promotion Agency (FFG).

References

1 Curty, J.-P., Declercq, M., Dehollain, C., and Joehl, N. (2007). *Design and Optimization of Passive UHF RFID Systems*. New York, NY: Springer Science+Business Media, LLC.

2 Finkenzeller, K. (2003). *RFID Handbook*. Chichester, GBR: Wiley.

3 Bhattacharyya, R., Floerkemeier, C., and Sarma, S. (2010). Low-cost, ubiquitous RFID-tag-antenna-based sensing. *Proceedings of the IEEE* 98 (9): 1593–1600.

4 Rida, A., Yang, L., and Tentzeris, M. (2010). *RFID-Enabled Sensor Design and Applications*. Norwood: Artech House.

5 Smith, J.R. (2013). *Wirelessly Powered Sensor Networks and Computational RFID*. New York, NY: Springer.

6 Colella, R., Tarricone, L., and Catarinucci, L. (2015). SPARTACUS: self-powered augmented RFID tag for autonomous computing and ubiquitous sensing. *IEEE Transactions on Antennas and Propagation* 63 (5): 2272–2281.

7 Correia, R., Boaventura, A., and Borges Carvalho, N. (2017). Quadrature amplitude backscatter modulator for passive wireless sensors in IoT applications. *IEEE Transactions on Microwave Theory and Techniques* 65 (4): 1103–1110.

8 Daskalakis, S.N., Kimionis, J., Collado, A. et al. (2017). Ambient backscatterers using FM broadcasting for low cost and low power wireless applications. *IEEE Transactions on Microwave Theory and Techniques* 65 (12): 5251–5262.

9 Sample, A., Yeager, D., Powledge, P. et al. (2008). Design of an RFID-based battery-free programmable sensing platform. *IEEE Transactions on Instrumentation and Measurement* 57 (11): 2608–2615.

10 Ussmueller, T., Brenk, D., Essel, J., et al. (2012). A multistandard HF/UHF-RFID-tag with integrated sensor interface and localization capability. *Proceedings of 2012 IEEE International Conference on RFID*, Orlando, FL, USA (3–5 April 2012). pp. 66–73.

11 Balanis, C.A. (2016). *Antenna Theory: Analysis and Design*. Wiley.

12 Babar, A. A., Manzari, S., Sydanheimo, L., et al. (2012). Passive UHF RFID tag for heat sensing applications. *IEEE Transactions on Antennas and Propagation*.

13 Capdevila, S., Jofre, L., Romeu, J., and Bolomey, J. (2011). Passive RFID based sensing. *Proceedings of 2011 IEEE International Conference on RFID Technologies and Applications*, pp. 507–512.

14 Occhiuzzi, C., Caizzone, S., and Marrocco, G. (2013). Passive UHF RFID antennas for sensing applications: principles, methods, and classifications. *IEEE Antennas and Propagation Magazine* 55 (6): 14–34.

15 Qiao, Q., Zhang, L., Yang, F. et al. (2013). Reconfigurable sensing antenna with novel HDPE-BST material for temperature monitoring. *IEEE Antennas and Wireless Propagation Letters* 12: 1420–1423.

16 Yang, F., Qiao, Q., Virtanen, J. et al. (2012). Reconfigurable sensing antenna: a slotted patch design with temperature sensation. *IEEE Antennas and Wireless Propagation Letters* 11: 632–635.

17 Zannas, K., Matbouly, H. E., Duroc, Y., and Tedjini, S. (2018). On the cooperative exploitation of antenna sensitivity and auto-tuning capability of UHF RFID chip. Application to temperature sensing. *Proc. IEEE International Microwave Symposium*, Philadelphia, PA, USA (10–15 June 2018).

18 Krigslund, R., Dosen, S., Popovski, P. et al. (2013). A novel technology for motion capture using passive UHF RFID tags. *IEEE Transactions on Bio-Medical Engineering* 60 (5): 1453–1457.

19 Krigslund, R., Popovski, P., and Pedersen, G.F. (2012). Orientation sensing using multiple passive RFID tags. *IEEE Antennas and Wireless Propagation Letters* 11: 176–179.

20 Paggi, C., Occhiuzzi, C., and Marrocco, G. (2014). Sub-millimeter displacement sensing by passive UHF RFID antennas. *IEEE Transactions on Antennas and Propagation* 62 (2): 905–912.

21 Caizzone, S., Giampaolo, E.D., and Marrocco, G. (2017). Setup-independent phase-based sensing by UHF RFID. *IEEE Antennas and Wireless Propagation Letters* 16: 2408–2411.

22 Goertschacher, L., Boesch, W., and Grosinger, J. UHF RFID sensor tag antenna concept for stable and distance independent remote monitoring. In: *2018 IEEE International Microwave Biomedical Conference (IMBioC)*, 16, 14.06.2018–18, 15.06.2018. IEEE.

23 Grosinger, J., Gortschacher, L., and Bosch, W. (2016). Passive RFID sensor tag concept and prototype exploiting a full control of amplitude and phase of the tag signal. *IEEE Transactions on Microwave Theory and Techniques* 64 (12): 4752–4762.

24 Jiang, Z., Fu, Z., and Yang, F. (2012). RFID tag antenna based wireless sensing method for medical transfusion applications. *Proceedings of 2012 IEEE International Conference on RFID Technologies and Applications*, pp. 126–130.

25 Chen, X., Ukkonen, L., and Bjorninen, T. (2016). Passive E-textile UHF RFID-based wireless strain sensors with integrated references. *IEEE Sensors Journal* 16 (22): 7835–7836.

26 Long, F., Zhang, X., Björninen, T., et al. Implementation and wireless readout of passive UHF RFID strain sensor tags based on electro-textile antennas. *Proceedings of 2015 9th European Conference on Antennas and Propagation*.

27 Yi, X., Cho, C., Cooper, J. et al. (2013). Passive wireless antenna sensor for strain and crack sensing – electromagnetic modeling, simulation, and testing. *Smart Materials and Structures* 22 (8).

28 Caccami, M., Manzari, S., and Marrocco, G. (2015). Phase-oriented sensing by means of loaded UHF RFID tags. *IEEE Transactions on Antennas and Propagation* 63 (10): 4512–4520.

29 Shuaib, D., Ukkonen, L., Virkki, J., and Merilampi, S. (2017). The possibilities of embroidered passive UHF RFID textile tags as wearable moisture sensors. *Proceedings of IEEE 5th International Conference on Serious Games and Applications for Health (SeGAH).*

30 Goncalves, R., Rima, S., Magueta, R. et al. (2015). RFID-based wireless passive sensors utilizing cork materials. *IEEE Sensors Journal.*

31 Manzari, S. and Marrocco, G. (2014). Modeling and applications of a chemical-loaded UHF RFID sensing antenna with tuning capability. *IEEE Transactions on Antennas and Propagation* 62 (1): 94–101.

32 Virtanen, J., Ukkonen, L., Björninen, T. et al. (2011). Inkjet-printed humidity sensor for passive UHF RFID systems. *IEEE Transactions on Instrumentation and Measurement* 60 (8): 2768–2777.

33 Kutty, A. A., Björninen, T., Sydänheimo, L., and Ukkonen, L. (2016). A novel carbon nanotube loaded passive UHF RFID sensor tag with built-in reference for wireless gas sensing. *Proceedings of IEEE MTT-S International Microwave Symposium (IMS).*

34 Occhiuzzi, C., Rida, A., Marrocco, G., and Tentzeris, M. (2011). RFID passive gas sensor integrating carbon nanotubes. *IEEE Transactions on Microwave Theory and Techniques* 59 (10): 2674–2684.

35 Vyas, R., Lakafosis, V., Lee, H. et al. (2011). Inkjet printed, self-powered, wireless sensors for environmental, gas, and authentication-based sensing. *IEEE Sensors Journal* 11 (12): 3139–3152.

36 Zhang, J. and Tian, G.Y. (2016). UHF RFID tag antenna-based sensing for corrosion detection & characterization using principal component analysis. *IEEE Transactions on Antennas and Propagation* 64 (10): 4405–4414.

37 Kim, S., Kawahara, Y., Georgiadis, A. et al. (2015). Low-cost inkjet-printed fully passive RFID tags for calibration-free capacitive/haptic sensor applications. *IEEE Sensors Journal* 15 (6): 3135–3145.

38 Nguyen, S.D., Dang, C.M., Pham, T.T. et al. (2013). Approach for quality detection of food by RFID-based wireless sensor tag. *Electronics Letters* 49 (25): 1588–1589.

39 Occhiuzzi, C., Ajovalasit, A., Sabatino, M. A., et al. RFID epidermal sensor including hydrogel membranes for wound monitoring and healing. *Proceedings of 2015 IEEE International Conference on RFID*, pp. 182–188.

40 Hao, Y. and Foster, R. (2008). Wireless body sensor networks for health-monitoring applications. *Physiological Measurement* 29 (11): R27–R56.

41 Patel, S., Park, H., Bonato, P. et al. (2012). A review of wearable sensors and systems with application in rehabilitation. *Journal of NeuroEngineering and Rehabilitation* 9 (21): 1–17.

42 Amendola, S., Bovesecchi, G., Palombi, A. et al. (2016). Design, calibration and experimentation of an epidermal RFID sensor for remote temperature monitoring. *IEEE Sensors Journal* 16 (19): 7250–7257.

43 Amendola, S., Palombi, A., and Marrocco, G. (2018). Inkjet printing of epidermal RFID antennas by self-sintering conductive ink. *IEEE Transactions on Microwave Theory and Techniques* 66 (3): 1561–1569.

44 Hall, P. and Hao, Y. (eds.) (2006). *Antennas and Propagation for Body-Centric Wireless Communications*. Norwood, MA: Artech House, INC.

45 Conway, G. and Scalon, W. (2009). Antennas for over-body-surface communication at 2.45 GHz. *IEEE Transactions on Antennas and Propagation* 57 (4): 844–855.

46 Gabriel, S., Lau, R., and Gabriel, C. (1996). The dielectric properties of biological tissue: III. parametric models for the dielectric spectrum of tissues. *Physics in Medicine and Biology* 41 (11): 2271–2293.

47 Durney, C., Massoudi, H., and Iskander, M. (1986). Radiofrequency Radiation Dosimetry Handbook. *Technical report, Brooks Air Force Base-USAFSAM-TR-85-73*.

48 ANSYS, Inc. (2008). Ansoft Corporation, HFSS Online Help.

49 Hall, P., Hao, Y., Nechayev, Y. et al. (June 2007). Antennas and propagation for on-body communication systems. *IEEE Antennas and Propagation Magazine* 49 (3): 41–58.

50 Occhiuzzi, C., Cippitelli, S., and Marrocco, G. (2010). Modeling, design and experimentation of wearable RFID sensor tag. *IEEE Transactions on Antennas and Propagation* 58 (8): 2490–2498.

51 Lea, A., Hui, P., Ollikainen, J., and Vaughan, R. (2009). Propagation between on-body antennas. *IEEE Transactions on Antennas and Propagation* 57 (11): 3619–3627.

52 Balanis, C. (2005). *Antenna Theory*. Hoboken, NJ: Wiley.

53 Griffin, J. (2009). High-frequency modulated-backscatter communication using multiple antennas. PhD thesis. Georgia Institute of Technology.

54 Mayer, L. (2009). Antenna design for future multi-standard and multi-frequency RFID systems. PhD thesis. Vienna University of Technology.

55 Nikitin, P. and Rao, K. (2006). Performance limitations of passive UHF RFID systems. *Proceedings of IEEE Antennas and Propagation Society International Symposium* (July 2006).

56 Molisch, A. (2005). *Wireless Communications*. Chichester, GBR: Wiley.

57 Icheln, C., Ollikainen, J., and Vainikainen, P. (1999). Reducing the influence of feed cables on small antenna measurements. *IET Electronics Letters* 35 (15): 1212–1214.

58 Grosinger, J. (2013). Feasibility of backscatter RFID systems on the human body. *EURASIP Journal on Embedded Systems* 2013 (1): 1–10.

59 Grosinger, J. and Scholtz, A. (2011). Antennas and wave propagation in novel wireless sensing applications based on passive UHF RFID. *e&i Elektrotechnik und Informationstechnik* 128 (11–12): 408–414.

60 Nikitin, P. and Rao, K. (2008). Antennas and propagation in UHF RFID systems. *Proceedings of IEEE International Conference on RFID* (April 2008).

61 Lasser, G., Langwieser, R., Xaver, F., and Mecklenbräuker, C. (2011). Dual-band channel gain statistics for dual-antenna tyre pressure monitoring RFID tags. *Proceedings of IEEE International Conference on RFID*, (April 2011).

62 Impinj, Inc. (2012). Monza® 5 Tag Chip Datasheet (IPJ-W1600).

63 Impinj, Inc. (2012). Indy® R1000 Reader Chip (IPJ-P1000).

64 IEEE Std C95.1-1991 (1992). *IEEE standard for safety levels with respect to human exposure to radio frequency electromagnetic fields, 3 kHz to 300 GHz.*

65 (1998). Guidelines for limiting exposure to time-varying electric, magnetic, and electromagnetic fields (up to 300 GHz). *Health Physics* 74 (4): 494–522.

66 Herndl, T. (2012). Chip sensitivities of semi-passive backscatter tags, Private Communication.

67 Rembold, B. (2009). Optimum modulation efficiency and sideband backscatter power response of RFID-tags. *Frequenz* 63 (1–2): 9–13.

68 Grosinger, J., Pachler, W., and Bösch, W. (2018). Tag size matters: Miniaturized RFID tags to connect smart objects to the Internet. *IEEE Microwave Magazine* 19 (6): 101–111.

69 Pachler, W., Pressel, K., Grosinger, J., et al. (2014). A novel 3D packaging concept for RF powered sensor grains. 2014 IEEE 64th Electronic Components and Technology Conference (ECTC), (May 2014), pp. 1183–1188.

70 Grosinger, J. and Griffin, J. (2014). Backscatter RFID Sensor with a Bend Transducer. US Patent US8917202 B2.

7

Robotics Meets RFID for Simultaneous Localization (of Robots and Objects) and Mapping (SLAM) – A Joined Problem

Antonis G. Dimitriou, Stavroula Siachalou, Emmanouil Tsardoulias, and Loukas Petrou

Faculty of Technology, School of Electrical and Computer Engineering, Aristotle University of Thessaloniki, Thessaloniki, Greece

7.1 Scope

This chapter introduces the joined problem of simultaneous localization (of robots and objects) and mapping (SLAM). The idea is to develop a system capable of creating a map of a previously unknown environment, to be able to navigate autonomously in that environment and to track all objects. In order to accomplish this target, we assume that all target-objects in the environment are tagged with a ultra-high frequency (UHF) (or higher frequency) radio frequency identification (RFID) tag (typically passive; i.e. battery-less). We have constructed a robot, carrying all necessary RFID equipment (reader and antennas) to interrogate all tags (hence the associated objects) in the environment, also carrying all necessary sensor equipment (lidar, depth camera, Red Green Blue (RGB) cameras, etc.), so as to be able to create a map of the environment and locate its position at any given time-stamp.

The aforementioned problem applies in logistics, mainly in controlling warehouses and retail stores in a continuous manner, though it could also be applied in any tracking problem (e.g. in an airport). Deploying a moving system, instead of a fixed network of RFID antennas and readers, has significant advantages:

- A single robot can cover any space, regardless of its dimensions. For larger areas, more time is needed. Considering the small range of passive RFID systems, the cost of a fixed installation is huge.
- The number of measured samples to perform localization can be infinitely large, by adjusting the speed or the navigation strategy of the robot. This implies that the accuracy of the system can be improved.

Wireless Power Transmission for Sustainable Electronics: COST WiPE - IC1301,
First Edition. Edited by Nuno Borges Carvalho and Apostolos Georgiadis.
© 2020 John Wiley & Sons, Inc. Published 2020 by John Wiley & Sons, Inc.

- It can be used in any unmapped (and unknown) environment and create a map of it.
- Can be used for active perception, i.e. on demand movement in areas of RFID tag localization uncertainty.

Due to the aforementioned advantages, such robotic systems have recently appeared both as commercial products and in recent bibliography. In this chapter, we will try to document prior-art, of at least what we have considered as the most representative techniques and finally present our experimental results.

In the next section, we briefly discuss on some properties of RFID technology and localization. Then, we present the most common methods for localization of RFID tags. In Section 7.4, we present the problems and prior-art related to robotic simultaneous localization and mapping (SLAM) and localization. Finally, we demonstrate our prototype and show some measurement-results from our robotic implementations.

7.2 Introduction

In principle, Radio Frequency IDentification (RFID) technology consists of an RFID reader (Figure 7.1), an RFID antenna (Figure 7.3), and an RFID tag (Figure 7.2). The reader is capable of interrogating and "reading" the unique information associated with the specific tag; namely, its ID and possible information stored in the tag. The tag is able to store such information in its internal memory and transmit it to the reader, when ordered to do so. Hence, it also includes some form of elementary intelligence, in order to define when it should transmit information to the reader.

Figure 7.1 "Impinj R420" RFID reader.

Figure 7.2 RFID tags.

Figure 7.3 UHF RFID antennas.

Communication between reader and tag takes place by means of back-scatter radio. The reader transmits a modulated signal and continues to transmit a carrier frequency. The tag switches its state (by means of a transistor), thus modulating the incident carrier wave on its antenna, in accordance to its stored information (unique ID and – if desired – information data). The reader demodulates the modulated back-scattered signal by the tag and hence "reads" the tag's ID and information. Several standards have emerged. In the remaining chapter, we will focus only on EPC UHF Gen2 Protocol.

The RFID tag may or may not include a battery. When a battery is included, the RFID tag is characterized as active or semi-passive. An active tag uses its battery to backscatter the information to the reader. Such a system could have a range in the order of 100 m. A semi-passive tag exploits its battery to provide the necessary "turn on" voltage to the diode at the front-end and thus improves the range of reader-tag-reader system, as the latter now depends on the sensitivity of the reader and not of the tag (the round-trip path-loss is of importance). Such systems could have a range in the order of tens of meters. Finally, the most popular RFID tag-type is the one without a battery, named "passive" RFID tag. Such tags transform the incident electromagnetic wave into the necessary DC to power up the transistor that modulates the desired back-scattered information. As a consequence, the range of the reader-tag-reader system (i.e. interrogation zone) is typically constrained by the ability of the tag to harvest the necessary power to communicate with the reader. The "electronics" component that introduces this limitation is the necessary rectification diode used at the front end of each passive RFID tag, which needs a minimum DC voltage to "turn on" [1]. State-of-the-art RFID chips report a minimum desired power in the order of −20 dBm, [2, 3]. Alternatives, to increase the range of passive RFID systems, have been reported and they typically rely on exploiting additional energy sources (scavenging) [4, 5], additional antenna elements (array) [6], or improving the shape of the incident signal to match the nonlinearity of the front-end of the tag (imagine a duty cycle of a signal with high energy peaks to activate a diode followed by zero transmissions, such that the total power, within a duty cycle, still agrees with the Federal Communications Commission (FCC) rules for maximum power transmission) [7, 8]. The aforementioned energy constraints, combined with the maximum allowed transmission power by the reader (in the order of 33 dBm equivalent isotropic radiated power [EIRP]), limit the interrogation zone in the order of a few meters from the reader (up to 13 m, under free-space propagation conditions, depending on the antenna of the tag).

Apart from the electronics, an RFID tag also includes an antenna, carefully tuned to the tag, in order to ensure maximum power transfer. As tuning of the tag-antenna is greatly influenced by the materials in proximity to the tag, [9], different tags should be selected depending on the application. Some commercial tags destined for different applications are illustrated in Figure 7.3, where size, material, and design differs.

Currently, the most common type of passive RFID tags costs as low as $0.02 per piece. From a financial perspective, this means that they can easily be deployed to "track" almost any object, without significantly "influencing" (if at all, considering the benefits in logistics) the cost of the product and its market value. The combinatorial advantages of (i) "low cost", (ii) different shaping/sizing, and (iii) customization for different applications, have enabled RFID technology to penetrate many areas of commercial applications, either replacing traditional technologies (e.g. barcode) or creating new ones. The main applications for RFID technology are in the fields of:

– Retail supply chain (passive UHF RFID tags). A tag "follows" the product from manufacturing to the end user, serving different purposes at each point of the supply chain.

- Livestock farms (high frequency (HF) and UHF passive RFID tags). Legislation forces adoption of the technology in many countries and is foreseen to become a global mandate.
- Precision agriculture.
- Financial, security, and safety (HF near-field passive RFID tags). This category includes plastic cards, e-passports, tickets, etc.
- Healthcare (UHF passive RFID tags). Tracking and management of personnel, patients, archiving, drugs for health.
- Land and sea logistics (can also be considered as part of the "wide" retail supply chain).
- Manufacturing, industry. Includes tagging of machines and robots, apart from the products, already covered before.
- Passenger, transport, automotive (UHF passive or UHF/2.4 GHz active RFID tags for highway-applications).
- Leisure activities, sports (all types of tags).
- Military (all types of tags, depending on the range of tracking target).
- Books, library, archiving (HF or UHF passive RFID tags).

In terms of volume of tags, passive UHF RFID tags hold the majority of the market. However, from a financial perspective, they are still behind HF tags (though estimated to surpass them when tagging of all commercial products will become standard), due to the massive deployment of the latter in "finance" (plastic cards). In general, growth is driven by "retail" and is further accelerated by the adoption of the "Internet of Things" (IoT), demanding low-cost devices to wirelessly interconnect objects.

Different analysts estimate an RFID market value of \$15–\$32 billion by 2023. Apparel tagging alone will demand 8 billion RFID tags in 2017, having penetrated only a small percentage of the related market. Ninety-nine percent of tags are expected to be passive (batteryless) by 2023, while active RFID tags are expected to operate in the 2.45 GHz band.

From an application perspective, localization of the tag is equivalent to locating the tagged object. Localization is inherently bound with RFID technology; i.e. an object is tagged in order to be "tracked" by the reader (this is an essential property of RFID technology). In principle, even if no algorithm is applied, we know (and utilize this information) that when a reader successfully interrogates a tag, the tag is located within the interrogation zone (range) of the reader at the specific moment in the time axis.

Localization is essential for many applications related to the aforementioned areas. Some representative cases are illustrated in Figures 7.4–7.9. For instance, tracking of goods is essential for productivity increase and error-control in many points of the supply chain management, including warehouse management (Figure 7.5), shipment tracking (Figure 7.6), inventory management, and

Figure 7.4 Tracking of passengers and luggage in an airport. Source: https://pixabay.com/photos/airport-gate-flight-1659008/

Figure 7.5 Tracking of goods in a warehouse. Source: https://pixabay.com/photos/forklift-warehouse-machine-worker-835340/

Figure 7.6 Tracking of cargo in a harbor. Source: https://pixabay.com/photos/business-cargo-containers-crate-1845350/

Figure 7.7 Tracking of books in a library. Source: https://pixabay.com/ photos/books-library- read-shelves-shelf- 1617327/

Figure 7.8 Tracking of products in retail. Source: https://pixabay.com/ photos/store-clothing- shop-bouique-984393/

Figure 7.9 Tracking of animals in livestock units. Source: https://pixabay .com/photos/cow-allg %C3%A4u-cows- ruminant-2788835/

real-time stock-out control in retail (Figure 7.8), avoidance of misplacement errors, etc. It is also essential for cost-reduction (reduction of flight delays) and safety issues in airports (tracking of passengers by tagging the boarding pass) as well as tracking luggage around the conveyor belts (Figure 7.4). Libraries (Figure 7.7) represent a typical application case, where misplaced books can be located. Similar needs are identified in healthcare and industrial applications. Other applications may include tracking of animals (Figure 7.9) or identifying the location of a plant in agriculture, where RFID tags can be used for communicating plant/soil/air-sensor data to the network (e.g. imagine a drone flying around the field to collect data and localize each sensor).

In this chapter, we present prior-art in algorithms that improve (in terms of accuracy) the inherent potential of readers to localize objects within their interrogation zone. Furthermore, we focus on robots, carrying all necessary equipment to perform localization and present the advantages of such an approach and the related market applications. Finally, we put forward our own prototype robot that performs such a task and present measurement results with state-of-the-art algorithms. Research findings confirm our expectations that accurate localization of passive RFID tags is strongly related with classic problems from robotics: localization or SLAM, of the robot itself.

7.3 Localization of RFID Tags – Prior Art

Localization of RFID tags is based on mapping a measurable quantity related to the tag into location. A first classification of RFID localization methods is derived from the measured physical quantity, namely:

- "Time-of-flight" – the round-trip time from reader-to-tag-to-reader of an RFID communication,
- "Received signal strength information" (RSSI) – the backscattered power received at the reader, and
- "Phase" of the backscattered modulated signal at the reader.

Depending on the way this information is processed to geometrically pinpoint the location of the tag, a different classification arises:

- "Distance" based methods, where tri (or multi)-lateration is involved in the next step,
- "Angle-of-Arrival" (AOA) base methods,
- "Exhaustive search" methods, where the location that maximizes a function based on observations (measurements) and expectations (physical model) is retrieved, and
- "Hybrid" methods, where a combination of the above is deployed.

Generally, there is no good or bad method. The "quality" of a localization method strongly depends on the application. In order to judge the suitability of a localization method, one should combinatorially consider the following criteria in conjunction with the application:

- *Accuracy* – mean error and standard deviation.
- *Speed* – how fast the algorithm decides on the locations.

- *Simplicity* – complexity of the deployment of the algorithm in actual applications.
- *Cost* – what is the cost of the necessary infrastructure for localization.

For example, "exhaustive" search methods typically ensure high accuracy, but involve many calculations, increasing the estimation-time for large problems, when many tags exist. On the other hand, "fingerprinting" methods (how much does an observation set of measurements "resembles" a set of measurements from a "reference" tag at known location) can be very fast, with smaller accuracy. If the discrimination-step of the problem (the desired accuracy) is larger than the expected accuracy of the simpler method, one could opt for the latter; e.g. imagine that we wish to discriminate large containers in the harbor, where the tags (the containers) are spaced by meters (due to their size) or imagine that the reader is placed next to a conveyor belt at a specific place in an airport, where measuring only the slope of the phase of each passing tag (on the suitcase) is more than enough to decide on the location of the suitcase (the slope will change sign exactly when the suitcase crosses the location of the reader). From an "expected accuracy" perspective, better accuracy is demanded in the following decreasing sequence: (i) Machine industry robotics (μm to mm), (ii) machine guidance, monitoring, indoor surveying (mm to 5 cm), (iii) ambient assisted living, law enforcement, pedestrian navigation, product tracking (5 cm to 10 m). The aforementioned classification is only indicative, since specific applications within a group are more demanding than others.

Historically, localization algorithms, currently applied to RFID technology, are influenced (based) on early works related to "Radars" and more recently on "wireless sensor networks" (WSNs) – in fact RFID can be considered as a communication technology to form a WSN. The similarity to the radar problem is obvious, as both problems involve backscattered information from a source. However, there exist many differences: in radars, the target is passive, while the tag modulates the incident wave, the bandwidth of the two systems is totally different (RFID is narrowband), influencing the resolution of time-based localization methods, the distance between source and target is different, affecting the expected accuracy of the method (e.g. a 2 m error is large in RFID technology). Differences with WSNs arise mainly because of the size of the problem. In WSNs, each node includes a power source and a transceiver. As a result, the range of such a system is (typically) much larger than the range of passive RFID systems. The latter observation seamlessly leads us to present the particularities of passive RFID technology that strongly affect localization algorithms and the expected accuracy. The main cause of inaccuracy in all localization algorithms is multipath. Bear in mind that most localization algorithms assume that we can measure a physical quantity that directly links the reader to the tag (in a straight-line vector).

7.3.1 Multipath in Passive RFID Systems

"Multipath" is the inherent phenomenon in all wireless systems that the same transmission arrives at the receiver following different paths around the propagation area. Due to the different length of these paths and the different inter-action with surrounding objects, each path, usually referred as "ray", arrives at the receiver at a different time, with different phase and different amplitude. These "rays" can be considered as rotating vectors with the same angular frequency equal to the carrier frequency with constant phase differences (for a fixed position) and constant (but different) amplitudes. These vectors are vectorially summed, and the resultant vector is the one typically measured by the receiver, [10], with the exception of ultra-wideband systems, where discrimination of paths can be performed – not the case of commercial RFID technology. As a consequence, the phased locked loop of the RFID reader measures the phase of the resultant vector (not the direct), the power corresponds to the power of the resultant vector (not the direct), and the time corresponds to the time-of-arrival of the resultant pulse; additionally, when AOA is deployed, the AOA corresponds to a vector that does not point to the direction of the actual source. A simple example is demonstrated in Figure 7.10, where the direct path that links a transmitter and a receiver is blocked and reception is carried out through a single reflection on a wall.

The "measured" distance corresponds to the distance traveled by the reflected path, the phase corresponds to that of the reflected ray, the angle of arrival points to the wall, and every time-measurement will be mapped to the distance traveled by the reflected path. The good news about batteryless/passive RFID technology is that this scenario is not as probable as in the case of long-range systems. The need for power in the order of at least −20 dBm, combined with the small power transmission limitation of the reader (~33 dBm EIRP), makes it very difficult to "read" a tag under non-line-of-sight (NLOS) conditions; i.e. when the direct path

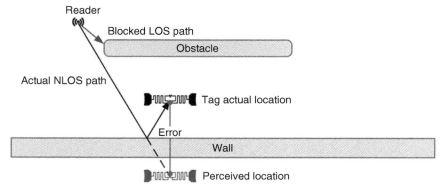

Figure 7.10 An obstructed case, where an RFID-reader measures the reflected path (LOS stands for Line-Of-Sight).

is blocked. Analysis and measurements of the expected backscatter channel can be found in [11, 12].

7.3.2 Representative Localization Techniques

Prior art in the field is extensive, due to the great interest for commercial applications. We will try to cover part of it and then focus on methods involving robots, similar to our approach.

7.3.2.1 Angle of Arrival

The most common way to estimate the AOA is to measure the phase of the backscattered signal at two antennas of the reader [13], as shown in Figure 7.11.

If $R \gg L$, i.e. vectors R_1, R_2 are approximated as being parallel, θ is:

$$\sin(\theta) \approx \frac{R_2 - R_1}{L} \tag{7.1}$$

The phase φ_i at antenna i of the backscattered EM wave is given by:

$$\varphi_i = \frac{2\pi}{\lambda} 2R_i + \varphi_{0i} \tag{7.2}$$

where two times the distance between antenna and tag has been replaced, which is valid for the monostatic RFID reader case. The term φ_{0i} denotes phase components due to transmit-receive leakage (hardware of the reader), static environment clutter, and differences due to the two antennas [14]. This term is constant for each transmit-receive setup and can be calculated. If the same antenna (of the same reader) is used for the two measurements (i.e. the reader is moving), then the term can be considered common for the two measurements (assuming the same environment). By substituting (7.2) in (7.1), we have:

$$\sin(\theta) = \lambda \frac{\varphi_2 - \varphi_1}{4\pi L} \tag{7.3}$$

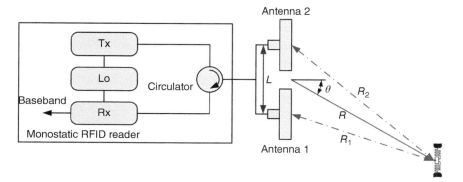

Figure 7.11 Phase measurement at two antennas connected to a monostatic RFID reader.

For the monostatic case, let $\Delta\varphi = \varphi_1 - \varphi_2$ and $\Delta R = R_1 - R_2$. Since the maximum phase difference that can be discriminated equals 2π, we have $\Delta\varphi \leq 2\pi \Rightarrow \frac{2\pi}{\lambda}\Delta R \leq 2\pi \Rightarrow R_1 - R_2 \leq \frac{\lambda}{2}$. The maximum difference is recorded along the line that connects the two antennas. Therefore,

$$L \leq \frac{\lambda}{2} \tag{7.4}$$

As the UHF RFID frequency band includes regions at 865—868 MHz and 902—928 MHz, the minimum requirement (at 928 MHz) states that $L \leq 16.1$ cm. Most UHF RFID antennas are usually planar microstrip, with dimensions from 11 cm up to 30 cm, depending on the desired gain; hence, only the smallest antennas can match the above constraint. However, within the main antenna lobe, which is smaller than $180°$ (typically from $60°$ to $75°$), constraint (7.4) is more relaxed; i.e. spacing between the antennas can be larger [13].

If the distance L between the two antennas is greater than $\lambda/2$, then the geometric locus of points that satisfy the desired inequality $\Delta R \leq \frac{\lambda}{2}$ is externally bounded by a hyperbola with foci at the two antennas. Let the distance L between the foci be:

$$L = k\frac{\lambda}{2} \tag{7.5}$$

and consider a Cartesian coordinate system centered between the two antennas, as shown in Figure 7.12. Assuming x to be the "antenna" axis and y the perpendicular one (pointing toward the antennas' main beam), the equation of the hyperbola is:

$$\frac{x^2}{\left(\frac{\lambda}{4}\right)^2} - \frac{y^2}{\left(\frac{\lambda}{4}\right)^2 (k^2 - 1)} = 1 \tag{7.6}$$

The asymptotes of the hyperbola are:

$$y = \pm\sqrt{k^2 - 1}\, x \tag{7.7}$$

The resultant asymptotes that geometrically bound the area, where the direction of a placed tag can be evaluated without phase ambiguity is demonstrated in Figure 7.12 for a monostatic reader configuration. This analysis can be easily extended for a bistatic case. As shown in Figure 7.12, the AOA estimation region matches a $70°$-beamwidth antenna main lobe for a maximum spacing of $L = 28$ cm for an RFID system operating at 928 MHz.

However, additional margins (thus reduced antenna-spacing) need to be considered to account for the phase measurement error, due to reader hardware. In [13], an experimental root mean square error (RMSE) equal to $1.7°$ is reported.

When additional elements are used to form an antenna array, well-known techniques can be applied in the phase measured by each antenna element, e.g. root MUSIC in [15], provided that the phase can be measured at each antenna element.

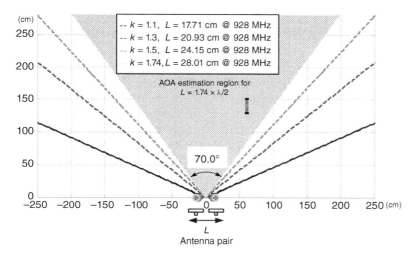

Figure 7.12 AOA estimation region for different antenna spacing L.

In [16], the authors use three downconverters to form an array of three elements and place three arrays to locate tags inside a 3 m × 3 m room. Antenna-element spacing is kept at 11.5 cm with a main beamwidth of 55°. Mean error is in the order of 21 cm with a standard deviation of 13 cm.

7.3.2.2 Received Signal Strength – Bayes' Theorem and Conditional Probability

The most common techniques for passive RFID localization are based on exploiting the measured backscattered power at the reader antenna. In its simplest concept, power is mapped to distance, assuming an injective function exists (e.g. path-loss formula) [17]; hence, by collecting measurements from several antennas, an estimation of the location of the tags can be performed either by multilateration or more typically by deploying a minimum mean square estimator. However, due to multipath [11, 12], such an injective function does not exist and most prior art exploits probability theory.

Most techniques are based on Bayes' theorem and conditional probability, properly formulated for the RSSI-measurements localization problem, e.g. [18]. Consider r to be the target position vector and $P = (P_1, P_2, \ldots, P_n)$ a set of n RSSI measurements related to the targeted object, i.e. measured at n different antenna locations. The conditional probability of r, given the observation set P is defined as:

$$p(r \mid P) = \frac{p(P \mid r)p(r)}{p(P)} \tag{7.8}$$

188 | 7 Robotics Meets RFID for Simultaneous Localization (of Robots and Objects) and Mapping

Note that $p(\boldsymbol{r})$ is a uniform distribution, and for a given set of observed data, $p(P)$ remains the same over all possible hypothetical values of $\boldsymbol{r}$, then:

$$\max(p(\boldsymbol{r} \mid \boldsymbol{P})) \equiv \max(p(\boldsymbol{P} \mid \boldsymbol{r})) \tag{7.9}$$

Thus, one can solve the inverse problem. $p(\boldsymbol{P}|\boldsymbol{r})$ is the likelihood the conditional probability of obtaining a set $\boldsymbol{P}$ of RSSI measurements at the n known antenna-positions, given that the true position of the targeted object is $\boldsymbol{r}$. Assuming that the n collected measurements are independent:

$$p(\boldsymbol{P} \mid \boldsymbol{r}) = p(P_1, P_2, \ldots, P_n \mid \boldsymbol{r}) = p(P_1 \mid \boldsymbol{r})p(P_2 \mid \boldsymbol{r}) \ldots p(P_n \mid \boldsymbol{r})$$

$$p(\boldsymbol{P} \mid \boldsymbol{r}) = \prod_{i=1}^{n} p(P_i \mid \boldsymbol{r}) \tag{7.10}$$

The maximum likelihood estimation of the position $\boldsymbol{r}$ of the targeted object is the value of $\boldsymbol{r}$ that maximizes the likelihood function:

$$\boldsymbol{r}^* = \operatorname{argmax}_{\boldsymbol{r}} p(\boldsymbol{P} \mid \boldsymbol{r}) = \operatorname{argmax}_{\boldsymbol{r}} \prod_{i=1}^{n} p(P_i \mid \boldsymbol{r}) \tag{7.11}$$

The most crucial parameters for successful application of probabilistic RSSI localization models is to (i) carefully define probability $p(P_i | \boldsymbol{r})$, such that the physical multipath channel is properly modeled, including uncertainties at the expected mean of the distribution, such as detuning of tag antenna due to interaction with attached material, possible blocking of line of sight (LOS) path, different chip-sensitivity, different overall tag design (small/large antenna, differently matched, etc.), polarization losses due to tag orientation, different tag-behavior versus frequency and (ii) collect a large number of measurements n, such that the effects of fading (around the expected value) are stochastically reduced; thus, the estimation is improved.

Based on the Bayes' theorem, different estimators can be applied. For instance, by modeling the system as a successive collection of measurements in time (like an RFID-reader-equipped moving robot), Bayesian filtering can be applied, like Kalman [19–21] (both phase and RSSI are exploited), [22, 23], or particle filters [24, 25].

The common problem with all estimators is to match the propagation-fading model with the actual scenario. The more uncertainties exist in the actual scenario, the greater the expected estimator's error. For instance, accurate results (less than 10 cm mean errors) are reported in [20, 23]. In both cases, the measurements' environment was fully controlled. For example, in [20], specific tags were mounted on the ceiling (for RFID-reader equipped robot navigation/localization), which were illuminated by a highly directive antenna. The fact that all tags "suffered" almost identical blocking and fading conditions (the directive antenna ensures less interaction with objects around the environment, while the location of the tag itself cancels the strongest reflected ray – i.e. the wall illuminated by the antenna, right behind the tag), led to such small errors. So again, the application (in this case robot navigation) allowed for the selection of this technique to be suitable.

Could one expect such accuracy, by applying Bayesian filtering at RSSI data, when tracking products in a retail store?

Definitely not as the uncertainty, related to propagation parameters is big.

The second problem with such estimators is the "estimation speed". The related algorithms "search" for the vector r that maximizes the expectation over a given set of observations. As a result, the estimation time grows with the possible search region multiplied with the population of tags (tracked items). The fingerprinting method, presented next, has the advantage of overcoming those limitations (speed and propagation uncertainties).

7.3.2.3 Fingerprinting – "Landmarc"

One of the most successful RSSI-based localization methods is "Landmarc" [26]. Passive RFID tags are placed at known locations around the region, where the unknown tags are "tracked". These will be used as reference "landmarks" to evaluate the unknown locations of target tags around them. The estimation is based on evaluating the similarity of measurements collected at an antenna grid between reference tags with each unknown tag.

Let n reader antennas collecting measurements of m "reference" tags and u "tracked" tags. Let $\mathbf{T}^j = (T_1^j, T_2^j, \ldots, T_n^j)$ be the signal-strength measurements' vector of "tracked" tag j by the n reader antennas, where $j \in [1, u]$ and

$$T_i^j = \begin{cases} E_i^j, \text{if tag } j \text{ is identified by antenna } i \\ \quad \text{null}, \quad \text{else} \end{cases} \tag{7.12}$$

where E_i^j is the measured backscattered signal strength of tag j at antenna i. Similarly, let $\mathbf{L}^l = (L_1^l, L_2^l, \ldots, L_n^l)$ be the corresponding collection of RSSI measurements of reference tag l, $l \in [1, m]$, from the same n antennas:

$$L_i^l = \begin{cases} E_i^l, \text{if tag } l \text{ is identified by antenna } i \\ \quad \text{null}, \quad \text{else} \end{cases} \tag{7.13}$$

For each "tracked" tag j and each reference tag l, we define the following distance-"resemblance" metric:

$$D_l^j = \begin{cases} \sqrt{\sum_{i=1}^n (T_i^j - L_i^l)^2} \quad, \quad \text{if } \left\{ (T_i^j \neq \text{null}) \cap (L_i^l \neq \text{null}) \right\} \\ \infty, \text{else} \end{cases} \tag{7.14}$$

For each "tracked" tag, we create a distance-resemblance vector:

$$\mathbf{D}^j = \left(D_1^j, D_2^j, \ldots, D_m^j \right) \tag{7.15}$$

The smallest element in (7.15) represents the reference tag, for which the measured RSSI values best fitted the corresponding measured RSSI values of the "tracked" tag. Hence, we expect the actual location of the tracked tag to be "closest" to that reference tag. Furthermore, the "resemblance" vector can be used as a distance indicator from each reference tag, thus "weighting" the distance of the "target" tag from each reference tag. Since the smallest elements in (7.15) are more significant, the corresponding resemblance metric can be inversed. In fact, one can select the kth smallest values in vector D^j, (quoted as k-nearest neighbors, abbreviation "k-nn") and estimate the coordinates of target tag j by the following two equations:

$$\left(x^j, y^j\right) = \sum_{i=1}^{k} w_i \left(x^i, y^i\right) \tag{7.16}$$

where the weights represent the squared inverse metric of Eq. (7.14), normalized as:

$$w_i = \frac{1/\left(D_i^j\right)^2}{\sum_{i=1}^{k} 1/\left(D_i^j\right)^2} \tag{7.17}$$

where only the finite values of D_i^j participate in the estimation. There are several "parameters" that need to be set and optimized in "Landmarc", e.g. the number k of considered "neighbors" in Eqs. (7.16)–(7.17), the weights in (7.16), the locations of the reference tags, the density of the corresponding locations, the number of antennas, etc. Furthermore, Eq. (7.14) should be carefully treated when the common measurement-samples of a given tracked tag with the reference tags are few (common non-null sample space). A minimum threshold of measurement-pairs between "reference" and "target" tag (Eq. (7.14)) must be defined.

There are two significant advantages of "Landmarc," compared to other algorithms: (i) estimation does not consider the locations of the antennas; it only depends on the locations of the reference tags, (ii) the algorithm is fast (in fact it can be considered real-time); estimation time depends on the number of the reference tags that have been "jointly" identified with each target tag (that number is very small for any typical computing hardware).

7.3.2.4 Holographic Localization

Perhaps one of the most promising methods that exploits only commercial-off-the-shelf (COTS) hardware is the "holographic" method. The method has been developed in the area of ultrasonic and radar and is applied in RFID in [27]. As discussed earlier, an RFID reader can measure the phase of the backscattered signal at the reader antenna. The method tries to answer the following question:

Given a set of phase-measurements of the same static RFID tag at n known antenna locations, what is the unknown location of the tag that best fits these observations?

The method represents a good candidate for application in an RFID-reader equipped mobile robot because a single moving antenna can provide the essential measurements' data. Furthermore, depending on the desired accuracy, one can collect the corresponding proper amount of data. The method is not suitable for static reader-antenna installations, as it would need an unreasonably large number of antennas for any practical implementation. Such properties of the method will be discussed later. Let's review the notation:

Consider θ_k the phase measurement collected at the kth out of n antenna-positions with known coordinates. The phase term θ_k also includes a "DC" phase term θ_{DC}, due to the reader's hardware [14] that does not depend on the Euclidian distance between tag and reader-antenna. If the same moving-reader-antenna (i.e. the same hardware) is used for the collection of the measurements, this phase term is common, i.e.

$$\theta_k = \theta_{DC} + \theta_k^{\text{Eucl.}} \tag{7.18}$$

Let a grid of m possible locations of the unknown tag be taken. For each possible location of the grid i, with coordinates $\{x_i, y_i\}$, we calculate the corresponding expected phase that would have been measured at the n known antenna locations as:

$$\varphi_{ik} = \frac{2\pi}{\lambda} 2d_{ik} \bmod 2\pi, k = 1, \dots, n, \tag{7.19}$$

where d_{ik} denotes the Euclidian distance between grid location i and antenna location k. For each grid location i, we calculate the following term:

$$P_i(x_i, y_i) = \left| \sum_{k=1}^{n} e^{j(\theta_k - \varphi_{ik})} \right|, i = 1, \dots, m \tag{7.20}$$

By replacing (7.18) in (7.20), we have:

$$P_i(x_i, y_i) = \left| e^{j\theta_{DC}} \right| \left| \sum_{k=1}^{n} e^{j\left(\theta_k^{\text{Eucl.}} - \varphi_{ik}\right)} \right| \Rightarrow$$

$$P_i(x_i, y_i) = \left| \sum_{k=1}^{n} e^{j\left(\theta_k^{\text{Eucl.}} - \varphi_{ik}\right)} \right|, \quad i = 1, \dots, m \tag{7.21}$$

At the actual tag location, represented at one of the grid-points, the different vectors in (7.21) are expected to add constructively. In all other grid-points, the vectors will add randomly. Therefore, Eq. (7.21) is expected to be maximized at the grid point i that best matches the phase-measurements. The plot of Eq. (7.21) is called "hologram," hence the name of the method. The coordinates that correspond to the grid-point of the maximum of the plot represent the estimated tag-position.

$$(x, y) = \text{argmax}_{(x_i, y_i)} P_i(x_i, y_i) \tag{7.22}$$

If different hardware is deployed, "phase-difference" samples are used instead; i.e. all measurements from a given antenna are subtracted from the first measurement-sample collected at the same antenna [28]. Of course, the same process is repeated for the theoretical values for each grid-point:

$$P_i(x_i, y_i) = \left| \sum_{k=1}^{n} e^{j[(\theta_k - \theta_1) - (\varphi_{ik} - \varphi_{1k})]} \right|, i = 1, \dots, m \tag{7.23}$$

This "subtraction" ensures that the common DC terms of each different hardware equipment (pair of transceiver with antenna) is eliminated. Again, it is reminded that in the mobile-robotic-platform, this method is useful when multiple antennas are connected to different antenna ports of the reader. Similarly, if multiple frequencies are involved, proper modifications are needed in the aforementioned equation to account for the different wavelength per hardware-antenna pair and per frequency.

One of the latest works, involving the "holographic" method is presented in [29], where comparisons versus "Tagoram," presented in [28] are reported. The algorithm named "MobiTagbot" deteriorates the effects of multipath in phase-measurement samples by repeating the measurements over multiple frequencies in the UHF RFID frequency band, ultimately achieving impressively high accuracy compared to prior art. The main idea is to suppress the importance of measurement-samples that suffer from strong multipath, with respect to samples suffering from "low" multipath (i.e. the direct component prevails over the received profile). They define an "entropy" metric to evaluate the effect of multipath, quantifying the variability of the phase-measurements over different frequencies. Depending on the metric, they weight each sample to be more or less important in the application of Eq. (7.21). In the experimental results, the authors report a median error of 3.5 cm, outperforming "Tagoram."

7.3.2.5 Other Methods

The previous sections summarize the most representative methods in prior-art. The majority of prior-art represents a modification of the presented algorithms. However, there are some additional approaches worth noticing.

One such method is called "PinIt," [30]. "PinIt" exploits the virtual antenna array (VAA) idea to estimate the AOA of the multipath profile for each tag. A VAA is formed when collecting measurements by a single antenna at discrete points in space (and at discrete time-slots); equivalently to an actual antenna-array with fixed antenna-elements at the same locations. Evidently, the RFID-equipped moving-robotic-platform represents in principle a VAA. The authors estimate the AOA profile of each tag, postprocessing the received backscattered signals, by applying complex phasors to each sample according to classic antenna-array theory:

$$B(\theta) = \left| \sum_{k=0}^{K-1} w(k, \theta) s(t_k) \right|^2$$

$$w(k, \theta) = e^{-j \frac{2\pi}{\lambda} t_k v \cos \theta} \tag{7.24}$$

where $B(\theta)$ is the AOA profile, K is the number of virtual antenna elements (K measurement-samples at discrete locations), $s(t_k)$ are the measured backscattered signals at time-slots t_k, and v is the speed of the moving antenna. The AOA profiles of target tags is compared to the corresponding AOA profile of reference tags, located at known locations (fingerprinting), applying dynamic time warping (DTW). Finally, weighting the coordinates of the k-nearest neighbors, based on the similarity of the fingerprinting algorithm gives an estimation of the location of the unknown tags.

As all fingerprinting techniques, success of PinIt relies on the similarity of multipath of the reference tags with target tags; hence the denser the grid, the better. However, as the population of tags increases (reference and unknown), the measurements per sample will decrease unless the robot moves at smaller speed. If the speed of the robot reduces, the total time for a collection of measurements will increase, and the channel variability will increase. This would cause different AOA multipath profiles among neighboring tags (reference and target), despite of the "efforts" of the DTW, due to large time-intervals of measurements samples. Still, "PinIt", represents a promising alternative to the localization problem.

Another approach, called "RFind", was presented in [31]. The authors "emulate" a wideband RFID system, in order to acquire high resolution. The authors transmit concurrently a "legitimate" UHF RFID signal and a low-power out-of-band carrier frequency (e.g. at 1 GHz). The in-band RFID signal causes the tag to power-up and start switching between its two states, thus modulating the in-band carrier. At the same time, the lower power carrier is also modulated since it also impinges on the powered-up RFID tag. As a result, a "custom" RFID reader (in this case universal software radio peripheral (USRP) N210 software radio transceiver) with sufficiently low sensitivity demodulates and stores the out-of-band backscattered signal. The same process is repeated for different carrier frequencies, thus ultimately capturing the response from a wide-bandwidth (200 MHz in [31]). The authors in [31] claim that the out-of-band transmission should be kept below −13.3 and 6.7 dBm for average and peak power, respectively, in order to comply with FCC rules for unlicensed low-power transmissions. As soon as the "wideband" data is collected, the authors apply the inverse fractional Fourier transform (IFRFT) to obtain a time-domain representation of the backscattered signal. From the power-delay received profile of the IFRFT, the authors identify the first peak to correspond to the LOS path. The time-resolution (multipath separability) depends on the bandwidth of the measurements $1/B$ (B is the bandwidth); e.g. a 220 MHz bandwidth corresponds to a necessary 4.5 ns time resolution. The estimated time-delay of the LOS path is better approximated in the next step, by matching the measured phase of the backscattered signal at all frequencies (in the 220 MHz bandwidth). The same process is repeated at three receivers at known locations (three USRPs), and 3D localization is accomplished by the cross-section of the three spheres. Impressive experimental results with 2D median accuracy of 0.91 cm are reported.

However, there are some important drawbacks of the proposed approach:

1. The results in [31] indicate that accuracy is poor (compared to prior-art) for a bandwidth less than 120 MHz.
2. In order to acquire the 220 MHz sampled frequencies at 10 MHz steps, a total time of 6.4 s(!) was needed (130 ms was the time for the USRP to change channel) for a single tag. This time-per-tag must be multiplied to the number of tags, which turns applicability of the algorithm for commercial problems (with many tags) to be impractical (even if half the bandwidth is used, accepting a lower accuracy).
3. RFind requires that the tag-antenna-chip front-end remains operational (even at reduced matching conditions) over a wide bandwidth – way much wider than the bandwidth that these tags are designed for. Such operational conditions are hardly ever expected to be met by commercial tags. This problem will be further aggravated, due to detuning of the tags, when attached to different materials [9].

7.3.3 Analysis of Prior Art

In the previous sections, we have presented the most representative localization techniques for RFID tags. The AOA methods are sensitive to the distance between successive samples. Fast localization is expected, while the accuracy depends on multipath (the direct LOS path should be detected).

Techniques based on conditional probability and RSSI are expected to be slow, since exhaustive search in a set of possible locations is required. Furthermore, uncertainties related to the environment, such as blocking of the tag, detuning of the tag antenna due to interaction with the attached material, mispolarization of the tag with respect to the incident field due to geometrical placement, etc. might result in large errors in the estimations.

Fingerprinting techniques based on RSSI will result in fast estimations. The accuracy depends on the similarity of the fading effects on reference tags with respect to the corresponding effects on the target tags. Hence, accuracy depends on the density of the grid of reference tags. A big advantage of the presented algorithm is that it does not depend on the location of the reader-antennas.

Holographic localization is expected to be accurate, since the measured phase does not depend on the orientation of the tag, its interaction with the attached object or any other parameter related to the magnitude of the backscattered signal. However, it is reminded that the measured phase is that of the resultant vector composed of all multiply scattered rays in the environment. Still, such multipath effects can be reduced when a large number of samples is available. Estimations are slow and depend on the size of the search-space; i.e. the size of the grid of possible locations. Yet, there are many possibilities to reduce the search-space.

We are investigating the potential to deploy an RFID-equipped robot to perform localization. The robot will be able to perform SLAM; i.e. create a map of the environment and track its position in that map. In the section, we present state-of-the-art SLAM techniques. Precision of SLAM strongly affects the accuracy of the localization of the tags. For instance, if the estimated location of the robot (hence the reader antenna) is inaccurate, localization of the tags will suffer from the same error (with the exception of "Landmarc," as discussed earlier).

7.4 A Brief Introduction in SLAM/Localization Techniques

If we were to provide a generalized description of what a robot is, we could state that a robot constitutes an agent equipped with sensors and effectors that can be programmed to autonomously present intelligence. In this context, sensors are hardware or software devices that transduce information to measurements, either concerning the interior of a robot (proprioceptive sensors such as velocity or battery sensors) or the outside world (exteroceptive sensors such as distance or vision sensors). On the other hand, effectors are electrical devices capable of directly affecting the state of the environment the robot is located in (e.g. motors or speakers). In contrast to purely teleoperated robots, autonomous robots form a special class of robotic vehicles, capable of intelligent motion and actions without requiring either a guide to follow or teleoperation control [32]. In order for a robotic agent to be autonomous, it must possess several functionalities that allow it to navigate the environment, such as gathering sensor input from its surroundings, interpreting these measurements toward accurately localizing itself in the environment, maintaining a correct environment representation, planning a path between two points, and ultimately safely traversing it.

It is understandable that the precise spatial localization of a robotic agent, as well as an accurate representation of the surrounding environment is of the utmost importance for a truly autonomous robot [33] and therefore essential to the autonomous RFID localization, if performed with the assistance of a robotic vehicle. In this section, an overview of Localization and SLAM techniques will be provided. Specifically, we will provide an introduction to the aforementioned techniques, defining their taxonomy according to the task at hand, and then present the core mathematical methodologies for describing and solving such problems, as well as different space representation techniques. Finally, a special mention will be made concerning what types of SLAM/Localization are suitable for RFID localization problems, including issues that may arise due to probabilistic uncertainty.

7.4.1 Introduction to Localization, Mapping, and SLAM

As aforementioned, SLAM stands for simultaneous localization and mapping and concerns the concurrent computation of the robot's poses and the map of the environment. SLAM is considered one of the hardest problems in robotics, since no single methodology exists that can correctly determine both the robot poses and the map of an environment, regardless of the robotic type, the environmental peculiarities, or the tasks the robot has to resolve. In fact, SLAM is the more generalized form of such problems, since mapping and localization are treated as different tasks.

Localization concerns the estimation of the robot's pose (or location), provided the map is a priori known. Therefore, the robot has knowledge of environmental morphology but not of its pose in it, which tries to determine based on sensory measurements. There are three main types of localization (presented from simpler to harder), namely position tracking, when the robot's initial pose is known and a method must exist to correctly track the future poses, global localization [34, 35] where the initial pose is unknown and the robot must progressively localize itself via motion and sensor measurements acquisition and the kidnapping localization problem where the robot must be globally localized but there is a chance of someone "kidnapping" it and transporting it to another location. In this case, the robot must detect that it has been kidnapped in order to relocalize itself [36, 37].

Mapping is the opposite problem, since the robot is constantly aware of its pose but has to construct the map of the environment. The most usual cases where the pose is known are when Global positioning system (GPS) or D-GPS sensors are utilized, by which the location of a moving vehicle (car, ground, or aerial robot) can be determined, always within a specific amount of error (some meters for GPS and centimeters for D-GPS).

The most general and complex problem is SLAM, where the robot tries to build a map without exactly knowing its location and at the same time localize itself in this map. This description resembles the chicken or egg problem, since an accurate map is needed for a correct localization, but in order to build an accurate map, you must first be correctly localized. There are many different types of SLAM, some of them being (among others):

- Volumetric versus feature-based SLAM, differentiating in the environmental representation.
- Topologic versus geometric maps, where in topologic maps the actual positions and distances between landmarks are of no importance, in contrast to the geometric representation
- Known versus unknown correspondence, where in known correspondence, the robot is able to detect unique features of the environment and associate them with already visited areas of the map
- Static versus dynamic environments, where in static environments, it is assumed that the only moving agent in the surrounding area is the robot, whereas in the dynamic environments, other active agents exist (people, other robots etc.)
- Single robot versus multi-robot SLAM, where in the latter case, several local maps must be correctly merged into one unified representation

Next, the mathematical formulation of the SLAM problem will be briefly presented.

7.4.2 Mathematical Formulation of SLAM

In SLAM, regardless of the different specialized approaches in existence, there are some known parameters:

- The **robot's controls** u, $u_{1:T} = \{u_1, u_2, ..., u_T\}$, where u_t denotes the control at time t, whereas $u_{1:T}$ the set of controls from the initial state to the current time T. Usually, the robot's controls are velocities, accelerations, or any other information resulting in alteration of the robot's pose in the environment.
- The **robot's observations** z, $z_{1:T} = \{z_1, z_2, ..., z_T\}$,, where z_t denotes the robot's observation at time t and $z_{1:T}$ the set of observations from the initial state to the current time T. The observations essentially are sensor-produced measurements via which the state or form of the environment can be inferred or calculated.

The unknown parameters of the SLAM problem are:

- The **map of the environment** m. The space is depicted in the form of a map and may have many representations, some of which are feature-based maps, metric/occupancy grid maps (OGMs), OctoMaps [38], etc.

- The **path of the robot**, comprising the robot poses x, $x_{1:T} = \{x_1, x_2, \dots, x_T\}$, where x_t denotes the position of the robot at time t, in the map m. Obviously, x_t is not a single number, since it is tightly correlated with the map representation and the sensor information, as well as the task at hand. For example, for the 2D paradigm (a ground vehicle traversing a plane), $x_t = \{x_t, y_t, \theta_t\}$, where x_t, y_t are the coordinates in a Cartesian plane and θ_t is the robot's orientation (or yaw). Similarly, in the 3D case (a drone traversing a 3D space), $x_t = \{x_t, y_t, z_t, roll_t, pitch_t, yaw_t\}$, where z_t is the robot's height, and *roll, pitch,* and *yaw* are the Tait–Bryan angles (a different convention on the classic Euler angles).

In SLAM, all the aforementioned parameters (controls, observations, maps, and poses) are considered to be multidimensional probability distributions, since all of them suffer from inherent uncertainty. Therefore, a verbal description of the SLAM problem is the definition of the probability of the robot's poses $x_{0:T}$ and the map m, provided the robot's controls $u_{1:T}$ and observations $z_{1:T}$, something that is described by the following equation:

$$p\left(x_{0:T}, m \mid z_{1:T}, u_{1:T}\right) \tag{7.25}$$

This mathematical formula describes the full SLAM problem, since all the robot's poses must be calculated, whereas the online SLAM problem requires the computation of only the latest pose:

$$p\left(x_T, m \mid z_{1:T}, u_{1:T}\right) \tag{7.26}$$

7.4.3 Probabilistically Solving SLAM

After generally describing the SLAM problem, some questions arise. How do we extract relevant information from raw sensor data and how do we represent the uncertainty of this information over time? Similarly, how do we model the control uncertainty and how does this affect the overall SLAM procedure? Finally, how can we determine if the robot is located in a previously visited area in the map based either on the sensor measurements or the map itself (also known as the data association or loop closing problems)? These questions bring forward the SLAM problem's complexity once more, as the robot path and the map are both unknown but the map and pose estimates are highly correlated. On top of that, the mapping between the observations and the map is usually unknown, and wrong data associations may lead to catastrophic consequences in the map quality [39]. Since the only known variables are the controls and the observations, their probabilistic models ought to be defined.

As far as controls are concerned, their model is usually denoted as the **motion model** and describes the relative motion of the robot. Specifically, its generalized mathematical formulation is $p(x_t \mid x_{t-1}, u_t)$, i.e. the probability distribution of the robot's pose, given the previous uncertain pose and the current control. This is usually modeled either by using closed form probabilistic distributions, such as Gaussian models, or by non-Gaussian models such as particle filters. It is understandable that the "larger" the control (e.g. higher speeds or application for a longer time), the more uncertainty will be introduced in the new pose assumption. Sources of uncertainty concerning the controls (e.g. in the case of ground vehicles equipped with wheels) may be mechanical differences in the wheels, slippages, or bumps.

On the other hand, the **observation model** relates the measurements with the robot's pose: $p(z_t | x_t)$, i.e. calculates the probability of receiving a measurement z_t, given that the robot is located at pose x_t. Again, the observation model may be formulated with a closed form (e.g. a Gaussian distribution) or in a non-Gaussian form such as particle filters.

The aforementioned equations derive by applying the Markov chain assumptions [37], i.e. that the world is static, the noise is independent and that our models are perfect with no approximation errors. Thus, the realistic motion and observation models result in a more manageable form:

$$p\left(z_t, x_{0:t}, z_{1:t-1}, u_{1:t}\right) = p\left(z_t \mid x_t\right) \tag{7.27}$$

$$p\left(x_t, x_{1:t-1}, z_{1:t-1}, u_{1:t}\right) = p\left(x_t \mid x_{t-1}, u_t\right) \tag{7.28}$$

Proceeding to the abstract solution of SLAM, since it essentially is a conditional probability problem ($p(x_{0:T}, m \mid z_{1:T}, u_{1:T})$), the most common way to resolve it is via employment of the Bayes theorem and the Bayes filter. The Bayes probabilistic theorem is:

$$P(A \mid B) \cdot P(B) = P(B \mid A) \cdot P(A) \tag{7.29}$$

On the other hand, the Bayes filter is a well-known tool for state estimation and comprises two iterative steps; the prediction and the correction. Concerning the SLAM problem, after applying a series of Bayes and total probability theorems, along with the Markov assumptions, the Bayes filter prediction and correction steps are

$$\text{Prediction step}: \overline{\text{bel}\left(x_t\right)} = \int p\left(x_t \mid u_t, x_{t-1}\right) \cdot \text{bel}\left(x_{t-1}\right) dx_{t-1} \tag{7.30}$$

$$\text{Correction step}: \text{bel}\left(x_t\right) = \eta \cdot p(z_t \mid x_t) \cdot \overline{\text{bel}\left(x_t\right)} \tag{7.31}$$

In simple words, initially a prediction about the robot pose is performed using the control model, along with the previous pose belief, and this belief is corrected via application of the observation model. There are four main families of SLAM solutions, namely SLAMs solved using Kalman filters, particle filters, graph theory (graph-based SLAM), and scan-matching techniques.

As far as **Kalman filter-based SLAMs** are concerned, all variables (pose, observations, motions, and the map) are modeled as Gaussian distributions [40]. Furthermore, linear transition and observation models are assumed, as well as zero mean Gaussian noise is added, resulting in the following equations:

$$\text{Pose prediction}: x_t = A_t \cdot \boldsymbol{x_{t-1}} + B_t \cdot \boldsymbol{u_t} + \varepsilon_t \tag{7.32}$$

$$\text{Observation prediction}: z_t = C_t \cdot \boldsymbol{x_t} + \delta_t \tag{7.33}$$

Here, A_t is the matrix describing how the state evolves from $t-1$ to t without controls or noise, B_t describes how the control alters the pose, C_t describes how to map the state $\boldsymbol{x_t}$ to an observation, and ε_t, δ_t are noise modeling variables. By applying these equations to the Bayes filter, we get the Kalman filter solution for the SLAM problem (which is not going to be presented here, as it is out of the scope of this chapter). It should be stated that in real life, state variables are not linearly connected or abide by Gaussian distributions; thus, several improvements over the classic Kalman filter algorithm have been proposed, including extended Kalman filter (EKF) [41] or the unscented Kalman filters (UKFs) [42].

Another way to deal with nonlinearities is to make use of a nonparametric representation and specifically **particle filters**. Particle filters constitute a numerical representation of Bayes filter, where a distribution is represented by particles, each of which is a district hypothesis of the problem's solution. Since our problem is SLAM, each particle must contain its assumption of the robot pose and the map, along with an importance weight that indicates how "good" is the specific particle's solution. Again, in the particle filter algorithm for SLAM, the two steps of Bayes filter are implemented, since initially the particle poses are updated using the control model, their important weights are calculated via the observation model (according to how well the real measurements were expected from each particle's pose and how well do they match each particle's map), and finally, the particles are sampled to probabilistically remove the ones with low importance weights [43, 44].

As far as **Graph-based SLAMs** are concerned, they follow a completely different approach than the aforementioned. Specifically, they treat SLAM as an overdetermined system, i.e. having more equations than unknowns, on which equations they try to minimize the sum of squared errors. The standard approach of a graph-based SLAM is to store, categorized into two types: based on the robot's motion, i.e. connecting successive robot poses using odometry and based on (allegedly) previously observed areas. Next, a graph is generated, whose nodes are robot poses, and an edge between two nodes is a spatial constrain. Finally, the problem is solved using least squares error minimization on the graph, generating a correct graph structure and therefore sound robot poses and measurement registrations [45].

Finally, a special category is the **scan matching SLAMs**. Usually, scan matching SLAMs reject odometry, since it agreeably is quite noisy, and try to utilize the sensor measurements, which are much more precise (e.g. scans from Lidars or Depth cameras). The main concept is that by spatially comparing two successive measurements (or scans in the case of distance-based sensors), one can calculate the geometric transformation between them, whose inverse is the robot's pose alteration, i.e. the successive poses transformation is not calculated by odometry or dead-reckoning but by geometric evaluation of sets of 2D or 3D points. There are quite a few approaches in the scan matching algorithmic family [46–48], but it is also common to use scan matching in combination with the aforementioned methods (e.g. particle filters), in order to minimize the accumulated error produced by the raw odometry [49].

7.4.4 Space Representation in SLAM

There are two major categorizations of SLAM algorithms, which differentiate on how they represent and therefore store the space; the **feature-based** or **landmark-based** SLAMs and the **metric or volumetric** SLAMs.

The concept behind **feature-based** SLAMs is that the robot must be able to process the abundant raw sensor measurements and generate higher level of information, which may be able to contribute toward the robot's correct localization. In human analogy, let's assume that you are blindfolded and your only sensor is your touch. If you are in a known space, once you grasp a door handle (i.e. process the touch information and conclude that this object is a handle based on a model you have trained and stored in your mind), you immediately lower your overall uncertainty of where you might be. In order for landmarks to be "good," i.e. to be able to be used in a SLAM scenario, they should be easily observable and reobservable, distinguishable from each other, plentiful in the environment and of course stationary. The most usual forms of landmark-based SLAM utilize cameras (V-SLAMs or Visual SLAMs [50–52]), since it is possible to generate a lot of easily distinguishable features in the environment via image processing (scale invariant feature transform (SIFT), speed up robust features (SURF), oriented FAST and rotated BRIEF (ORB), etc.). Finally, the usual way of dealing with landmarks is to model each one of them using a Gaussian probability distribution.

On the other hand, **metric maps** are usually stored in grids which actually represent the world discretization into cells. This representation does not rely on a feature detector, usually the grid structure is rigid (all cells are of the same, constant size) and each cell contains a probability of it being occupied. Therefore, these maps are often referred to as **occupancy grid maps** or **OGMs**. Since each cell is a random variable that models occupancy, the cells are assumed to be

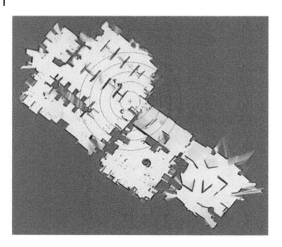

Figure 7.13 OGM produced by CRSM SLAM.

independent of each other (although in reality they are not), in order to simplify the procedure of probabilistically solving the problem, i.e. the probability of having a map m becomes $p(\boldsymbol{m}) = \prod_i p(m_i)$. The most common way to update the cells of an OGM is via the log odds notation by using the previous occupancy value of a cell and the inverse observation model [53]. Finally, metric SLAMs utilize distance information; therefore, the robot must be equipped with appropriate sensors like Lidars, sonars, depth cameras, etc. An example of a metric OGM, calculated using critical rays scan matching (CRSM) SLAM can be seen in Figure 7.13 [48].

As understandable, Kalman filter-based SLAMs mainly prefer the landmark-based space representation, since the robot pose and each landmark can be modeled using a Gaussian distribution, whereas the particle filter SLAMs may use either landmarks or metric maps.

7.4.5 SLAM Algorithm Selection

As described in the previous two sections, numerous different approaches to the SLAM problem exist; therefore, the selection of the proper algorithm for a specific scenario is not an easy task by an inexperienced researcher. Usually, there are three questions that must be answered in order to choose the best SLAM algorithm: (i) what and where are the robot's sensors, (ii) which is the environmental morphology, and (iii) how will the generated map be utilized?

7.4.5.1 What Are the Robot's Sensors?

Since SLAM is a procedure that generates new structures of data, its inputs must be acquired from active sensors, i.e. devices that measure properties of the environment. The two sensor categories used in the SLAM scenarios are distance and optical sensors. Some sensors that fall under the distance sensors category are Lidars, Sonars, IRs, Depth, and Stereo cameras, whereas optical sensors are different types of cameras, such as RGB, gray scale, thermal, IR, depth, etc. Since metric SLAMs represent the environment as a spatial decomposition with specific dimensions, they can mainly be deployed using distance sensors, i.e. devices that provide a measure of range, whereas optical sensors are mostly used for feature-based SLAMs. Of course this distinction has exceptions, since with methods like optical flow camera-derived information can be transformed into 3D points (therefore it can produce metric maps if a distance reference exists) and distance sensors can produce spatial features in order to be used in feature-based SLAMs. Finally, the topology of the sensors on the robot's chassis is of upmost importance, since an erroneous sensor pose can lead to a malformed map.

7.4.5.2 Which Is the Environmental Morphology?

One of the most important information in the processes of picking a SLAM algorithm is the type of environment the robot operates in. If metric maps are concerned (i.e. if distance sensors are mainly utilized), the density of obstacles in the environment must be in accordance with the sensors nominal range of operation, since if the world is too sparse, the input will be sparse as well, leading to loss of reference (e.g. in a scan-matching SLAM). Also, if the environment's obstacle density is satisfactory, one must make sure that the obstacles possess distinct spatial features (e.g. corners), in order for scan matching or metric feature-based approaches to operate. On the contrary, if the environment has sufficient obstacle density but lacks spatial features (e.g. mazes or long corridors), a SLAM that utilizes odometry must be selected, since this is the only way to maintain a valid robot pose assumption. Moving on to feature-based methods which utilize cameras (i.e. visual SLAMs), it is obvious that the environment must be rich in color or depth features in order for maintaining correct reference. Finally, several usual environmental culprits exist that can completely destroy the result of a SLAM algorithm, such as glass walls (where IR-based sensors cannot operate and RGB cameras fail), operating in dark conditions (where RGB cameras obviously fail), moving obstacles (where a dynamic SLAM algorithm must be chosen), or large circles/loops (where scan-matching fails and SLAMs with loop-closing abilities must be used) [54].

7.4.5.3 How Will the Generated Map Be Utilized?

Finally, a vital question is how the researcher will utilize the map. The two prominent answers fall into the aforementioned "metric or not" dilemma. If the maps are to be utilized as a research outcome or as a visualization artifact, the SLAM type is of no importance, as far as the generated map is precise (therefore the selection is performed based on the two previous questions), thus both metric and feature-based approaches can be utilized. When the generated map comprises the input of another algorithm, it makes sense that the SLAM selection is critical. For example, if the generated map will be used for navigation, it must have a metric representation, since almost all path planning methods operate in cell-based representations (A*, Probabilistic Roadmaps – PRMs, Rapidly Exploring Random Trees – RRTs, etc.). The same decision is made for applications like Monte-Carlo localization, where a metric map must exist in order to evaluate the sensor measurements against the possible robot poses.

7.4.6 SLAM/Localization and RFID Localization Issues

Since the task at hand is the precise RFID tags localization, it is reasonable that the accurate knowledge of the reader's position at all times is of utmost importance. In fact, most of the research around this field assumes a perfect knowledge of the reader's pose, aiming to focus in the tag localization problem itself. Nevertheless, the attachment of the RFID readers and antennas to a mobile robot introduces high levels of uncertainty, as no SLAM or localization algorithm can ensure robot pose and map correctness, regardless of the environmental peculiarities. In the localization domain, where the map a priori known, the pose uncertainty may be introduced by uncertainty in the control or observation models (Figure 7.14), by similar areas in the environment (Figure 7.15), or even from inherent errors in the map depiction.

When the robot does not have knowledge of its surroundings, therefore a SLAM algorithm must be applied, the uncertainty becomes even larger. Some of the most common errors in metric SLAM algorithms are usually presented via malformed maps, nevertheless since the SLAM procedure computes the robot poses based on the current map, these errors are included in the RFID reader's pose as well. For example, known SLAM issues are erroneous loop closing (Figure 7.16) resulting in doubling specific areas of the map, loss of reference in specific areas (Figure 7.17),

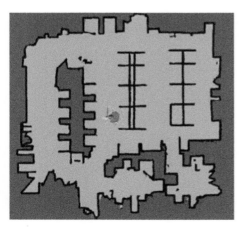

Figure 7.14 AMCL (http://wiki.ros.org/amcl) algorithm in ROS – particles dispersion.

Figure 7.15 AMCL algorithm in player – ambiguity due to symmetry.

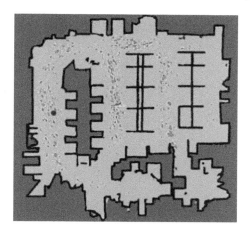

Figure 7.16 Loop closing problem in SLAM.

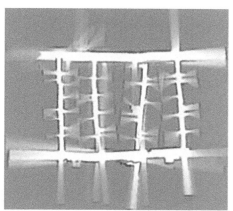

Figure 7.17 Loss of reference at a long corridor in a scan-matching SLAM.

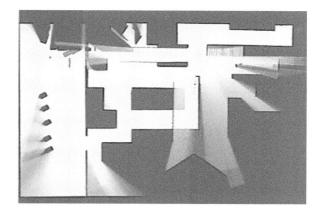

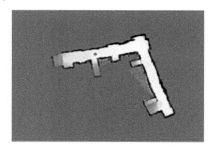

Figure 7.18 Nonlinearity error introduced by robot slippage and/or SLAM algorithm.

Figure 7.19 Total loss of robot pose and map reference.

nonlinearities in otherwise linear corridors (Figure 7.18) or even total failure to maintain correct pose and map estimation (Figure 7.19) [46, 48, 49].

In addition, special attention must be paid into selecting the appropriate SLAM algorithm for the type of robot that carries the RFID readers and antennas, as well as for the type of environment the robot exists in (i.e. outdoors, indoors with large open areas or indoors with rooms, corridors, etc.), in order for the error introduced by the SLAM inherent uncertainty to be minimized. Finally, it should be stated that since our task is to accurately calculate the RFID tags pose in the produced map, this means that the utilization of metric SLAMs is more favorable, as metric maps model both the existence and lack of occupancy, in contrast to the feature-based SLAMs where only information about the detected landmarks exists.

7.5 Prototype – Experimental Results

7.5.1 Equipment

The joined experiments were performed using a "Turtlebot 2" robot, [55], of the R4A (Robotics 4 All) robotics team, School of Electrical and Computer

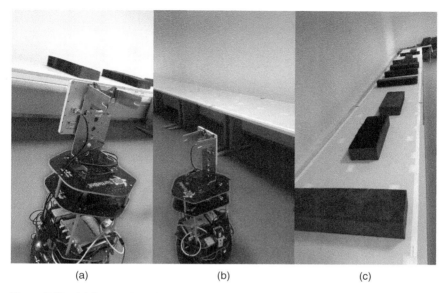

(a)   (b)   (c)

Figure 7.20   (a) Photo of the RFID-equipped robot, during measurements (with dielectrics placed on top of RFID tags). (b) Photo of the RFID-equipped robot, during measurements (unobstructed measurements). (c) The tags are attached to a long millimeter paper on top of a bench. During some of the measurements, dielectric materials are placed on top of some of the tags, blocking the LOS path.

Engineering (ECE), AUTH. This robot is modified in order to serve the research strands of the team, which mainly are autonomous SLAM, exploration/coverage, enhanced perception, and social applications (Figure 7.20).

The robot is equipped with an Intel i7 CPU attached on a Mini-ITX motherboard (MI980 from IBase), which serves as the main computational unit, and all required data are stored in an solid state drive (SSD) drive. The platform's main motion is supported by the Kobuki mobile base [56], which operates in a differential motion model fashion. As far as sensors are concerned, Turtlebot is equipped with an RPL-idar A1 [57], responsible for the SLAM operations, as well as with an Xtion Live Pro depth camera [58], mounted on a pan-tilt mechanism built with Dynamixel servos [59]. In order to support the RFID localization application, the "MT-242032/NRH" circularly polarized antenna from "MTI Wireless Edge" was mounted in a custom rack. The antenna has a 72° and 73° half power beamwidth along the azimuth and elevation planes, respectively and 7.5 dBic gain [60]. The antenna was connected to the Speedway Revolution R420 RFID reader, manufactured by IMPINJ [61]. All algorithms were executed in-robot and were implemented in the form of ROS nodes, grouped in ROS packages [62].

The experiments took place in a long corridor-type room inside the Campus. Eighty-seven EPC Gen2 "ALN-9662 Squiggle-SH Inlay" UHF-RFID tags, [63], were attached to a long millimeter paper, laid on a laboratory bench. In the experiments, a single antenna was used, which faced toward the area of the tags.

7.5.2 Methodology

The experimental formulation dictates that the environment is a priori unknown, and no blueprints of the area are available. Since we want to compare the estimated with the actual locations of the tags, we need to be able to transform the coordinates of the tag system to the local coordinates of the ROS system. Some reference optical landmarks are considered, in order to perform the transformation. The coordinates of the bench-legs were marked and will be used to account for proper transformation among the coordinate systems.

Since no map is available beforehand, the experiments were divided in two phases: (i) The first phase utilizes a SLAM algorithm to create a metric map of the environment, while (ii) the second applies the localization algorithm on the RFID tags.

7.5.2.1 Phase 1

During the first phase, the robot was manually steered in the entire unoccupied area, in order to produce a metric, OGM. The selected SLAM algorithm is Cartographer from Google, a real-time solution for indoor mapping [34]. Cartographer utilizes submaps (maps that represent a small percentage of the environment – usually the area around the robot), where Lidar scans are registered using a scan-to-map matching procedure. In order to be lightweight, it does not employ a particle filter, but when a submap's construction is finished, a pose optimization procedure is performed on the existent submaps, in order to solve the loop-closing problem. The generated OGM is presented in Figure 7.21 with a length of 14 m and width of 5 m, where the table with the RFID tags exists at the top side of the corridor. As evident, the mapping procedure was not perfect, since a minor rotational deformation can be detected along the corridor. This is the rule and not the exception in the SLAM domain, as it is impossible for a SLAM algorithm to achieve a blueprint quality, even in the simplest of cases.

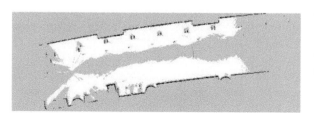

Figure 7.21 OGM of the Microwaves laboratory, ECE, utilizing Cartographer as SLAM algorithm.

7.5.2.2 Phase 2

The robot traverses the currently known environment, and at the same time captures the necessary RFID information. Since the robot pose plays a crucial role in the tag localization, the AMCL (Adaptive Monte-Carlo Localization) algorithm was utilized for calculating the vehicle position in the known OGM. AMCL uses a particle filter to represent the robot pose's multimodal probabilistic distribution, where each particle contains an assumption of the robot's pose. The particles are filtered based on the degree of alignment of the distance measurements input to the known map, thus maintaining a "correct" distribution. As evident, this procedure introduces another source of noise in the system, since it is quite improbable for a robot localization technique to compute the robot's pose robustly, correctly, and with precision at all times.

As soon as the map of the environment and localization of the robot is finished, one can apply any of the state-of-the-art algorithms, presented in Section 7.3. Initially, we must connect each measurement related to the tags with the estimated pose of the robot; i.e. the location and direction of the RFID reader-antenna, where each measurement-sample was collected. It is reminded that the robotic-localization system is independent from the RFID reader measurement system. However, both are controlled by the same processor and the same operating system (ROS). Hence, measurements (of tags) and estimations (of the robot's pose) from the two systems can be matched via the common timing of the system.

As the robot moves inside the room, the RFID-reader continuously interrogates every tag within range. For each tag response, the reader collects the following information:

Timestamp,   EPC-CODE,   Backscattered Power (RSSI),   Phase

From "Phase 2", described previously, the robot's pose (position and orientation) is also time-stamped. Hence, from the robot's localization system, the following data are collected with respect to the robot's (hence also the antenna's) pose:

Timestamp, x-coordinate, y-coordinate,   direction (along the x-y plane)

Evidently, each RFID-tag measurement can be associated with the location (and direction) of the RFID-reader antenna through the timestamp of each measurement. The denser the timestamps of the robot's pose, the better for the RFID-localization system. The robot estimates its pose every 20 ms. Depending on the speed of the robot, the desired spatial density of samples can be adjusted. We have performed experiments at three different speeds: 5, 10, and 20 cm/s. They correspond to 4, 8, and 16 mm space-intervals for the estimated poses of the robot, respectively. A characteristic result is demonstrated in Figure 7.22. The estimated route of the robot for the 5 cm/s experiment is shown (in dark gray).

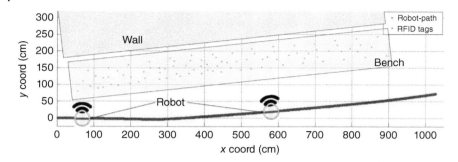

Figure 7.22 Estimated path of the robot (gray), actual locations of RFID tags ("x") and geometry of the environment.

The actual locations of the tags are also shown with an "x" marker. The location of the wall (opposite to the RFID-antenna) is demonstrated, as it introduces a strong reflection in the measurements.

When the reader interrogates the tags, they respond according to a slotted ALOHA random access protocol. As a consequence, the timing between successive tag measurements is random. It depends on the population of tags (which affect the collision-probability), the time-window (for the random back-off timer), etc. In our experimental setup, $10\,\text{tags/m}^2$ were placed on the bench, approximately. Each tag can be identified only when the power at its antenna is greater than a given threshold. Given the radiation pattern and the distance of the antenna from the tag population, the reader actually powers-up approximately 35 tags (out of the 87) from each location.

When a robot passes at the vicinity of the tag (under excellent link conditions), each tag (specifically) is identified at an average time interval among 48–100 ms (depends on the actual distance of the tag from the robot's path). Given the three speeds of the reader, the wavelength at 865.7 MHz (34.63 cm), and the 100 ms measurement-periodicity, the reader collects 70, 35, and 18 measurement-samples per tag within a wavelength of the robot's displacement for the 5, 10, and 20 cm/s, respectively.

Such a characteristic result is shown in Figure 7.23, where the measured phase, the backscattered power, and the path of the reader (that corresponds to these measurements) are demonstrated for the 5 cm/s experiment.

Notice that the phase-curve repeats in 2π intervals and changes slope when the reader antenna is located opposite of the tag (because this point of the path represents the location where the distance between tag and reader-antenna is minimized). The specific point is marked with a red-dotted line in Figure 7.23. It is not a surprise that the estimated location of the specific tag is $x_{\text{coord}} = 91.5\,\text{cm}$, $y_{\text{coord}} = 74\,\text{cm}$ (shown with a "+", while the actual location is shown with a "x"),

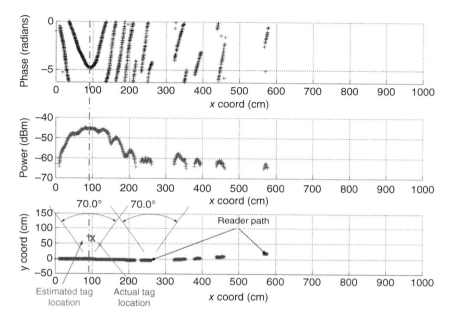

Figure 7.23 Measured phase and power for one of the tags, as the robot moves in the room. The locations of the robot are shown in the bottom plot.

as will be shown later. Also, notice that the reader antenna continues to measure the tag, even when the tag is no longer illuminated by the main radiation lobe of the antenna (see the 70° angle, when the reader is at 250 cm). In fact, the reader continues to identify the tag even when the robot is as far-away as 580 cm from its initial position.

The phase-curve changes smoothly with respect to the movement of the robot. On the contrary, the power curve may increase, at increasing distance (due to multipath as explained earlier). Furthermore, the power curve remains relatively stable for a large displacement of the robot (from 70 to 130 cm). This is due to the nonlinearity of the tag's chip, which is optimally designed for the minimum reception threshold. This power-flatness in the back-scattered power response reveals much less information with respect to the corresponding phase-curve.

Figure 7.23 represents a "good" measurement, implying that there is a lot of information stored to apply the RFID-localization algorithms presented earlier. However, there are also measurements, where such information is not so "rich". Such an example is demonstrated in Figure 7.24.

In the measured tag of Figure 7.24, the reader was unable to identify the tag when the robot crossed the path opposite to the tag, even though the distance was minimized. The most probable reason to justify this finding is that the strong

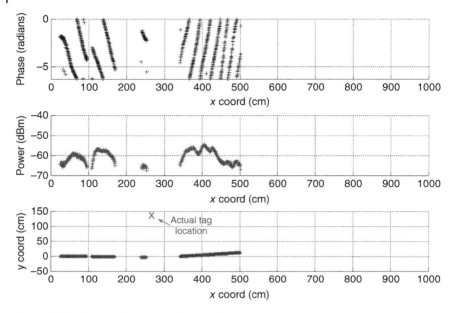

Figure 7.24 Measured phase and power for a tag.

reflection at the wall opposite to the antenna added destructively to the direct path (the LOS contribution). As a consequence, the resultant field (the phase-sum of the two vectors) was smaller than the required threshold to power-up the tag. The actual location of this tag further supports our hypothesis, since the tag is located at the upper end of the bench, next to the wall. Localization-accuracy of the specific tag is expected to be smaller, compared to the sample of Figure 7.23.

7.5.3 Results

Our purpose is not to present an analytical comparison between the methods, presented in Section 7.3. In fact, it seems that each of the methods might be advantageous over the others, depending on the propagation environment. Our goal is to demonstrate how the localization and mapping of the robot itself affects the performance of the localization of the tags.

Keeping that in mind, we have decided to present the performance results of the "holographic" localization method, presented in paragraph 7.3.2.4, without the improvements suggested in "Tagoram" [28] (sharpening the estimation of the holographic method) or "MobiTagbot" [29] (reducing fast-fading effects, which were not encountered in our "static" setup). In fact, we believe that insignificant improvements were to be expected, because our main sources of error are related to localization and mapping of the robot. We have decided to work with the phase (over the power information), because of the rich information of the measured data.

We have performed six experiments with three different speeds as explained in the following:

– The robot moves at 5 cm/s at approximately 1 m distance from the bench. All tags are unblocked.
– The robot moves at 5 cm/s at approximately 50 cm distance from the bench. All tags are unblocked.
– The robot moves at 5 cm/s at approximately 1 m distance from the bench. Some tags are blocked by a dielectric material, typically used in an anechoic chamber.
– The robot moves at 10 cm/s at approximately 1 m distance from the bench. All tags are unblocked.
– The robot moves at 10 cm/s at approximately 1 m distance from the bench. Some tags are blocked by a dielectric material, typically used in an anechoic chamber.
– The robot moves at 20 cm/s at approximately 1 m distance from the bench. All tags are unblocked.

We have applied Eqs. (7.21) and (7.22) and verified that (7.23) (only useful when different hardware is deployed) gives the exact same results. One characteristic result is demonstrated in Figures 7.25 and 7.26. Figure 7.25 shows the holographic function (Eq. (7.21)) for the tag that corresponds to the measurements of Figure 7.23. The maximum of the surface corresponds to the location that best fits the measured samples, according to Eq. (7.22).

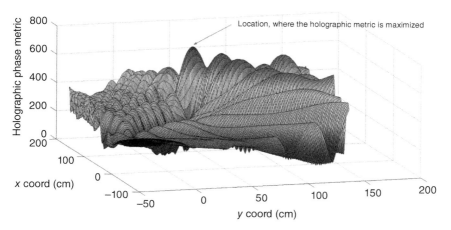

Figure 7.25 E-dimensional plot of the holographic metric, applied for a given tag.

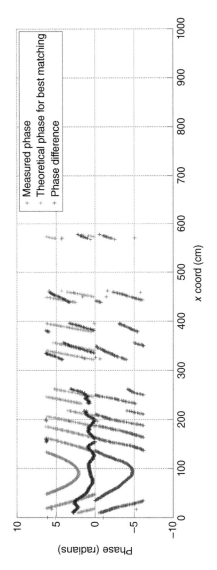

Figure 7.26 Measured versus best "matched" phase for the "holographic" localization method.

Table 7.1 Localization error, applying the "holographic" algorithm for the six experiments.

Speed (cm/s)	Obstacles	Distance from tags (m)	Mean error (cm)	Standard deviation (cm)
5	No	1	39	25.1
5	No	0.5	38	21.1
5	Yes	1	39.5	37.8
10	No	1	37.5	21.6
10	Yes	1	37.8	30.8
20	No	1	33.6	19.8

The result of Figure 7.26 shows the measured phase versus the theoretical phase for the coordinates that maximized Eq. (7.22) for the same tag (of Figures 7.25 and 7.23). The "dark gray" marks represent the measured phase (same measurement as in Figure 7.23). The "light gray" marks represent the theoretical phase for the location in the area that best matches these measurements, according to Eqs. (7.21), (7.22). The black marks represent the difference between those phase samples; i.e. the exponents in Eq. (7.21). Ideally (if no multipath exists), the black marks should have formed a straight line, parallel to the x-axis; suggesting that there is a point for which the measured phase is identical and differs only by a constant. Evidently, the black marks do not form a straight line due to multipath; the phase measured is the phase of the resultant vector comprising direct and multiply scattered contributions. The algorithm accomplished to "locate" the curvature of the measured phase at the region, where the slope changes sign (when the robot is opposite to the tag), which is a desired characteristic. The mean and the standard deviation of the error for the six measurements are given in Table 7.1.

The mean error ranges from 33 to 39 cm and seems stable in all experiments. The standard deviation is increased in both experiments with the dielectrics. The reason is that there are some samples with few data (due to blocking) that result in large errors. Such data affect the mean and the standard deviation of the experiment. Such an example is shown in Figure 7.27, where a poorly sampled measurement results in a large estimation error (214 cm).

In Figure 7.28, the estimated versus the actual locations of the tags are demonstrated for the experiment, when the reader moved at 20 cm/s. Notice that the estimated locations of the tags are always near the actual location; as was verified by the small standard deviation. If the application is in a warehouse or a retail store and that the robot informs a human of the locations of all products, this result dictates that the robot will always send the human to the correct shelf.

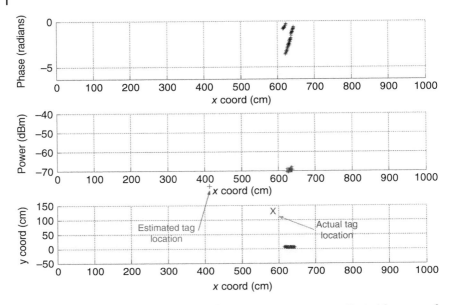

Figure 7.27 Under blocking conditions, few measurements were collected for some of the tags, which resulted in large localization error.

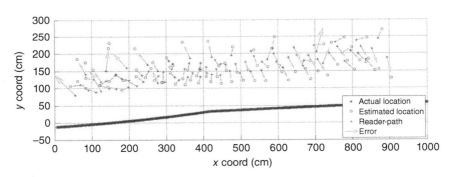

Figure 7.28 Estimated versus actual location of RFID tags for the 20 cm/s reader speed.

7.6 Discussion

In this chapter we have presented the most representative methods for simultaneous localization (of a moving robot in any unknown area) and mapping of the area. Furthermore, we have also presented the most representative methods for localization of UHF RFID tags. We have constructed a prototype robot, capable to perform both and have presented characteristic experimental results in Section 7.5.

To the best of our knowledge, this is the first time that experimental localization results are presented, considering both problems: (i) the possible error because of the localization and mapping of the robot and (ii) the error, due to localization of the tags.

State-of-the-art tag-localization algorithms report accuracies below 10 cm. However, in those experiments, the precise locations of the reader-antennas are known. As a consequence, the only source of error is in the tag-localization algorithm. In the real-world problem, the locations of the reader-antennas are not known, as treated herein. Despite the fact that a large number of measurements per tag are collected, the mean localization error ranges between 30 and 40 cm. We believe that this is partly due to the localization error of the tested "holographic" method and partly due to the error on the estimation of the map of the area and the positioning of the robot inside it.

Errors in the constructed map of the area, lead to erroneous interpretations of the measurements; i.e. the measured quantities no longer correspond to the visual representation of the environment. Errors in the estimated location of the robot lead to errors in the locations of the RFID tags, since the locations of the tags are estimated with reference to the locations of the reader-antenna (hence the poses of the robot).

As far as the RFID localization methods are concerned, the main source of uncertainty appears to be multipath. Strong reflections may significantly affect the performance of a method, since the reader will measure characteristics (magnitude and phase) of the resultant field vector at the receiver, while the direct path is only of interest. Multipath needs to be deteriorated, by applying intelligent techniques, during the collection of measurements (if commercial-off-the-shelf – COTS) equipment is used, or by applying conjugate-weighting in the collected samples if custom-RFID readers are deployed. For example, performing measurements in different frequencies and from different antennas is expected to limit multipath effects. Different frequencies result in different wavelengths and thus different phasors in the resultant multipath vectors for the same antenna-tag geometry; such property however cannot be exploited in Europe, where the UHF RFID bandwidth is very narrow (the wavelength differences are very small). Different antennas (on top of the same robot) imply that the geometrical relationships of the direct with the multiply scattered field components are different. Another possibility is the use of diversity in the tags. If the size of the tagged object allows it, one can attach more multiple tags on each object; even better performance is expected if those are orthogonally attached (interact with a different polarization of the incident field).

Even if such techniques are deployed and the potential accuracy is increased, still the presented "holographic" method is very slow (unsuitable for real-time tracking applications). The reason is that an exhaustive search on all possible locations of the environment is carried out in order to decide on the positions that best matches the observations. Acceleration techniques must be applied to reduce the search-space.

In conclusion, there is still room for significant improvements in this promising robotic-RFID application field. The advantages of the system, compared to static installations, combined with the increased commercial interest on such technology have gained a lot of attendance in the scientific community recently. A single system can perform mapping, navigation, inventorying, and localization at selective speeds, accuracy, and periodicity in any unknown space. The reported accuracy from our range of experiments is already good-enough for many commercial applications, including the main target applications, namely warehouse and retail-store inventorying as well as libraries.

Acknowledgments

This work has been co-financed by the European Union and Greek national funds through the Operational Program Competitiveness, Entrepreneurship and Innovation, under the call RESEARCH – CREATE – INNOVATE (project code:T1EDK-03032).

References

1 Dobkin, D. (2012). *The RF in RFID*, 2e. Elsevier.
2 Alien (2018). Higgs Chip – Alien. http://www.alientechnology.com/products/ic/higgs-ec/ (accessed 02 April 2018).
3 IMPINJ (2018). Monza chip – IMPINJ. https://www.impinj.com/platform/endpoints/ (accessed 02 April 2018).
4 Kampianakis, E., Kimionis, J., Tountas, K. et al. (2014). Wireless environmental sensor networking with analog scatter radio & timer principles. *IEEE Sensors Journal (SENSORS)* 14 (10): 3365–3376.
5 Vougioukas, G. and Bletsas, A. (2017). 24 µW 26 m range batteryless backscatter sensors with FM remodulation and selection diversity. *IEEE International Conference on RFID-Technology and Applications (RFID-TA)*, Warsaw, Poland (20–22 September 2017).
6 Griffin, J.D. and Durgin, G.D. (2008). Gains for RF tags using multiple antennas. *IEEE Transactions on Antennas and Propagation* 56 (2): 563–570.
7 Collado, A. and Georgiadis, A. (2014). Optimal waveforms for efficient wireless power transmission. *IEEE Microwave and Wireless Components Letters* 24 (5): 354–356.

8 Kimionis, J. and Tentzeris, M. (2016). Pulse shaping: the missing piece of backscatter radio and RFID. *IEEE Transactions on Microwave Theory and Techniques* 64 (12): 4774–4788.

9 Griffin, J.D., Durgin, G.D., Haldi, A., and Kippelen, B. (2016). RF tag antenna performance on various materials using radio link budgets. *IEEE Antennas and Wireless Propagation Letters* 5 (1): 247–250.

10 Faseth, T., Winkler, M., Arthaber, H., and Magerl, G. (2011). The influence of multipath propagation on phase-based narrowband positioning principles in UHF RFID. *2011 IEEE-APS Topical Conference on Antennas and Propagation in Wireless Communications (APWC)*, Torino, Italy (12–16 September 2011).

11 Dimitriou, A.G., Bletsas, A., and Sahalos, J.N. (2011). Room coverage improvements of UHF RFID with commodity hardware. *IEEE Antennas and Propagation Magazine* 53 (1): 175–194.

12 Dimitriou, A.G., Siachalou, S., Bletsas, A., and Sahalos, J.N. (2014). A site-specific stochastic propagation model for passive UHF RFID. *IEEE Antennas and Wireless Propagation Letters* 13: 623–626.

13 Zhou, J., Zhang, H., and Mo, L. (2011). Two-dimension localization of passive RFID tags using AOA estimation. *2011 IEEE Instrumentation and Measurement Technology Conference (I2MTC)*, Binjiang, China (10–12 May 2011).

14 Nikitin, P. V., Martinez, R., Ramamurthy, S. et al. (2010). Phase based spatial identification of UHF RFID tags. *2010 IEEE International Conference on RFID*, Orlando, FL (14–16 April 2010).

15 Zhang, Y., Amin, M.B., and Kaushik, S. (2007). Localization and tracking of passive RFID tags based on direction estimation. *International Journal of Antennas and Propagation* 2007: 1–9. (ID 17426).

16 Azzouzi, S., Cremer, M., Dettmar, U. et al. (2011). New measurement results for the localization of UHF RFID transponders using an angle of arrival (AoA) approach. *2011 IEEE International Conference on RFID*, Orlando, FL (12–14 April 2011).

17 Chawla, K., McFarland, C., Robins, G., and Shope, C. (2013). Real-time RFID localization using RSS. *2013 International Conference on Localization and GNSS (ICL-GNSS)*, Turin, Italy (25–27 June 2013).

18 Siachalou, S., Bletsas, A., Sahalos J. N., and Dimitriou A. G. (2016). RSSI-based maximum likelihood localization of passive RFID tags using a mobile cart. *2016 IEEE Wireless Power Transfer Conference (WPTC)*, Aveiro, Portugal (5–6 May 2016).

19 DiGiampaolo, E. and Martinelli, F. (2014). Mobile robot localization using the phase of passive UHF RFID signals. *IEEE Transactions on Industrial Electronics* 61 (1): 365–376.

20 Martinelli, F. (2015). A robot localization system combining RSSI and phase shift in UHF-RFID signals. *IEEE Transactions on Control Systems Technology* 23 (5): 1782–1796.

21 Subedi, S., Pauls, E., and Zhang, Y.D. (2017). Accurate localization and tracking of a passive RFID reader based on RSSI measurements. *IEEE Journal of Radio Frequency Identification* 1 (2): 144–154.

22 Bekkali, A., Sanson, H., and Matsumoto, M. (2007). RFID indoor positioning based on probabilistic RFID map and Kalman filtering. *Proceedings of 3rd International Conference on Wireless and Mobile Computing, Networking and Communications*, White Plains, NY, USA (8–10 October 2007).

23 Zhang, J., Lyu, Y., Patton, J. et al. (2018). BFVP: A probabilistic UHF RFID tag localization algorithm using Bayesian filter and a variable power RFID model. *IEEE Transactions on Industrial Electronics* 65 (10): 8250–8259.

24 Hahnel, D., Burgard, W., Fox, D. et al. (2004). Mapping and localization with RFID technology. *2004 IEEE International Conference on Robotics and Automation*, New Orleans, LA, USA (26 April–1 May 2004).

25 Yang, P. and Wu, W. (2014). Efficient particle filter localization algorithm in dense passive RFID tag environment. *IEEE Transactions on Industrial Electronics* 61 (10): 5641–5651.

26 Ni, L.M. and Liu, Y. (2004). LANDMARC: indoor location sensing using active RFID. *Wireless Networks* 10 (6): 701–710.

27 Miesen, R., Kirsch, F., and Vossiek, M. (2011). Holographic localization of passive UHF RFID transponders. In: *2011 IEEE International Conference on RFID*, 32–37. Orlando, FL: IEEE.

28 Yang, L., Chen, Y., Li, X.-Y. et al. (2014). Tagoram: real-time tracking of mobile RFID tags to high precision using cots devices. In: *Proceedings of the 20th Annual International Conference on Mobile Computing and Networking*, 237–248. New York, NY: ACM.

29 Shangguan, L. and Jamieson, K. (2016). The design and implementation of a mobile RFID tag sorting robot. In: *Proceedings of the 14th Annual International Conference on Mobile Systems, Applications, and Services*, 31–42. New York, NY: ACM.

30 Wang, J. and Katabi, D. (2013). Dude, where's my card? RFID positioning that works with multipath and non-line of sight. In: *Proceedings of the ACM SIGCOMM 2013 Conference on SIGCOMM, Hong Kong, China*, 51–62. New York, NY: ACM.

31 Ma, Y., Selby, N., and Adib, F. (2017). Minding the billions: ultra-wideband localization for deployed RFID tags. *MobiCom 2017, 23rd Annual Conference on Mobile Computing and Networking*, Utah, USA (16–20 October 2017).

32 Lozano-Perez, T. (2012). *Autonomous Robot Vehicles*. Springer Science & Business Media.

33 Cox, I.J. (1991). Blanche-an experiment in guidance and navigation of an autonomous robot vehicle. *IEEE Transactions on robotics and automation* 7 (2): 193–111.

34 Dellaert, F., Fox, D., Burgard, W., and Thrun, S. (1999). Monte Carlo localization for mobile robots. In: *Proceedings 1999 IEEE International Conference on Robotics and Automation*, vol. 2, 1322–1326. IEEE.

35 Thrun, S., Fox, D., Burgard, W., and Dellaert, F. (2001). Robust monte carlo localization for mobile robots. *Artificial Intelligence* 128 (1–2): 99–141.

36 Engelson, S.P. and McDermott, D.V. (1992). Error correction in mobile robot map learning. In: *Proceedings 1992 IEEE International Conference on Robotics and Automation*, 2555–2555. IEEE.

37 Thrun, S., Burgard, W., and Fox, D. (2005). *Probabilistic Robotics*. MIT Press.

38 Hornung, A., Wurm, K.M., Bennewitz, M. et al. (2013). OctoMap: an efficient probabilistic 3D mapping framework based on octrees. *Autonomous Robots* 34 (3): 189–117.

39 Csorba, M. (1997). Simultaneous Localization and Map Building. PhD dissertation. University of Oxford.

40 Cheeseman, P., Smith, R., and Self, M. (1987). A stochastic map for uncertain spatial relationships. In: *4th International Symposium on Robotic Research*, 467–467. Cambridge: MIT Press.

41 Bailey, T., Nieto, J., Guivant, J. et al. (2006). Consistency of the EKF-SLAM algorithm. In: *2006 IEEE/RSJ International Conference on Intelligent Robots and Systems*, 3562–3566. IEEE.

42 Julier, S., Uhlmann, J., and Durrant-Whyte, H.F. (2000). A new method for the nonlinear transformation of means and covariances in filters and estimators. *IEEE Transactions on Automatic Control* 45 (3): 477–475.

43 Montemerlo, M. and Thrun, S. (2007). *FastSLAM: A Scalable Method for the Simultaneous Localization and Mapping Problem in Robotics*, 63–27. Springer.

44 Montemerlo, M., Thrun, S., Koller, D., and Wegbreit, B. (2002). FastSLAM: a factored solution to the simultaneous localization and mapping problem. In: *AAAi*, 593–595. Menlo Park, CA: American Association for Artificial Intelligence.

45 Thrun, S. and Montemerlo, M. (2006). The graph SLAM algorithm with applications to large-scale mapping of urban structures. *The International Journal of Robotics Research* 25 (5–6): 403–426.

46 Mingas, G., Tsardoulias, E., and Petrou, L. (2012). An FPGA implementation of the SMG-SLAM algorithm. *Microprocessors and Microsystems* 36 (3): 190–204.

47 Nieto, J., Bailey, T., and Nebot, E. (2007). Recursive scan-matching SLAM. *Robotics and Autonomous Systems* 55 (1): 39–10.

48 Tsardoulias, E. and Petrou, L. (2013). Critical rays scan match SLAM. *Journal of Intelligent & Robotic Systems* 72 (3-4): 441–462.

49 Thallas, A.G., Tsardoulias, E.G., and Petrou, L. (2017). Topological based scan matching–odometry posterior sampling in RBPF under kinematic model failures. *Journal of Intelligent & Robotic Systems* 91: 1–26.

50 Davison, A.J. (2003). Real-time simultaneous localisation and mapping with a single camera. In: *Proceedings of the Ninth IEEE International Conference on Computer Vision*, vol. 2, 1403. IEEE.

51 Eade, E. and Drummond, T. (2006). Scalable monocular SLAM. In: *Proceedings of the 2006 IEEE Computer Society Conference on Computer Vision and Pattern Recognition*, vol. 1, 469–467. IEEE.

52 Mur-Artal, R., Montiel, J.M.M., and Tardos, J.D. (2015). ORB-SLAM: a versatile and accurate monocular SLAM system. *IEEE Transactions on Robotics* 31 (5): 1147–1163.

53 Montemerlo, M. and Thrun, S. (2003). Simultaneous localization and mapping with unknown data association using FastSLAM. *IEEE International Conference on Robotics and Automation* 2: 1985–1986.

54 Hess, W., Kohler, D., Rapp, H., and Andor, D. (2016). Real-time loop closure in 2D LIDAR SLAM. In: *2016 IEEE International Conference on Robotics and Automation*, 1271–1278. IEEE.

55 TurtleBot 2 (2018). Turtlebot 2 Platform. http://www.turtlebot.com/turtlebot2/ (accessed 02 April 2018).

56 Kobuki (2018). Kobuki Mobile Base. http://kobuki.yujinrobot.com/ (accessed 02 April 2018).

57 SLAMTEC (2018). RPLidar sensor. https://www.slamtec.com/en/Lidar/A1 (accessed 02 April 2018).

58 ASUS (2018). Xtion Live Pro Depth Camera. https://www.asus.com/gr/3D-Sensor/Xtion_PRO_LIVE/ (accessed 02 April 2018).

59 Dynamixel Servos (2018). http://www.robotis.us/dynamixel/ (accessed 02 April 2018).

60 MTI Wireless Edge LTD.. (2018). UHF RFID reader antennas by MTI wireless edge Ltd.. http://www.mtiwe.com/ (accessed 02 April 2018).

61 IMPINJ (2018). R420 RFID Reader. https://www.impinj.com/platform/connectivity/speedway-r420/ (accessed 02 April 2018).

62 ROS (2018). The Robot Operating System. http://www.ros.org/ (accessed 02 April 2018).

63 Alien (2018). Squiggle RFID tag – Alien. http://www.alientechnology.com/products/tags/short/ (accessed 02 April 2018).

8

From Identification to Sensing: Augmented RFID Tags

Konstantinos Zannas[1], Hatem El Matbouly[1], Yvan Duroc[2], and Smail Tedjini[1]

[1] *University of Grenoble Alpes, Grenoble INP, LCIS, Valence, France*
[2] *University of Lyon, University Claude Bernard Lyon 1, Ampere Laboratory, Villeurbanne, France*

8.1 Introduction

Nowadays, the concept of Internet-of-Things (IoTs) is getting more and more popular as it is becoming real and present in our everyday life. A universal network interconnecting multiple devices and integrating multiple functionalities will be the backbone toward the realization of various "smart" environments (smart cities, smart houses, smart agriculture, and smart industries). With the IoT concept, the users will have access and will be able to monitor and control many devices remotely through their smart phones, tablets, or laptops. Many domains such as healthcare, smart grids, environment monitoring, and traffic control will find great advantages from the IoT, enabling a vast amount of information to be distributed and shared among the end users. The IoT has the potential to become a key service for the society and specific areas of interest, such as industrial environments, where it will be able to achieve better monitoring and controlling of the production process. This could eventually result in a higher productivity with a more effective line of production [1]. In addition, the cost of the implementation of IoT is certainly one of the most important features as the expected numbers of devices to be connected is in the order of tens of billions.

Moreover, the IoT concept becomes even more appealing when it is connected with the upcoming 5G technology [2]. The parameters of 5G cellular networks are expected to be flexible and able to target specific applications and services with IoT being one of the top priorities [3]. Considering this flexibility of 5G technology, it will be very advantageous for the implementation of wireless sensor networks (WSNs) that will be the backbone of IoT.

In order to realize such networks effectively, several attributes are very substantial. The IoT networks should be robust and reliable, especially when sensitive and

Wireless Power Transmission for Sustainable Electronics: COST WiPE - IC1301,
First Edition. Edited by Nuno Borges Carvalho and Apostolos Georgiadis.
© 2020 John Wiley & Sons, Inc. Published 2020 by John Wiley & Sons, Inc.

impactful processes should be controlled and monitored (in healthcare or industry areas for instance); secure and resilient to malicious interactions; sustain their functionality with low-power requirements; offer good accuracy; and present a low cost in order to be easily affordable. All these attributes have created the need for cheap, easy deployed sensors, and actuators, which are capable of operating with very low-energy consumption. These augmented devices will be able to transmit the sensed information from the environment, as well as, interact with it.

A very good candidate for realizing the sensing functionalities for such an ambitious plan is the radio frequency identification (RFID) technology [4]. The RFID technology is a well-established identification technology that has become popular during the past two decades and it is still expanding. It offers reliable wireless identification solutions for many applications. This is also indicated by the current trend of the RFID market that is the expected to grow in revenues in the near future, as shown in Figure 8.1 [5].

Due to the mass production printing techniques and the advances in fabrication of integrated circuits (ICs) the cost of an RFID tag can be very low, less than $0.10 for passive tags intended for general use [6]. However, there are also RFID tags with more capabilities and complexity that can cost more than $100. These tags are usually active tags with very long read range (more than 100 m) and more than just identification capabilities [7]. The RFID tags can work without a conventional energy source, in a passive mode, acquiring theoretically lifelong operation. The advances in IC technology allowed the development of very sensitive RFID chips, which offer a very sufficient read range that enables several applications and is leading the implementation of the last few meters of the IoT. Table 8.1 shows the evolution of passive RFID chip sensitivity and its impact on the read range. More recently, RFID chips can also be designed for sensing a variety of physical parameters, such as temperature, pressure, acceleration, humidity, light,

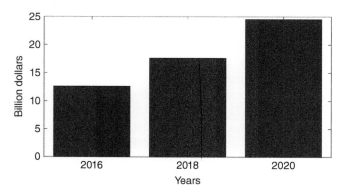

Figure 8.1 RFID market growth forecast. Source: Statista.org.

Table 8.1 RFID chip sensitivity and read range for ETSI regulation during the past 20 years.

Year	Sensitivity (dBm)	Read range (m)[a]
1997	−8	5.07
1999	−10	6.38
2005	−12	8.03
2007	−13	9.01
2008	−15	11.34
2010	−18	16.02
2011	−20	20.16
2014	−22	25.38

a) Estimated theoretical read range considering: operating frequency at 868 MHz, reader power 2 W ERP (effective radiated power), reader antenna gain 2 dBi, tag antenna is ideal dipole and perfect matching between RFID chip-tag antenna.

inclination, strain, and location to name a few, with adjustable characteristics, such as sensitivity, read range, and accuracy.

Furthermore, the RFID technology operating especially in the ultra-high frequency (UHF) band is very promising. It offers connectivity to a widely used frequency band (860–960 MHz) according to the local regulations (mainly the European Telecommunication Standards Institute – ETSI [EU] or the Federal Communication Commission – FCC [USA]) with well-defined protocols (ISO, EPCglobal), achieving reading ranges in some cases above 20 m for passive tags. Moreover, the RFID has a unique characteristic in comparison with other technologies used for establishing WSN: the RFID sensor is actually a "fusion" between a communication protocol and a sensor, while in the other technologies, there is an actual sensor connected to a communication module in order to transmit the information.

Also, the RFID technology allows WSN to consist of vast number of sensors due to the 32-bit Electronic Product Code (EPC) code used for identification, whereas in other technologies, this number is considerably lower. Because of this particularity of RFID technology, any new features should be capable of being implemented in the communication protocol. In addition, the RFID technology is mainly operating passively lowering substantially the cost of the devices, prolonging their lifetime, while reducing the carbon footprint [8]. All these aspects are presenting a clear view of why during the last years much effort has been given in designing and developing efficient UHF RFID sensors for a wide range of applications.

Several classification attempts have been made in the past regarding the integration of RFID sensor systems in WSN [9, 10]. Also, proposals about RFID sensor classification either in terms of "transducer" type RFID tags have been proposed in [11, 12] or in a more general context, concerning RFID tags including circuits and chips with processing capabilities have been proposed in [13–15]. In [16], there is a classification between analog and digital sensor, and in [17], is presented an RFID classification specialized in food industry.

This chapter aims to present primarily a review, mainly focusing on the part of the RFID sensor tag, presenting a comprehensive classification concept based on previously published scientific articles, trying in parallel to encapsulate the different techniques and prototypes of UHF RFID sensor tags that have been presented in the literature. This classification is associated with a comparison of the different approaches and also provides several examples serving as an illustration and state-of-the-art of RFID sensors.

In Section 8.2, a description of the different topologies and attributes of RFID sensors is presented with a detailed diagram. In Section 8.3, paradigms of RFID sensors found in the corresponding literature are being presented, compared, and classified according to the proposed reasoning.

8.2 Generic RFID Communication Chain

In Figure 8.2, a general and synthetic diagram describes every block of the UHF RFID sensor communication chain (integrating identification and sensing functionalities) with ultimate goal of underlining the properties of each block. This diagram also serves as a summary of all the met cases and highlights the different links in each step of radio communication, including the transmission of sensing data and identification (ID). The communication chain between RFID sensor tag and RFID reader is presented in six levels: the RFID sensor tag, the RFID data capture level, the RFID tag process level, the RFID communication channel, the RFID reader process level, and the RFID reader.

8.2.1 RFID Sensor Tag

The RFID sensor tag exceeds the usual RFID tag functionality by offering sensing capabilities on top of the identification functionality. This is why these types of RFID tags are also known as augmented RFID sensor tags [18]. One important parameter that must be taken into account is the energy requirements of the RFID sensor tag, Figure 8.2 level I. The optimum design should be capable of sustaining its functionality without any additional external power source (passive mode), using only the electromagnetic field of an RFID reader and the rectifying capabilities of the RFID chip. However, in many situations, this is not feasible, thus the RFID tag should have a complementary external power source

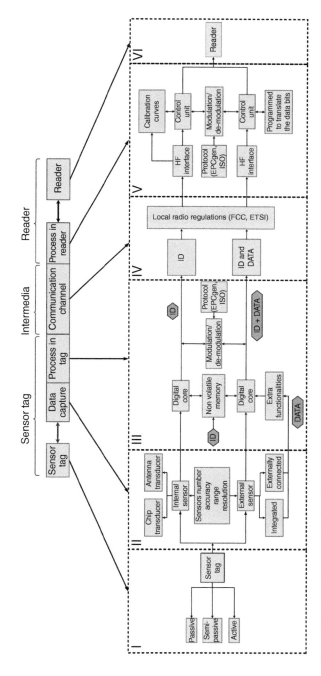

Figure 8.2 Description of the classification format. Each Latin number represents a part of communication between an RFID reader and an RFID sensor tag.

(semipassive mode) or it is powered exclusively by an external power source (active mode). For instance, in many cases, additional power source is required either to power up sensors or an extra microcontroller, while the communication with the reader is supported by the rectified RF power. In any case, the energy requirement is an application-oriented characteristic, and it is dependent on the operating environment and the desired read range.

8.2.2 RFID Data Capture Level

Initially, the state of the sensor must be clarified in terms of how the sensing procedure is being performed at the sensor tag. The desired physical parameter can be sensed either using an internal sensor (Figure 8.3a) or an external sensor (Figure 8.3b).

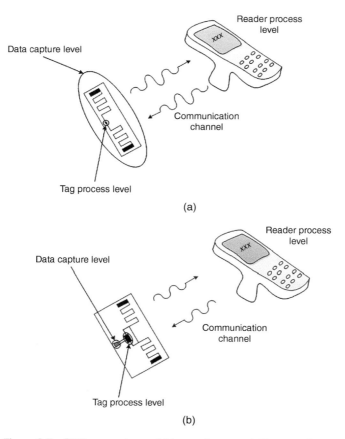

Figure 8.3 RFID sensor types: (a) internal type and (b) external type.

The concept of internal sensor refers to designs using partially or even the whole RFID tag structure to perform the physical parameter detection. The associated technique consists of either frequency shifts due to the effective permittivity changes or the polarization conditions (i.e. where the antenna is used as a transducer) or matching condition alterations due to changes to the input impedance of the chip (i.e. where the chip is used as a transducer). Sensor tags without a chip (so-called chipless RFID tags) can also be classified in this category. These sensing capabilities can be detected in the reader as backscattered signal alterations: power variations, shift of frequency, and phase variations.

On the other hand, external sensor is referred to RFID sensor tags that use sensors to enhance the sensing mechanism. These sensors are connected to the RFID chip either with an external port through a microcontroller or they are integrated within. There is a variety of sensors that can be used in such cases depending on the application. Usually, the types of sensors used in these cases are resistive, capacitive, or even voltage sources.

Besides the characteristics of each sensor type, regarding the RFID communication chain, there are some more considerations that should be taken into account.

The number of sensors that each type of RFID sensor tag can accommodate is one interesting parameter. For instance, for the internal type sensor, the simultaneous measurement of different physical parameters is a challenge, since the behavior of different parameters is very difficult to be separated and recognized. On the other hand for the external type sensor RFID tag, the existence of external ports can allow the simultaneous usage of sensors for different physical parameters. The limitation of the number of sensors for the external type sensor tag is due to the energy requirements since each sensor consumes more power in order to be operational. In addition, the range, resolution, and accuracy of the measured parameter is an important factor, where the internal type sensor tag usually relies heavily on the capability of the RFID reader to detect the corresponding changes, while in the external type sensor these attributes depend on the used sensor and the resolution of the analog to digital converter (ADC) used for the measurement.

A graphical representation of this level can be seen in Figure 8.2 level II.

8.2.3   RFID Tag Process Level

The processing capability of the ICs used in sensor tags is a very important feature for their overall operation. The ICs used in RFID sensor tags can be divided into two categories: common off-the-shelf ICs with their functionality limited to identification and ICs with upgraded functionalities including micro-processors, analog to digital (A/D) converters, serial peripheral interface (SPI) ports or inter-integrated circuit (I^{2}C) ports for more processing capabilities and extra functionalities. Currently, there are a series of RFID ICs with these kinds of

Table 8.2 RFID IC's with extra functionalities.

	RFID IC's				
Name	AMS SL900A [19]	FARSENS ANDY100 [20]	EM –Micro-electronics EM4325 [21]	RF Micron Magnus S3 [22]	Impinj Monza X – 8K Dura [23]
Read sensitivity in passive mode	−6.9 dBm	−2 dBm	−8.3 dBm	−16.6 dBm	−21.6 dBm (dual antennas)
Analog to digital converter	Yes	No	No	No	No
SPI or I²C connectivity	SPI	SPI	SPI	No	I²C
Integrated temperature sensor	Yes	No	Yes	Yes	No
Temperature resolution	0.029 to 0.232 °C	—	0.25 °C	—	—
Temperature accuracy	±1 °C	—	±0.5 °C	±1 °C	—
Temperature range	−40 to 125 °C	—	−40 to 64 °C	−40 to 85 °C	—

capabilities from different manufacturers and with different specifications (see Table 8.2). This separation is related to the internal/external sensor concept. Internal type sensor tags can be realized with any types of ICs. On the other hand, in order to realize a sensor tag with an external type sensor, the usage of an IC or a circuitry with additional processing capabilities either with an integrated sensor or external ports that can accommodate various sensors is mandatory. However, the RFID chips with additional capabilities usually have a low sensitivity that reflects to limited read range, while the off-the-shelf RFID chips have higher sensitivity and thus superior read ranges.

For the internal type sensor tag, the physical parameter can be evaluated when the RFID reader receives the interrogation response and according to the received power level or frequency or phase, the state of the physical parameter can be determined according to preliminary calibration curves. In this case, the RFID sensor tag transmits the unique ID code that is stored in the nonvolatile memory of chip and the useful information is acquired by the reader prior to decoding process of the received signal.

For the external type sensor tag, a portion of the generated data from the sensor is digitalized and processed in the tag (since the RFID tag in this case has extra functionalities) and embedded in the binary signal transmitted to the reader. A great example for the case of the external type RFID sensor tag can be seen in [24]. In this case, the information about the physical parameter is acquired after the decoding process of the received signal. The graphical representation of these processes can be seen in Figure 8.2 level III.

8.2.4 RFID Communication Channel

The concept of internal/external sensor is closely related to the communication channel (Figure 8.2 level IV) and to its role at the sensing procedure.

In the case of internal sensor concept, the communication channel is being altered in connection to the sensed physical parameter. These alterations include the back-scattered signal power level, the center frequency, and the phase. In this way, the variations of physical parameters are linked to measurable quantities in the reader by calibration curves that correspond to the status of the sensor to the measured physical parameter. Moreover, the most advanced off-the-shelf RFID chips include "auto-tuning" capabilities. In this case, the RFID chips are capable of changing their input impedance according to environmental de-tuning effects. These changes can then be detected by the RFID reader by simply acquiring the value of the "auto-tune" parameter from the memory bank of the chip. If these changes can be related to certain effects linked to a physical parameter, then the RFID tag can work as an internal type sensor communicating digitally.

In the case of the external sensor concept, the communication channel for an RFID tag sensor in a fixed position is independent from the status of the physical parameter measured. The sensing information is being processed in the RFID chip, and it is enhanced in the bit stream received by the reader. The signal contains all the information about the status of the sensor in binary form acquired by the reader after the decoding process of the received signal.

Another consideration is the read range for each type of RFID sensor, especially in a passive mode case. The internal type sensor depends on phase/frequency shift and impedance mismatches to transmit information. This fact has an immediate impact on the read range due to the condition of maximum power transfer to the RFID IC, since the perfect match condition cannot be maintained continuously in most of the cases. This issue is discussed in [25], where a different approach is suggested. For the external type RFID sensor, an initial clarification to the meaning of read range should be made. When the desired information is included in the binary stream, a separation between the "power on" and "read" range should be included. The fact that the RFID tag can be identified by the reader and there is a received power indication does not ensure that the sensor tag can transmit the necessary power so that the bit steam with the information of the sensor to be read. Thus, in this case, the "read" range corresponds to the actual read range.

From this prospective, there can also be a distinction of internal/external sensor to a distinction between analog/digital information transmissions over the communication link.

8.2.5 RFID Reader Process Level and RFID Reader

In the RFID technology, the main processing load is carried out by the RFID reader. The main parts of a typical RFID reader are the high-frequency (HF) interface and the control unit, responsible for the radio communication and the data processing respectively [26]. The reader according to the RFID sensor tag should have different attributes. In the case of the internal sensor type tags, the reader should be first calibrated with appropriate curves in order to translate correctly the back-scattered power, frequency shift, or phase alterations received to match them to the correct physical parameter status. In this part, it is very important to take into consideration the regulations, since for ETSI the available bandwidth is 2 MHz and for the FCC is 26 MHz. Furthermore, if the polarization of the signal is used as an indication parameter, the reader's antenna should be carefully selected to match the criteria.

In the case of external type sensors, since the communication channel, for a fixed-position RFID tag sensor, remains relatively stable, the reader has to be programmed to match the binary code, which is included to the received bit stream, to the appropriate physical parameter magnitude. In the case of external sensor type, the antenna polarization is not as significant a factor as in the case of internal type sensor. The RFID reader and RFID reader process block corresponds to levels V and VI.

Furthermore, a very important attribute is the existence of the same protocol in both sensor types. For the RFID tag sensor to be identified by the majority of commercial readers, the existence of a communication protocol is crucial. In the literature, designs without the existence of protocol can be found, which mainly focus on the proof of a certain concept. On the other hand, since there are globally used predefined protocols like EPCglobal and ISO 18000-6, the majority of designs in the literature are compatible with either of these protocols.

One more characteristic that applies to both categories is the local regulations of the RFID communication. The regulations used for the RFID communication can be distinguished, mainly between the regulations in USA (FCC) and the regulations in Europe (ETSI). The main differences can be found at the frequency bands that are being used for each territory (865–868 MHz for ETSI band and 902–928 MHz for FCC band), at the receiver transmitter technique (four-channel plan for ETSI and frequency hopping spread spectrum [FHSS] for FCC) as well

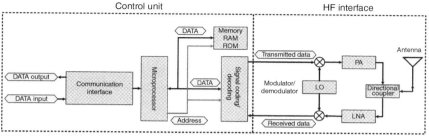

LO, local oscillator; LNA, low noise amplifier; PA, power amplifier; RAM, random access memory; ROM, read only memory.

Figure 8.4 Architecture of an RFID reader.

Table 8.3 Summary of RFID sensor types.

	Type of sensors	
RFID level	**Internal**	**External**
Data capture (level II)	Via tag structure	Via dedicated sensors
Tag process (level III)	Basic RFID functionalities	Advanced functionalities
Communication channel (level IV)	Varying/stable	Stable
Reader process (level V)	Predefined curves/ memory bank	Accordingly programmed

as the maximum allowed transmitted power from the reader, using the effective isotropic radiated power (EIRP) indication (3.2 W for ETSI and 4 W for FCC). Finally, it is worth noting that new UHF spectrum, in addition to the 865–868 MHz band, is planned to be available in Europe (around 915 MHz) demonstrating the dynamic of UHF RFID evolution [27].

A typical RFID reader architecture can be seen in Figure 8.4. In Table 8.3, a summary of the attributes of each RFID sensor type is presented.

8.3 RFID Sensor Tags: Examples from Literature or Commercially Available

Significant examples of prototypes of UHF RFID sensors proposed in the literature are presented according to the sensed parameter(s) and described according to the general diagram proposed in Figure 8.2. In Table 8.4, the literature examples, as well as some commercially available UHF RFID sensors, are summarized, highlighting their characteristics according to the introduced classification. Finally, based on RFID temperature sensing that is one of the most researched areas regarding RFID sensors (i.e. [28–30]), an example of illustrative comparison is made for sensors of both types aiming to underline the performance in terms of read range, accuracy, resolution, and temperature range.

8.3.1 Examples from Literature

Orientation sensing: In [31], the orientation of an object is determined by the usage of two passive UHF RFID tags attached on an object. These two tags have a known angle difference between them. The communication channel is varying according to the orientation and the status of the orientation parameter is calculated by a statistical estimator based on the received signal strength (RSS) variation, which is being caused by the polarization angle between the tags and the antenna of the reader. This is an example of an RFID tag with an internal type sensor, since the sensing data capture is due to antenna's characteristics. The chip used in this case is an off-the-self chip (ALN-9740) with basic identification capabilities.

Motion and temperature sensing: In [32], both motion and temperature semipassive UHF RFID sensor tag is presented. The communication channel in this case is independent of the physical parameters that are being sensed and an appropriate programmed reader is needed to evaluate the status of these parameters. According to the inclination of the object toward the earth's magnetic field, the sensor can monitor and log the motion of the object and at the same time, it can monitor the temperature of the object as well. This is an external type RFID sensor able to detect the motion and the temperature of an object, using a chip with processing capabilities (AMS SL-900A), accommodating an integrated temperature sensor and an externally connected anisotropic magneto-resistive sensor.

Temperature sensing: In the example of [33], two temperature passive UHF RFID sensors are presented, one external type Figure 8.5 and one internal type Figure 8.6. For the external type sensor, an RFID chip (EM Microelectronic – EM 4325) with an integrated temperature sensor and extra functionalities is being used. The communication channel is independent of the physical parameter sensed and the reader must be programmed to make the temperature readout. In this case, a double-folded patch antenna is used. For the internal type RFID sensor, a meander dipole antenna is used with a thermistor connected in parallel to the chip. According to the ambient temperature, the resistance of thermistor changes which affects the reflecting impedance of the RFID chip. The chip used is an off-the-shelf chip (Impinj Monza-R6), the communication

Table 8.4 Comparison between examples found in literature.

Example/ References	1 [31]	2 [32]	3a [33]	3b [33]	4 [13]	5 [34]	6 [13]	7 [35]	8 [36]	9 [37]	10 [38]
Energy (level I)	Passive	Semi-Passive	Passive	Passive	Semi-Passive	Passive	Passive	Passive	Passive	Passive	Passive
Physical parameter	Orientation	Motion and temperature	Temperature	Temperature	Pressure	Gas	Light and temperature	Temperature	Permittivity of meat	Temperature	Temperature and pressure
Number of sensors	1	2	1	1	1	1	2	1	1	1	2
Data capture (level II)	Internal (via antenna)	External	External	Internal (via chip)	External	Internal (via chip)	External	Internal (via antenna)	Internal	Internal (via antenna)	External
Tag Process (level III)	ALN 9640	SL – 900A	EM – 4325	Monza R6	SL – 900A	Chip – less	SL – 900A	ALNH2	AK3 tagsys	Magnus S3	ANDY100
Chip sensitivity (Passive mode) (dBm)	-20	-6.9	-8.3	-21	-6.9	—	-6.9	-14	-15	-16.6	-2
Communication channel (level IV)	Varying	Stable	Stable	Varying	Stable	Varying	Stable	Varying	Varying	Stable	Stable
Reader process (level V)	Calibration curves	Programmed	Programmed	Calibration curves	Programmed	Calibration curves	Programmed	Calibration curves	Calibration curves	Through memory bank	Programmed
Maximum read range (m)	2	7	15	13	1.5	—	1.6	7	1.4	19	1.5
Protocol	EPCglobal Gen2/ ISO18000 – 6C	EPCglobal Gen2	EPCglobal Gen2/ ISO18000 – 63/64	EPCglobal Gen2/ 18000 – 63	EPCglobal Gen2	No Protocol	EPCglobal Gen2	EPCglobal Gen2/ ISO 18000 – 6C	EPCglobal Gen2	EPCglobal Gen2/ ISO18000 – 6C	EPCglobal Gen2
Regulations	ETSI/FCC	ETSI/FCC	ETSI/FCC	ETSI/FCC	ETSI/FCC	ETSI	ETSI/FCC	ETSI/FCC	ETSI	ETSI/FCC	ETSI/FCC
Dimensions (mm^3)	–	$102 \times 22 \times$ NaN	$52 \times 74 \times 3$	$80 \times 34 \times$ NaN	$135.7 \times 22.2 \times 2$	$120 \times 30 \times 0.2$	$79 \times 5.5 \times 1.5$	$44 \times 30 \times 1$	$63 \times 67 \times 0.2$	$50 \times 52.5 \times 3.55$	$76 \times 34 \times 4$

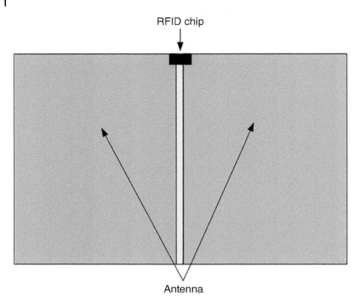

Figure 8.5 External type RFID temperature sensor. Source: Adapted from [33].

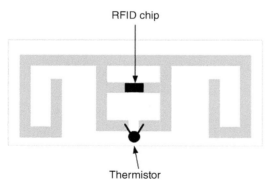

Figure 8.6 Internal type RFID temperature sensor. Source: Adapted from [33].

channel is varying according to the physical parameter and a reader will be able to make the measurement of the temperature using calibration curves.

Pressure sensing: In the example of [13], a passive RFID external sensor is presented. The pressure sensor is externally connected to the RFID tag and a chip with processing capabilities is used (AMS SL900A), connected with a micro controller in SPI mode. The communication channel is independent from the physical parameter and the reader must be programmed in order to have the readout of the pressure. A dipole antenna is used in this case.

Meandered dipole antenna

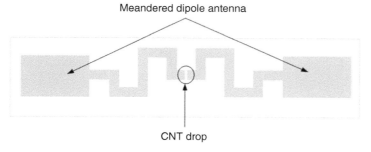

CNT drop

Figure 8.7 Chipless gas sensor. Source: Adapted from [34].

Gas sensing: In the example of Figure 8.7, [34], a passive internal RFID sensor for gas sensing (ammonia) is designed utilizing carbon nanotubes (CNTs). The CNT drop is connected with a meander dipole antenna. The existence of ammonia in the vicinity of the CNT is causing an impedance change and this change affects the back-scattered power. This is an example of internal RFID sensor. Even though this example is chipless, it can be classified as an internal type passive RFID sensor with the usage of the chip as transducer since the sensing mechanism is due to matching alterations caused by the physical parameter sensed. The communication channel changes according to the physical parameter and a reader with calibration curves should be used in order to measure the quantity of ammonia.

Light sensing: In the example of Figure 8.8 [13], an external type passive UHF RFID light sensor is being realized using a chip with processing capabilities (AMS-SL900A). The communication channel is independent of the physical parameter variations and the reader must be programmed to translate the data bits. Since this chip includes an integrated temperature sensor, it is capable of sensing temperature and luminosity at the same time.

Threshold temperature sensing: The internal type RFID sensors can be also used as threshold sensors. In the example of [35], a passive UHF RFID sensor with an internal type temperature sensor, realized on appropriate paraffin substrate,

Figure 8.8 RFID light sensing tag. Source: Adapted from [35].

Dipole antenna

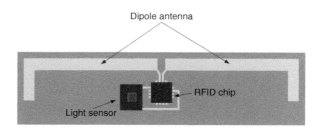

RFID chip

Light sensor

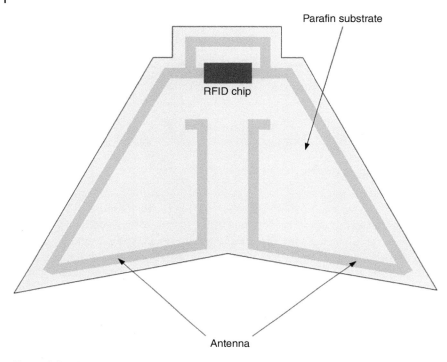

Figure 8.9 Threshold temperature RFID sensor tag. Source: Adapted from [36].

can sense the temperature rise over a certain threshold. An off-the-shelf chip is used (Alien Higgs 2), and the communication channel is varying along with the temperature. From the reader's part, calibration curves should exist. When the temperature exceeds a certain threshold, the RFID sensor changes its proprieties permanently (Figure 8.9).

Meat quality sensing: Another method of sensing is to exploit the permittivity variations of the substrate. It is well known that foods, like meat, contain a large percentage of water (more than 60%) that will decrease over time due to the evaporation process. Exploiting this feature can be used to measure the freshness of meat. This process has been exploited in [36] to design an RFID sensor of meat quality that provides back-scattered signal linked to the freshness of the meat. The communication channel is varying in this case, and a reader with predefined curves should be used in order to evaluate the condition of the meat (Figure 8.10).

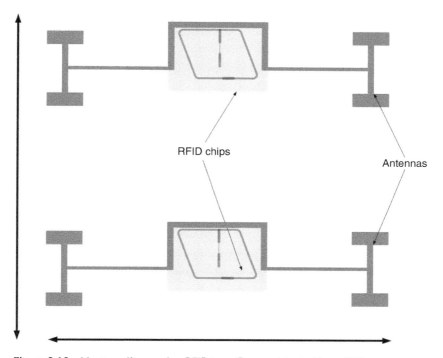

Figure 8.10   Meat-quality sensing RFID tags. Source: Adapted from [36].

8.3.2   Examples Commercially Available

Temperature sensing: A commercially available UHF RFID sensor is presented
in Figure 8.11 utilizing the RFM3240 of RF Micron company [37] It is a passive temperature sensor dedicated for on-metal applications with a temperature
range from −40 °C to +85 °C. This sensor utilizes the Magnus S3 RFID chip that
has the ability to change its internal impedance to de-tuning effects. It can be
classified to internal sensors, but in this case, the communication channel is
not varying since the information is in digital form. Based on the value of the
variable input impedance of the chip, this sensor can be calibrated to transmit
information about the temperature that is a detuning factor. The accuracy of
this sensor is less than ±2 °C.

Pressure and temperature sensing: Another commercially available UHF RFID sensor is the CYCLON-14BA from the company Farsens [38]. This is a passive pressure and temperature sensor utilizing a circuit with an ANDY100 RFID chip, a

Figure 8.11 Temperature sensor from RFMicron. Source: Adapted from [37].

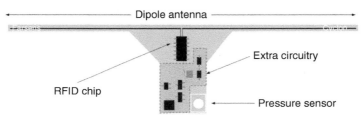

Figure 8.12 Temperature and pressure sensing tag from Farsens. Source: Adapted from [38].

microcontroller, a pressure, and a temperature sensor. It has a pressure range from 0 to 14 bars with an accuracy of ±140 mbar and resolution of 14 mbar. The temperature range is from −30 °C to +85 °C, with an accuracy of ±1 °C and resolution 0.1 °C. This is an external type RFID sensor (Figure 8.12).

8.4 Comparison of Different Types of RFID Temperature Sensors

RFID sensor tags capable of measuring the temperature are selected from Table 8.4. Based on the previous discussion, the type of RFID sensor tag (internal/external) is expected to have an important role in the performance of the temperature RFID sensor tag.

In the case of the internal RFID temperature sensor tag type, the sensing mechanism is realized through the variation of the reflection impedance and thus, the read range is expected to be affected due to mismatch between antenna and RFID

IC. The other attributes, such as resolution, accuracy, and temperature range are mostly results from the designing process, the interacting method with temperature, and the calibration curves used in the reader to quantify the measured parameter.

In the case of the external RFID temperature sensor tag, the actual sensor used for the measurement of the temperature presents a specific resolution, accuracy, and temperature range. When an integrated temperature sensor is used in a chip with processing capabilities, the aforementioned attributes are fixed according to its initial characteristics. When the temperature sensor such as a thermistor or a thermocouple is interconnected externally through a microcontroller, the resolution, accuracy, and temperature range are solely dependable on the sensor used. Finally, the fact that the communication channel is independent from the measured physical parameter makes the reading range dependent upon the chip sensitivity, the polarization mismatch, the antenna design, the operating environment, and the power of the RFID reader. In general, the comparison of the examples of RFID temperature sensor tags can lead to remarks concerning in general the RFID sensor tags. In both types, internal and external, pros and cons can be observed (see Table 8.5).

For the case of external type RFID temperature sensors, the usage of RFID ICs with low sensitivity results in limited read range and also the processing functions of the chip are adding a time delay that is important, especially when there are involved in real-time sensing applications [39] On the other hand the connectivity option of these chips with a variety of sensors using even the SPI mode offers great advantages in terms of design flexibility, ease of realization, and reliability in the communication.

Table 8.5 Comparison between temperature sensing RFID tags.

| Attributes | Examples | | |
| | 3a [33] | 3b [33] | 6 [13] |
Sensor type	External	Internal	External
RFID chip	EM 4325	Monza R6	SL 900A
EIRP of reader (W)	0.47	0.13	0.4
Chip sensitivity (dBm)	−8	−22	−15
Read range (m)	1.5	1.3	1.5
Temperature resolution (°C)	0.25	0.5 to 3	0.15
Temperature accuracy (°C)	±0.5	±1	±1
Temperature range (°C)	−40 to 64	28 to 200	−40 to 125

For the case of internal type RFID sensor tags, the usage of chips with high sensitivity leads to higher read ranges even though the techniques used involve impedance mismatching conditions that affect partially the performance of the sensor tag. The physical phenomena detected by such sensor tags can be seen to the reader faster, without additional delay due to processing functionalities at the side of the tag. The internal type RFID sensor lacks the ease of realization and usually requires high expertise in antenna design, since in every case the tags should be carefully designed and matched to the sensing structure or mechanism. Also, effects in the communication channel can lead to ambiguous results. The fact that calibration curves should preexist makes these type of tag sensors highly customized according to the operating environment and application.

8.5 Conclusion

In this chapter, a review and a comprehensive classification concerning the UHF RFID sensor tags were presented. This chapter aimed to offer a global vision of the RFID sensing technology and to underline the main advances in RFID sensor tag design. In the meantime, the main differences in each part of the RFID communication chain of the current state-of-the-art UHF RFID sensor designs are discussed. Furthermore, this chapter could serve as a design guide for RFID sensor tags. The readers can find a complete and synthetic view on the components and the associated strategies to integrate sensing capabilities to UHF RFID tags.

The introduction of RFID chips with even higher sensitivity and complexity is very likely to happen in the near future, expanding the read range of a passive RFID tag more than 26 m that is currently available. Moreover, the RFID sensor tags in many cases are exploiting the antenna structure, opening the door for new concepts related to different antenna designs or the usage of different materials for their realization. In addition, the architecture of the RFID technology allows a vast number of sensors to be part of a WSN, exceeding the number of sensors that implement other technologies. This attribute can enable future applications, containing millions of small sensors. In addition, insisting on the standardized character of RFID communication through a protocol is on the top of any vision related to WSN.

The 5G and IoT implementation and the extended research on RFID sensor tags worldwide are very likely to enable new and unprecedented applications. These milestones can be the breakthrough to move a step closer to the realization of the IoT, and all these reasons are making the RFID sensor technology without any doubt the cornerstone of WSN and consecutively of IoT.

References

1 Da Xu, L., Wu, H., and Li, S. (2014). Internet of things in industries: a survey. *IEEE Transactions on Industrial Informatics* 10 (4): 2233–2243. https://doi.org/10.1109/TII.2014.2300753.

2 Gupta, A. and Jha, R.K. (2015). A survey of 5G network: architecture and emerging technologies. *IEEE Access* 3: 1206–1232. https://doi.org/10.1109/ACCESS.2015.2461602.

3 Ijaz, A. et al. (2016). Enabling massive IoT in 5G and beyond systems: PHY radio frame design considerations. *IEEE Access* 4: 3322–3339. https://doi.org/10.1109/ACCESS.2016.2584178.

4 Want, R. (2004). Enabling ubiquitous sensing with RFID. *IEEE Computer* 37 (4): 84–86. https://doi.org/10.1109/MC.2004.1297315.

5 Statista. (2015). RFID tags – Global Market Size Forecast 2016–2020. http://www.statista.com/statistics/299966/size-of-the-global-rfid-market/ (accessed 04 September 2019).

6 AtlasRFIDstore. (2016). Avery Dennison AD-229r6 UHF RFID wet inlay (Monza R6). https://www.atlasrfidstore.com/avery-dennison-ad-229r6-uhf-rfid-wet-inlay-monza-r6/ (accessed on 16 October 2017).

7 AtlasRFIDstore (2012). Active RFID. https://www.atlasrfidstore.com/omni-id-view-4-rfid-visual-display-tag/ (accessed 16 October 2017).

8 NXP. (2013). RFID white papers. https://nxp-rfid.com/wp-content/uploads/2013/04/Carbon-Footprint-White-Paper.pdf (accessed 17 October 2017)

9 Liu, H., Bolic, M., Nayak, A., and Stojmenovic, I. (2008). Taxonomy and challenges of integration of RFID and wireless sensor networks. *IEEE Network* 22 (6): 26–35. https://doi.org/10.1109/MNET.2008.4694171.

10 Mitsugi, J. et al. (2007). Architecture Development for Sensor Integration in the EPC-global Network. Auto-ID Labs, white paper.

11 Occhiuzzi, C., Caizzone, S., and Marrocco, G. (2013). Passive UHF RFID antennas for sensing applications: principles, methods, and classifications. *IEEE Antennas and Propagation Magazine* 55 (6): 14–34. https://doi.org/10.1109/MAP.2013.6781700.

12 Grosinger, J., Gortschacher, L., and Bosch, W. (2016). Sensor add-on for batteryless UHF RFID tags enabling a low cost IoT infrastructure. *IEEE MTT-S International Microwave Symposium Digest*: 1–4. https://doi.org/10.1109/MWSYM.2016.7540331.

13 Salmeron, J.F., Molina-Lopez, F., Rivadeneyra, A. et al. (2014). Design and development of sensing RFID tags on flexible foil compatible with EPC Gen 2. *IEEE Sensors Journal* 14 (12): 4361–4371. https://doi.org/10.1109/JSEN.2014.2335417.

14 Colella, R., Tarricone, L., and Catarinucci, L. (2015). SPARTACUS: self-powered augmented RFID tag for autonomous computing and ubiquitous sensing. *IEEE Transactions on Antennas and Propagation* 63 (5): 2272–2281. https://doi.org/10.1109/TAP.2015.2407908.

15 Virtanen, J., Ukkonen, L., and Björninen, T. (2011). Temperature sensor tag for passive UHF RFID systems. *Proceeding of 2011 IEEE Sensors Applications Symposium*: 312–317. https://doi.org/10.1109/SAS.2011.5739788.

16 Islam, M.M., Viikari, V., Nikunen, J., and Reinikainen, M. (2016). UHF RFID sensors based on frequency modulation. *2016 IEEE Sensors*: 1–3. https://doi.org/10.1109/ICSENS.2016.7808807.

17 Bibi, F., Guillaume, C., Gontard, N., and Sorli, B. (2017). A review: RFID technology having sensing aptitudes for food industry and their contribution to tracking and monitoring of food products. *Trends in Food Science & Technology* 62 (Suppl C): 91–103. https://doi.org/10.1016/j.tifs.2017.01.013.

18 Tedjini, S., Andia-Vera, G., Zurita, M. et al. (2016). Augmented RFID tags. *2016 IEEE Topical Conference on Wireless Sensors and Sensor Networks (WiSNet)*: 67–70. https://doi.org/10.1109/WISNET.2016.7444324.

19 AMS (2010). SL900A EPC sensor tag. http://ams.com/eng/Products/Wireless-Connectivity/Sensor-Tags-Interfaces/SL900A (accessed 16 May 2016).

20 Farsens. ANDY100. http://www.farsens.com/en/products/andy100/ (accessed 26 June 2016).

21 EM Microelectronics. EM4325. http://www.emmicroelectronic.com/products/rf-identification-security/epc-and-uhf-ics/em4325 (accessed 23 June 2016).

22 RFMICRON. Magnus S product family. http://rfmicron.com/magnus-family/ (accessed 05 June 2017)

23 Impinj. Monza X-8K Dura datasheet. https://support.impinj.com/hc/en-us/articles/202756868-Monza-X-8K-Dura-Datasheet (accessed 05 March 2017).

24 Sample, A.P., Yeager, D.J., Powledge, P.S. et al. (2008). Design of an RFID-based battery-free programmable sensing platform. *IEEE Transactions on Instrumentation and Measurement* 57 (11): 2608–2615. https://doi.org/10.1109/TIM.2008.925019.

25 Caizzone, S., DiGiampaolo, E., and Marrocco, G. (2016). Constrained pole-zero synthesis of phase-oriented RFID sensor antennas. *IEEE Transactions on Antennas and Propagation* 64 (2): 496–503.

26 Finkenzeller, K. (2003). Readers. In: *RFID Handbook: Fundamentals and Applications in Contactless Smart Cards and Identification*, 2e, 309–328. Munich, Germany: Wiley.

27 European Telecommunications Standards Institute. (2008). Radio frequency identification equipment operating in the band 865 MHz to 868 MHz with power levels up to 2 W and in the band 915 MHz to 921 MHz with power levels up to 4 W; Harmonized Standard covering the essential requirements of article 3.2 of the Directive 2014/53/EU. *European Telecommunications Standards Institute*, ETSI EN-302-208-v3.1.0, 2016. http://www.etsi.org.

28 Rima, S., Georgiadis, A., Collado, A. et al. (2014, 2014). Passive UHF RFID enabled temperature sensor tag on cork substrate. *2014 IEEE RFID Technology and Applications Conference, RFID-TA*: 82–85. https://doi.org/10.1109/RFID-TA.2014.6934205.

29 Bjorninen, T. and Yang, F. (2015). Signal strength readout and miniaturised antenna for metal-mountable UHF RFID threshold temperature sensor tag. *Electronic Letters* 51 (22): 1734–1736. https://doi.org/10.1049/el.2015.2661.

30 Bhattacharyya, R., Floerkemeier, C., Sarma, S., and Deavours, D. (2011). RFID tag antenna based temperature sensing in the frequency domain. *2011 IEEE International Conference on RFID* https://doi.org/10.1109/RFID.2011.5764639.

31 Krigslund, R., Popovski, P., and Pedersen, G.F. (2012). Orientation sensing using multiple passive RFID tags. *IEEE Antennas and Wireless Propagation Letters* 11: 176–179. https://doi.org/10.1109/LAWP.2012.2185918.

32 Vena, A., Sorli, B., Foucaran, A., and Belaizi, Y. (2014). A RFID-enabled sensor platform for pervasive monitoring. *2014 9th International Symposium on Reconfigurable and Communication-Centric Systems-on-Chip (ReCoSoC)*: 1–4. https://doi.org/10.1109/ReCoSoC.2014.6861358.

33 Manzari, S., Caizzone, S., Rubini, C., and Marrocco, G. (2014). Feasibility of wireless temperature sensing by passive UHF-RFID tags in ground satellite test beds. *2014 IEEE International Conference on Wireless for Space and Extreme Environments (WiSEE)*: 1–6. https://doi.org/10.1109/WiSEE.2014.6973074.

34 Tentzeris, M.M. and Nikolaou, S. (2009). RFID-enabled ultrasensitive wireless sensors utilizing inkjet-printed antennas and carbon nanotubes for gas detection applications. *IEEE International Conference on Microwaves, Communications, Antennas and Electronic Systems, COMCAS*: 1–5. https://doi.org/10.1109/COMCAS.2009.5385940.

35 Babar, A.A., Manzari, S., Sydanheimo, L. et al. (2012). Passive UHF RFID tag for heat sensing applications. *IEEE Transactions on Antennas and Propagation* 60 (9): 4056–4064. https://doi.org/10.1109/TAP.2012.2207045.

36 Nguyen, S.D., Pham, T.T., Blanc, E.F. et al. (2013). Approach for quality detection of food by RFID-based wireless sensor tag. *Electronics Letters* 49 (25): 1588–1589. https://doi.org/10.1049/el.2013.3328.

37 RFMICRON. RFM3240 long-range wireless temperature sensor. http://rfmicron .com/rfm3240-long-range-temperature-sensor/ (accessed 15 September 2017).

38 Farsens. Cyclon-14BA. http://www.farsens.com/en/products/cyclon-14ba/ (accessed 12 October 2017).

39 El Matbouly, H., Zannas, K., Duroc, Y., and Tedjini, S. (2017). Analysis and assessments of time delay constrains for passive RFID tag-sensor communication link: application for rotation speed sensing. *IEEE Sensors Journal* 17 (7): 2174–2181. https://doi.org/10.1109/JSEN.2017.2662058.

9

Autonomous System of Wireless Power Distribution for Static and Moving Nodes of Wireless Sensor Networks

Przemyslaw Kant, Karol Dobrzyniewicz, and Jerzy Julian Michalski

SpaceForest, Pomeranian Science and Technology Park, Gdynia, Poland

9.1 Introduction

The problem of power transfer in extreme environments is well known and, in general, quite difficult to solve. In many real situations, the use of traditional power transmitting means, i.e. cables, is not possible due to their limited resistance to environmental threats, for example temperature, fire, or physical features like mass. These reasons are the main motivation for designing and constructing a WPT system. The use of electromagnetic waves as the means to transfer power should allow for achievement of very high resistance against multiple threads, such as extreme temperature variations, medium interruption, or vibrations [1].

WSNs are sets of tens or even hundreds of autonomous network nodes capable of measuring certain physical quantities and performing two-way wireless communication with other nodes or data sink – a central node where all the data are collected. These networks are gaining recognition due to their flexibility and reliability, i.e. ability to perform autoconfiguration and self-healing. The main drawback of the WSN is the necessity to periodically change or charge the batteries powering up the WSN nodes. In some cases, especially in extreme environments, like in the outer space or in other hazardous conditions, there is no possibility to recharge or exchange the batteries. This problem can be overcome with an appropriately designed WPT system used for powering the WSN nodes. Examples and applications of WPT for WSNs can be found in literature, positions [2, 3]. Apart from the increased redundancy and flexibility, since the WPT/WSN system introduces significant mass reduction of e.g. flying objects, due to extrusion of the powering and data transmission cables, the use of such a system can introduce significant savings, e.g. increase of the payload mass for high altitude/orbital rockets [4]. A great example is the Ariane 5 launcher rocket, which has more than

Wireless Power Transmission for Sustainable Electronics: COST WiPE - IC1301,
First Edition. Edited by Nuno Borges Carvalho and Apostolos Georgiadis.
© 2020 John Wiley & Sons, Inc. Published 2020 by John Wiley & Sons, Inc.

500 sensors measuring various physical quantities, like for example accelerations, temperatures, pressures, etc. The sensors are placed in various parts of the rocket and connected by cables. The total weight of the sensor network including cables is about 70% of the total mass of the rocket's avionics. The use of wireless power distribution system should also result in increasing the sensor network's reliability due to a decreased risk of wire breakage. This is a very big advantage of the wireless solution because of rapid temperature changes and high accelerations, which the rocket experiences during the flight to an orbit. The same applies to other high reliability applications.

In Chapter 2, we describe a concept of a WSN based on a novel routing approach. In this approach, a concept of multiple spanning trees is used for routing data between the nodes of the WSN. Such approach significantly increases the reliability of data transfer by allowing for very fast reconfiguration of the network, even in case of multiple node failures, preventing the network from massive data losses. Chapter 3 describes a proof-of-concept of a WPT system used to propagate the microwave energy toward WSN nodes located on a single 2D plane. In Chapter 4, considerations on development of a WPT system, allowing for propagation of microwave energy toward WSN nodes located in 3D space are presented. To supply microwave energy to individual nodes, in general distributed freely in 3D space, an autonomous tracking system is proposed in Chapter 5. The system was originally designed to track a research sounding rocket, as described in [4].

9.2 Data Routing in WSN Based on Multiple Spanning Trees Concept

In this chapter, the main area of interest is the use of WPT in order to power up a WSN. The amount of energy required by the WSN [5] for optimal operation depends on multiple factors, such as power consumption of a single network node in Tx (transmitting) and Rx (receiving) mode, number and size of data packets to be sent within a time unit, etc. A typical network node uses tens of milliwatts of power during transmission of data packets and single milliwatts or less during data reception. The node can also be put into a deep sleep mode, in which it uses only single microwatts of power. This obviously shows that in order to minimize the amount of power which needs to be sent to the WSN nodes, the number of transmissions within a time unit should be minimized.

In the simplest case, all WSN nodes could be configured in a way which allows for sending the data straight into a central WSN node called a network coordinator or a data sink. In this case, each data packet is sent only once, which minimizes the total power consumption, making wireless powering easier.

Because of node visibility limitations, caused by e.g. long distances between the nodes or obstacles preventing data transmission, multi-hop data transmission including router network nodes forwarding the data packets from the source node to the data sink could be used as a more effective and reliable solution. Multi-hop data transmission can also be used to limit the total transmitter power, and consequently the power consumption.

In a WSN containing hundreds of nodes, assigning a role to each node is a complex task. In general, three basic roles can be defined: end node, which only generates and transmits the data, router node, which generates, transmits and forwards the data from other nodes, and data sink, which receives the data from other network nodes. The role of the network node role assignment is done by routing algorithms, which optimize network parameters and make it possible to keep a direct or indirect connectivity between each node generating the data and the data sink during network operation, as a result increasing the lifetime of the network [6–10]. The routing protocols can optimize multiple parameters of the WSN, such as energy efficiency, medium latency of a data packet or network reliability, which describes resistance of the network to node failure due to physical damage or lack of power. In an unfavorable case of high throughput router failure, multiple network nodes using the router to forward data to the data sink would be completely disconnected from the data sink until the end of network reconfiguration, causing a significant data loss. In case of, for example, a sounding rocket equipped with a WSN, such data loss could contribute to the loss of the rocket.

To improve network reliability, a concept of a routing protocol based on multiple spanning trees will be described.

9.2.1 Multiple Spanning Trees Routing Protocol

The main idea behind the described routing protocol was to allow for instant reconfiguration of the network in case of a node failure. In a typical routing protocol, in case of router node failure, a network reconfiguration process would be started, taking typically from a few dozen seconds to minutes, depending on the network size. To minimize the duration of the network reconfiguration process, static routing tables, each assigned to a single node and including a list of addresses of the target routers/data sink were prepared. The use of static routing tables allows for instantaneous reconfiguration of a sector of the network influenced by the failed node. Each routing table includes a list of target nodes ranked by the target priority (Figure 9.1). The target priority is defined by the routing algorithm. In case of damage to a target node with the highest priority, the source node sends the data to the node with the second highest priority, thus making it possible to maintain connectivity with the data sink. The switch can be instantaneous after finding no response from the failed network node.

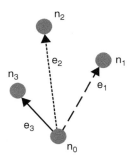

Figure 9.1 Priority connection between nodes. Node n_0 is a data sender. If node n_3 is accessible (highest priority), data is sent exclusively to it. If node n_3 is not accessible, data is sent exclusively to node n_2 and so on.

In order to find the static routing tables for each network node, a multiple spanning tree computation algorithm was developed. In graph theory, a spanning tree is an undirected graph connecting all the vertices (network nodes) in an unambiguous way, i.e. only a single path connects each pair of the vertices in such a graph representing the physical network. Examples of different spanning trees can be found in Figure 9.2a–c. A particular case of a spanning tree is a minimal spanning

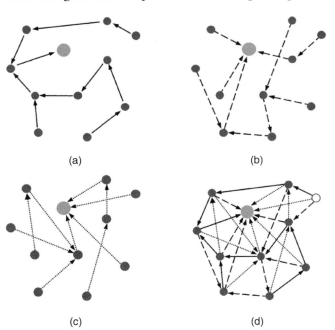

Figure 9.2 (a–c) Three exemplary spanning tress, with disjointed set of edges – connections between WSN nodes. (a) Spanning tree with highest priority. (b) Spanning tree with secondary priority. (c) Spanning tree with lowest priority. (d) Proposed WSN topology. Small light gray circles – WSN nodes sending and routing data, small not coloured circle – WSN nodes sending data only, large circle – data sink.

tree. In this case, the spanning tree is chosen to minimize the total value of edges (edge weights) connecting the vertices of the spanning tree. The weight of an edge can describe different parameters of a network, for example the distance between the nodes. In this case, the minimal spanning tree would describe a tree (network), in which all the vertices are connected unambiguously and by the shortest possible distances, i.e. the total sum of edge weights is minimal. Many minimal spanning tree generation algorithms exist, from among which the most commonly used ones were introduced by Kruskal [11], Prim and Borůvka.

The proposed algorithm is based on Kruskal's minimal spanning tree algorithm. The algorithm is used to generate "n" disjoint spanning trees based on a complex graph edge weighting factors including such information as the distance between the connected nodes, the amount of data generated within a time unit, and the distance to the data sink and others. An example of a network including three trees generated with the use of the developed algorithm can be seen in Figure 9.2d.

The starting point for generation of static routing tables is the generation of a full graph in which all nodes are mutually connected. The trees can be generated sequentially – with the first spanning tree achieving the lowest total sum of weights and with the sum of weights increasing with each subsequently generated tree, excluding the edges assigned to the trees which have already been generated. Alternatively, in the parallel generation of the trees, n trees, each characterized by a comparable total edge weight sum, are generated. In this case, a single edge is assigned to each tree in a single iteration of the algorithm. The trees are generated until no unconnected node is left. Using Kruskal's minimal spanning tree algorithm results in the confidence that no cycles are generated within the end graph for each tree, i.e. only straight (unambiguous) connections between each network node and the data sink will be generated in the network, with no possibility to pass more than once through each router included in the path.

With the spanning tree calculations finished, a static routing table can be assigned to each node in the network, each including a list of n routers to which the data should be transmitted, structured in order from the highest to the lowest priority. During the network operation, each node should periodically send data to the router with the highest priority. Due to problems connected with too many data sent using the router or with the router failure, reception of the packet may not be confirmed. In such case, the node will try to retransmit the data after a short back-off period. After a specified number of subsequent failed retransmissions (in the case implemented in the physical test network the number of subsequent failed retransmissions was 4), the nodes switch to the router with the second highest priority, thus making it possible to send the data to the data sink with low information losses and a very short delay. An exemplary behavior of the network during multiple node failure can be found in Figure 9.3.

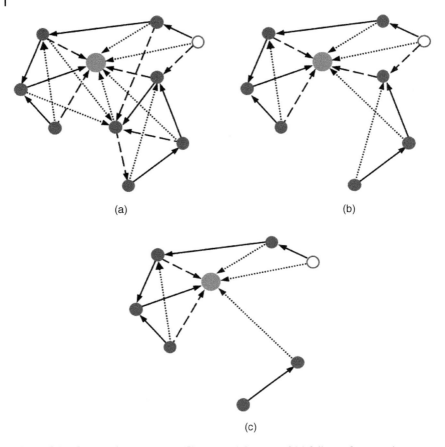

(a)   (b)

(c)

Figure 9.3 Routing in proposed WSN network in case of (a) failure of one node, (b) failure of two nodes, and (c) failure of three nodes. Small light gray circles – WSN nodes sending and routing data, small not coloured circle – WSN nodes sending data only, large circle – data sink.

9.2.2 Software WSN Simulator

The algorithms were implemented in a dedicated software simulator. The simulator was prepared in order to model WSN network traffic and to optimize the proposed WSN routing algorithm. The simulator is a piece of software, which utilizes a set of input data, such as the number of nodes and their location, the amount of generated data, the duration of the main events during node operation, the visibility of two nodes, etc. in order to generate a set of static routing tables based on the aforementioned routing algorithm and to obtain network operation statistics, which can be used to evaluate routing quality and to further optimize the network topology. The initial simulator algorithm was prepared for time-step-based simulation, i.e. the next step of the simulation was calculated after addition of a constant time step to the time counter. In the newest version, however, the software is event-based. The event-based nature of the software means that during the next steps of the simulation, changes of states of WSN nodes (sleeping, transmitting, waiting for packets, etc.) are simulated, which makes the whole simulation incomparably faster. The maximum number of nodes, which can be simulated, depends theoretically only on the amount of memory space in the computer and can exceed a few hundred. For a higher number of network nodes, a possibility to use a clustering algorithm (K-means algorithm) was implemented in order to allow for division of the entire network into a set of subnetworks, which makes it possible to divide the total traffic between a specified number of subnetworks. The subnetworks can be separated using different channels of the unlicensed ISM 2.4 GHz band used for data transmission within the described network.

After network traffic simulation, network operation statistics, including the information about the number of packets successfully sent and lost, can be displayed. Because the number of lost packets is the decisive factor for reliability of the network, these network statistics can also be generated for a single and multiple node failure scenarios. In case of one node failure simulation, a full simulation is repeated for the whole network with only one node (which is different in each simulation) failing during a single simulation. The simulation can be repeated for two node failure, where failures of all combinations of two different nodes are taken into account. Following all the simulations, statistics of the network, including the reliability data, can be obtained. In Figure 9.4, the main view of the time-based simulation software can be seen. Small dark gray circles represent routers, small light gray circles – end nodes, and the large circle in a upper left corner – a coordinator node. Arrows between the nodes represent packet flow in the network. In Figure 9.5, an example of multi-tree (two trees) calculated topology can be seen. The small dark gray arrows belong to the primary tree, whereas the small light gray arrows represent connections of the secondary tree (the connection is activated if a node failure occurs).

9.2.3 Experimental Verification

Operation of the network routing algorithm was tested using an experimental physical network consisting of 30 network nodes located on a test board. In the experiment, routings using ZigBee and the proposed algorithm were compared. The comparison was performed using 10, 20, and 30 network nodes operating at the same time, with a variable packet generation rate from 1 packet (measurement) per second to 50 data packets per second.

Because the ZigBee routing is calculated autonomously, no control over routing was assured. In high stress situation, the network topology was most probably

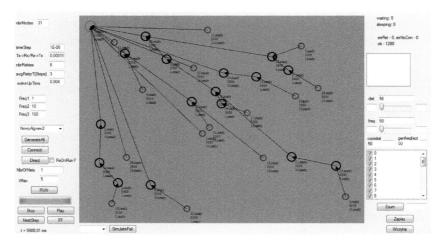

Figure 9.4 View of main window of network simulator. Nodes are indicated with circles (large circle in upper left corner – coordinator, small dark gray circle – routers, small light gray circle – end nodes; operation modes defined in accordance with routing tables).

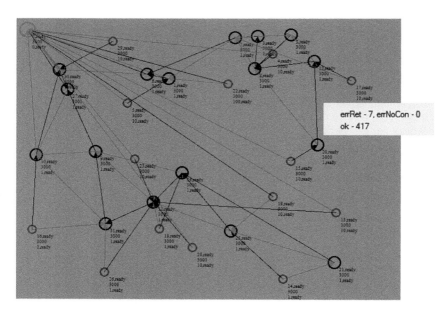

Figure 9.5 Example of multi-tree (two trees) topology calculated for redundant network operation. Alternative path is provided for each node. Network statistics can be seen (ok – packet delivered to coordinator, errRet – packet delayed, and errNoCon – packet not delivered).

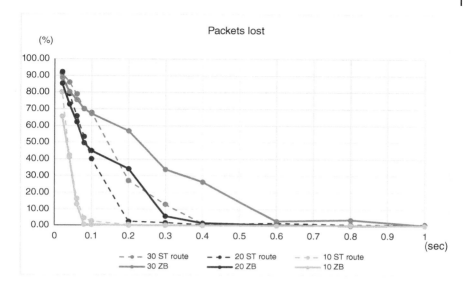

Figure 9.6 Comparison of percentage of packets lost for proposed routing algorithm, and routing based on ZigBee protocol. Series are presented for 10, 20, and 30 network nodes configured using ZigBee algorithm (ZB) and spanning tree algorithm (ST route).

switched to star configuration, i.e. no routing was performed. In case of the routing based on multiple spanning trees, multiple hopping was enforced on the network, based on network optimization using the WSN network simulator.

The experiment results show that the proposed routing algorithm works better than ZigBee algorithm up to 10 packets per second, especially in dense networks. This could also be influenced by a smaller amount of data included in a single measurement packet. Above 10 packets per second, the proposed routing algorithm generated a higher amount of lost packets, mostly due to the imposed multi-hop routing within the network. In case of ZigBee, switch to a star configuration made it possible to improve a little the statistics regarding the number of lost packets. Still, the results in this packet generation rate range are comparable. A chart showing the percentage of lost packets during the test period versus the number of network nodes can be seen in Figure 9.6.

In a physical situation where the network nodes would have to be distributed in a full-scale model of a sounding rocket or other flying object (especially made of metal or carbon fiber), where direct communication between the end node and the data sink would not always be possible, a high advantage of the proposed routing algorithm can be noticed.

During a different test, random nodes were turned off during the network operation. The network was able to successfully and quickly reroute the data, making it possible to achieve a high level of reliability of the network, necessary for its use in flying objects.

In conclusion, the proposed algorithm shows the potential to allow for reliable operation of the WSN in applications requiring efficient reconfiguration in case of node failure. The proposed routing algorithm was tested in laboratory conditions, showing advantages over an industry standard protocol – ZigBee.

9.3 WPT System for 2D Distributed WSN

To meet the requirements of most applications utilizing WPT, considerations regarding flexibility of the entire system need to be carried out. Among many different WPT features, one of the most important is the ability to change microwave energy propagation direction in order to maximize the energy efficiency of the system and to maximize the amount of energy on the receiver side [1, 12, 13].

This chapter shows a concept and physical realization of a WPT system capable of setting the power distribution direction to one of four predefined areas in 2D space. In each of the four predefined areas, a WSN ZigBee node based on TI2530 transceiver including a WPT receiver was used in order to verify the correctness of the system operation.

In order to obtain positive energy balance of the WSN node including a WPT receiver, the WPT system needs to be designed to meet the requirement that more energy needs to be delivered to the WSN node then is consumed during its operation. The main equation, which is the starting point for the design of the WPT system, is the Friis' transmission formula, which describes the power level P_R at the receiver with the transmitter located at the distance of R from the receiver:

$$P_R = \frac{P_T G_T G_R c^2}{(4\pi R f)^2} \tag{9.1}$$

P_T describes the transmitted power level, f describes the WPT signal frequency, G_T is the gain of the transmitter antenna, and G_R is the gain of the receiver antenna, c is the speed of light.

One of the most important factors in this equation is the operation frequency f. In the version of the system presented in this chapter, the operation frequency is located within the 2.45 GHz unlicensed ISM band. The choice of this frequency range allows for easy access to a wide variety of components (transceivers, sensors) and for integration of the WPT system with WSN using a user-friendly ZigBee protocol. The main drawback of this frequency range is that the system components at this relatively low frequency are rather large.

9.3.1 System Concept

The Friis' equation shows that at constant frequency and distance between the transmitter and the receiver, the most important factors influencing the received power level are the antenna gains and the transmitter power level. The gains of the antennas can be maximized using antenna arrays, which consist of multiple radiators, all connected to a single antenna port. By increasing the number of radiators, the antenna becomes more directional (the narrower the antenna pattern is, the more the maximum gain increases), and thus more power is received by the receiver antenna. Due to reciprocity, the same rule applies to the receiver antenna. The rule of thumb says that the gain increases by 3 dB with the doubled number of radiators. The transmitted power level can be increased using a high output power (high P1dB) amplifier placed before the transmitted antenna. After meeting the aforementioned conditions, an optimal power level should be achieved at the receiver antenna.

The main conclusion arising from the Friis' formula is that in order to transfer the radio-frequency (RF) power over long distances in a desired direction, antennas with narrow radiation patterns must be used, which results in smaller spatial coverage of the WPT system. To increase the spatial coverage and maintain the high antenna gain, a system with steered radiation direction can be used. An antenna array powered with the use of a Butler matrix is proposed as physical implementation of such a system. A block diagram of the proposed WPT/WSN system can be seen in Figure 9.7.

Butler matrix is a planar structure utilizing a combination of three basic types of microwave structures in order to equally divide the input power and to produce equal phase differences between the outputs of the structure [14]. Equal power

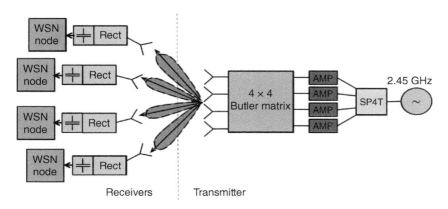

Figure 9.7 Block diagram of 3D WPT system used for powering WSN nodes.

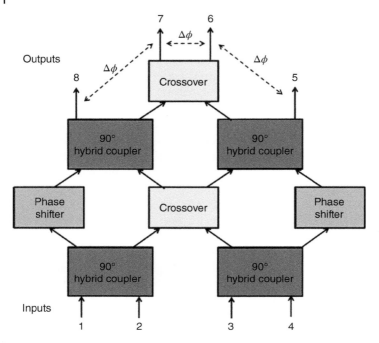

Figure 9.8 Block diagram of 4×4 Butler matrix.

levels and phase shifts between the subsequent radiators make it possible to tilt the radiation direction of the antenna array. The structures used in the Butler matrix are as follows (see Figure 9.8):

- 90° hybrid coupler – used to obtain equal division of the input power and to introduce 90° phase shift between the outputs.
- Phase shifter – used to define the appropriate phase shift between the subsequent outputs of the Butler matrix.
- Crossovers – used to allow for safe crossing of signals, without the need to use multilayer planar structures.

By supplying microwave power to the right input of the Butler matrix, the beam, i.e. the maximum radiation direction, can be tilted in one of n directions, where n is the number of inputs/outputs of the Butler matrix [4, 15]. In the presented system, a 4×4 Butler matrix is used.

By using a Butler matrix, multiple spatially separated WSN nodes can be powered up just by switching the main radiation direction of the transmitter. In this implementation, no mechanical parts are involved in directing the transmitter antenna array toward the target, which makes it possible to switch the radiation

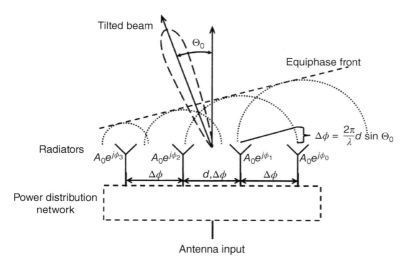

Figure 9.9 Generation of equiphase wave front using equal phase differences between Butler *matrix outputs (Power distribution network) resulting in tilt of radiation pattern.*

direction even within nanoseconds. The mechanism of tilting the beam (generation of an equiphase wavefront) using a 4×4 Butler matrix and a 4×1 antenna array is shown in Figure 9.9. In order to realize this feature in the proposed solution, a 1-to-4 nonreflective low insertion loss SP4T switch is used. This component can direct the input power toward the right input of the Butler matrix, thus steering the power flow (radiation) direction of the transmitter. Because the switch used in the system cannot handle high power levels (maximum +24 dBm), a high power amplifier (approximately 1 W of output power level, AMP in Figure 9.7) had to be added between each switch output – Butler matrix input pair.

All the aforementioned components are used to deliver a desired power level to the output of the receiver antenna. In order to power up the WSN nodes attached to the receiver antennas, a rectifier capable of converting RF power into DC voltage must be used. The rectifier is used to supply the energy to a supercapacitor, which is finally used to power up the WSN nodes.

In this solution, the rectifier (Rect in Figure 9.7) is based on zero-bias Schottky diodes, connected in a voltage multiplier configuration. The reason for this is that, for low input powers and low impedance loads, the output voltages are much too low to enable powering of ZigBee nodes (the currently used nodes require at least 2 V supply). By combining multiple diode pairs (as shown in Figure 9.10), the output voltage can be increased [17]. In theory, the output voltage should be m times higher than in the single diode pair case, where m is the number of diodes used. However, in reality, the output DC voltage is lower than the theoretical one, mainly due to the voltage drop on the diodes and due to circuit losses (Figure 9.10).

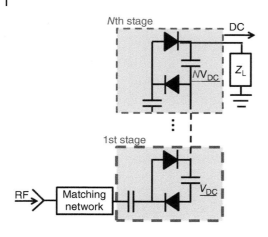

Figure 9.10 Rectifier with Villard-bridge-based voltage multiplier [16].

The DC voltage prepared in this way is then used to charge up a supercapacitor, which powers a single node of the WSN. As the currents flowing in the rectifier circuits are rather low, a supercapacitor with moderate capacitance should be used to store energy for the WSN nodes.

9.3.2 Physical Realization of 2D WPT System

The WPT/WSN system elements were prepared based exclusively on off-the-shelf and low-cost components. The first design assumed the use of standard ZigBee nodes based on TI2530 transceiver. The nodes were configured to perform periodic voltage-level measurements and send the obtained data to the data sink within a short period of time, which allowed for supervision of the supercapacitor charges in all WSN nodes. After each measurement and transmission, the nodes were set into sleep mode in order to minimize energy consumption. The communication channel chosen for ZigBee WSN was 2.405 GHz channel, which ensured a high level of isolation between the WSN channel and the 2.45 GHz WPT channel. Each node built into this proof-of-concept was connected to a 0.47 F supercapacitor, which was recognized as the optimal value for charging (relatively short charging time) and discharging times (relatively long WSN operation time of the node).

The signal source for the WPT system was based on the transceiver used in the WSN system design. The output from the signal generator consisted of a single sine (2.45 GHz) signal with a maximum power level of +10 dBm. This signal was then switched using Skyworks AS208 SP4T. Both microwave switches and power amplifiers can be seen in Figure 9.11. Such switch provides a very high level of isolation between the outputs (unused ports are 50 Ω matched), which is of high importance due to the necessity of powering up only one input port of Butler matrix at a time to obtain the correct output signals from the Butler matrix.

Figure 9.11 Microwave switch and power amplifiers to switch Butler matrix inputs and provide suitable power level for wireless powering.

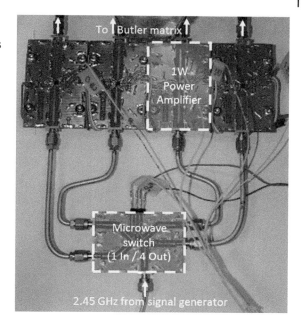

The Butler matrix was designed in Advanced Design Systems (ADS) and manufactured on 0.6-mm-thick FR-4 laminate. The presented version of the Butler matrix is narrow-band because of the fact that only a single sine signal is pumped into the system. There is a possibility to widen the bandwidth of the structure with the use of multiple section components [1]. The assumed phase shift introduced by the phase shifter in the Butler matrix was set to approximately 45° in order to achieve wide scanning/radiation direction tilt ability. The measurement results for the Butler matrix showed maximum deviation from the assumed output port phase differences, namely 8° for the outer Butler matrix input ports (average deviation 4.5°) and 3.8° for the inner input ports (average deviation 1.7°). The maximum amplitude variation recorded during the measurements was 1.7 dB (1.2 dB average).

The outputs of the Butler matrix were connected to four patch antennas optimized for 2.45 GHz and placed on a single 1-mm-thick FR-4 printed circuit board (PCB). The whole structure can be seen in Figure 9.12. The measured radiation patterns generated by supplying subsequent inputs of the Butler matrix connected to the antenna array can be seen in Figure 9.11.

In antennas with steerable radiation pattern, the higher the tilt of the main radiation direction, the higher the side lobes and the lower the maximum gain of the antenna array. In the case presented in Figure 9.13, because of high phase shifts between the outputs of the Butler matrix and relatively short distance between the

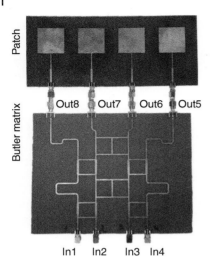

Figure 9.12 2.45 GHz Butler matrix connected to 4 × 1 microstrip patch antenna array.

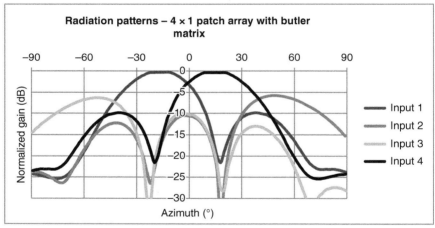

Figure 9.13 Radiation patterns of 4 × 1 patch antenna connected to Butler matrix. Inputs are labeled in accordance with Figure 9.8.

antennas, a very high tilt of the antenna was achieved (approximately ±50° from the radiation direction without beam scanning), which resulted in the decrease of the antenna gain by approximately 6 dB in comparison with the remaining radiation directions (approximately ±17° from nonscanning configuration). These gain differences can be lowered by decreasing the phase shift introduced by the phase shifter in the Butler matrix, i.e. by decreasing the maximum antenna sweep angle.

The antennas used for receivers were realized as 2 × 2 patch antenna arrays manufactured using 1 mm FR-4 substrate (Figure 9.14). The approximate maximum

Figure 9.14 Manufactured 2×2 antenna array.

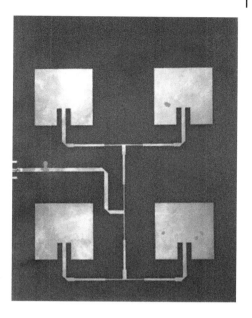

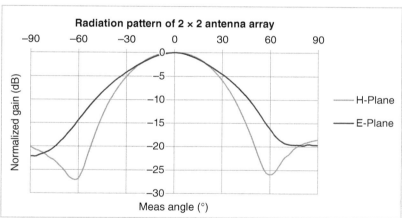

Figure 9.15 Radiation patterns of manufactured 2×2 antenna array.

measured gain was 11.6 dBi. The measured radiation patterns in both electric and magnetic planes of the antenna array can be found in Figure 9.15.

The RF power received by the receiver antenna must be converted into DC power using a rectifier. The rectifier was built using a five-stage Villard cascade voltage multiplier to achieve high DC voltage levels at the rectifier's output [16]. The circuit was optimized using an impedance matching circuit which, after

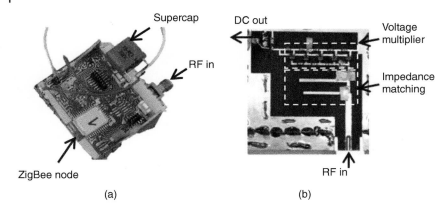

Figure 9.16 Components used in receiving parts of WPT system. Rectifier (b) is placed on bottom layer of WSN node PCB (a).

manufacturing, was tuned with the use of copper strips to achieve the highest possible output DC voltage for the input RF power of 0 dBm and the resistive load of 100 kΩ. To eliminate the influence of external interferences on the operation of the rectifiers, each of them was put in a metal housing. The optimized rectifiers were placed on the bottom side of the ZigBee PCB, which can be seen in Figure 9.16. The ZigBee module PCB was prepared to allow for low-energy operation of the Anaren A2530 with an integrated antenna.

The components prepared in such a way were integrated into a single WPT system. The following chapters summarize the measured performance of the system with few variations of each of the components.

9.3.3   Experimental Verification of the 2D WPT System

The measurement setup for a proof-of-concept WPT system consisted of the transmitter subsystem including a 2.45 GHz signal generator, a microwave switch, amplifiers, a Butler matrix, and a transmitter antenna array, as well as four receiver subsystems, each consisting of a rectenna (an antenna connected to an optimized/matched rectifier), a 0.47 F supercapacitor, and a ZigBee node. All the receivers/ZigBee nodes were located on a circle of a constant radius of 65 cm, with the transmitter placed in the middle. Each receiver antenna was set in the maximum radiation direction of one of the four possible radiation beams. The experimental setup can be seen in Figure 9.17.

The signal generator and the amplifiers were set to provide the maximum power on all the antennas of 0 dBm. Because of the high tilt of the radiation pattern for the two inner Butler matrix inputs, the maximum gain of the transmitter antenna for those settings was approximately 6 dB lower than for the remaining inputs, giving

Figure 9.17 Final experimental setup of WPT system.

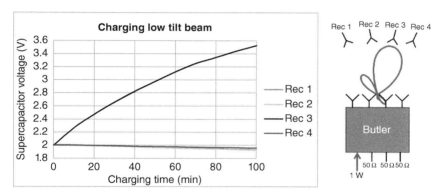

Figure 9.18 Results of charging experiment with low tilt beams (outer inputs of Butler matrix).

the received power level of −6 dBm at the two outer receivers. In order to verify the performance of the WPT system, the voltage levels of supercapacitors connected to the rectifier were periodically read out.

The results of the experiment are shown in plots of supercapacitor voltage levels in function of time, which can be found in Figure 9.18 (for the lower tilts of the radiation pattern) and Figure 9.19 (for the higher tilts of the radiation patterns).

Results of the experiment show that, for all the radiation patterns which can be obtained with the use of the presented system, constant charging of the supercapacitor (i.e. increase of the supercapacitor voltage) powering the WSN nodes can be obtained. Due to higher maximum gains of the less tilted radiation patterns (Figure 9.18), the charging is much faster, and much higher voltages can be obtained than in the second case (Figure 9.19). The obtained performance allows

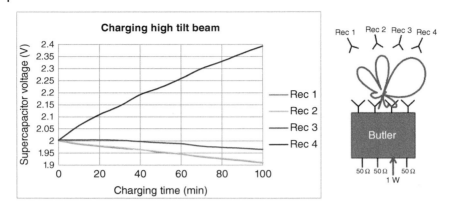

Figure 9.19 Results of charging experiment with high tilt beams (outer inputs of Butler matrix).

for sending approximately 30 measurement results per minute, using a standard off-the-shelf WSN node for the lower tilts of the radiation pattern, and approximately five measurements per minute for higher tilts of the radiation pattern without obtaining negative energy balance. With more and more efficient WSN nodes available on the market, the possible number of measurements per minute can be increased. Alternatively, with the same number of measurements per minute, a lower transmitter power could be used to power up the network nodes. The difference in the power levels on the receivers located at different spatial angles in relation to the transmitter antenna can be potentially compensated by using a signal source with an adjustable output power level and higher gain amplifiers.

The proof-of-concept of the WPT system makes it possible to wirelessly charge supercapacitors powering WSN nodes, located at a maximum spatial angle of approximately 120°. The obtained performance makes it possible to achieve minimum five measurements per minute with standard ZigBee WSN nodes placed in the maximum radiation directions of the obtained radiation patterns, at the distance of 65 cm from the transmitter. By using less lossy laminates with lower permittivity, the range of the system should be significantly increased, and the improvement of the system efficiency could be achieved as a result of obtaining higher radiation efficiencies and lower dielectric losses. However, lower permittivity in antennas would result in increased size of the radiators.

9.3.4 Tests of 2D WPT System with Implemented Switching Algorithm

During operation of the WSN node, measurements of certain physical quantities can be collected and sent to the data sink (WSN receiver in Figure 9.7).

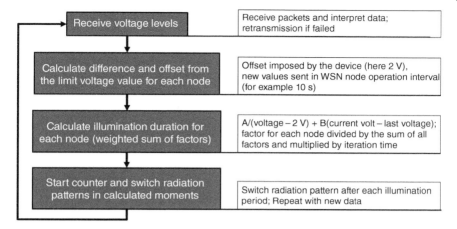

Figure 9.20 Schematic diagram of algorithm.

Apart from that, each WSN node can measure and send the voltage level of the supercapacitor to which it is connected. Voltage levels can be collected periodically from each WSN node and sent to the data sink located at the transmitter. Basing on responses of the WSN nodes, the transmitter's radiation pattern can be autonomously switched in time, in order to cover current energy demand of each WSN node and thus allow for the network to remain operational for a long time.

Using the received WSN node voltage levels, the transmitter can switch radiation patterns according to the following algorithm (shown in Figure 9.20):

(1) Receive voltage levels from all nodes.
(2) Calculate difference and offset from the limit voltage value for each node.
(3) Calculate illumination duration for each node (weighted sum of factors from point (2)).
(4) Start counter and switch radiation patterns in calculated moments.
(5) Go to point (1).

The algorithm, although simple, makes it possible to efficiently charge multiple network nodes and to maintain (in an ideal situation) the same voltage level at each supercapacitor (network node). In case of a high number of powered network nodes or a higher number of data samples generated in a time unit, a lower average voltage level of supercapacitor voltages can be expected.

The system prepared in order to test the automated switching of power propagation directions, utilizes a 4 × 1 patch array transmitter with a switched radiation beam in order to ensure wide spatial coverage. The configuration of the transmitter is similar to the configuration shown in Figure 9.7.

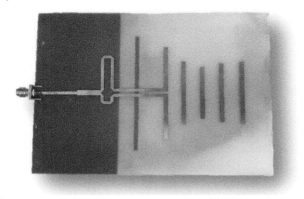

Figure 9.21 Highly directional planar Yagi–Uda antenna for 2.45 GHz.

Supercapacitor

Rectifier

WSN node

Figure 9.22 Rectifier with WSN node.

On the receiving side, the RF power was collected using a highly directional planar Yagi–Uda antenna (Figure 9.21). The antenna was connected to a rectifier including a five-stage voltage multiplier and a matching circuit for high DC output level. The output of the rectifier (shown in Figure 9.22) was connected to the 0.47 F supercapacitor gathering energy for powering one WSN node. In total, four receivers were used in the proposed system, each placed in the maximum gain direction of radiation patterns of each of the four transmitters.

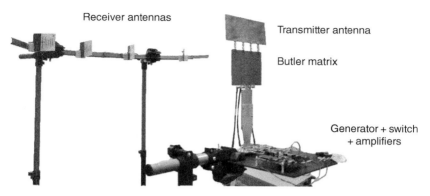

Receiver antennas

Transmitter antenna

Butler matrix

Generator + switch
+ amplifiers

Figure 9.23 Autonomous WPT system realized for supplying power to WSN nodes distributed in space.

In Figure 9.23, the WPT system for automated charging of WSN nodes can be seen. To verify performance of the WPT system, a charging experiment has been performed. In the experiment, all WSN nodes were measuring and sending the data to the data sink while being charged in accordance with the algorithm. Voltage levels measured by all WSN nodes versus time are presented in Figure 9.24.

It can be seen that the node with the lowest voltage level obtains the highest amount of energy from the transmitter. This is done by extending the illumination time for the network node requiring the highest amount of energy. As a consequence, the illumination time for other nodes decreases, causing a slight decrease of the supercapacitor voltages. In the case of high energy loss (sudden voltage drop of N3 in Figure 9.24), the system directs the energy flow to the node which needs energy the most, in this case the N3, reducing to a minimum the illumination time for other network nodes in order to extend the lifetime of the entire network. This ensures achievement of optimal voltage levels for all the network nodes and quick response of the WPT in case of high activity of a single (or multiple) node included in the network, or in case of an unforeseen voltage drop at a single (or multiple) nodes.

9.4 WPT System for 3D Distributed WSN

In this chapter, a fully autonomous wireless power transmission system, able to transmit microwave energy in predefined directions in 3D space, is proposed. The possibility of power distribution in any direction in 3D space is really important because it increases the scope of applications.

In the concept described in this chapter, four antenna arrays forming a tetrahedron are used. The power is delivered to the proper antenna using a series of

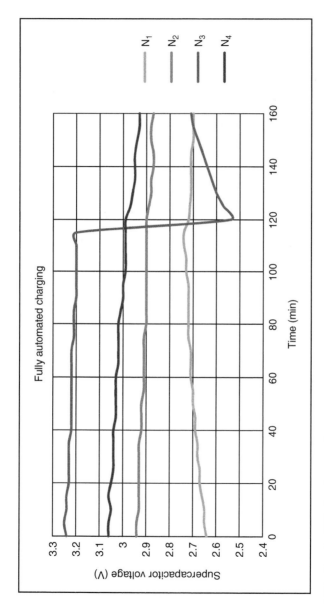

Figure 9.24 Result of 2D charging experiment.

(a)   (b)

Figure 9.25 (a) Four antenna arrays placed in tetrahedral configuration. (b) All possible radiation directions.

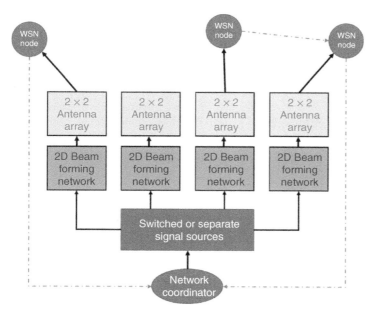

Figure 9.26 A simplified block diagram of the WPT system for 3D powering WSN nodes.

microwave switches. With four 4-element antenna arrays, coverage of nearly the entire 3D space can be obtained. The proposed electromagnetic propagation concept (configuration) is presented in Figure 9.25a,b. A simplified block diagram of the system can be found in Figure 9.26.

The maximization of gain is obtained with the use of a switched-beam antenna array. In the presented case, high directivity in different spatial directions can be obtained, which, together with high transmitter power, significantly increases the range. A simple and cost-effective solution is to use a planar microstrip antenna array, where spatial coverage of approximately 120°, in both electric and magnetic planes of the array, can be obtained. As a result, effective back radiation practically does not exist. Radiation in other directions can be made possible with the use of the remaining antennas.

In order to maximize the efficiency of the system, the transmitter should be able to adjust the receiver illumination times to current voltage levels at the receivers, i.e. for fast dropping voltage levels, the transmitter should illuminate the receiver with higher voltage drop longer than all the others. This can be obtained using feedback from receivers to the transmitter, as shown in Section 9.4. Feedback from WSN nodes to the data coordinator includes current voltage level at a particular receiver, which is used to calculate the required illumination times for all the WPT system receivers (WSN nodes). With a well-adjusted switching algorithm, an average voltage level should be present at each receiver after a longer operation time of the system.

In the proposed concept, the switching algorithm based on a weighted sum of voltage drops in a time unit and current voltage levels is implemented. For low voltage levels, a risk of node failure is increased; therefore, a longer illumination time is necessary. Also for fast voltage drops, which can be caused, for example, by a temporarily or constantly increased measurement frequency in comparison with other WSN nodes, the illumination time is increased. With voltage levels delivered as few extra data bits attached to the main measured value by the sensor, fully autonomous and intelligent WPT System for WSNs is obtained.

9.4.1 Design of Components of the 3D WPT System

The structure of the proposed 3D WPT system is quite similar to the 2D WPT system developed and presented previously. In this case, the transmitter antenna has to be able to tilt the beam in both horizontal and elevation planes. Also, the beam switching circuit has to be redesigned. A block diagram of the proposed 3D WPT system is shown in Figure 9.27.

In order to allow for beam steering, a four-element-patch antenna array can be used. In contrast with the transmitter patch array used in the 2D WPT solution, in this case, a 2×2 antenna array with separate feeds is used. The proposed 2×2 patch antenna array is shown on Figure 9.28. The array was simulated in CST environment.

With appropriate phase shifts between inputs of antennas ($\Delta\varphi x$ in magnetic plane and $\Delta\varphi y$ in electric plane), an appropriate beam tilt can be obtained. It is

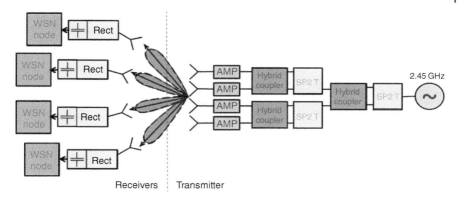

Figure 9.27 Block diagram of 3D WPT system used for powering WSN nodes.

Figure 9.28 Structure of 2×2 patch antenna array.

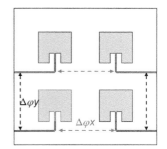

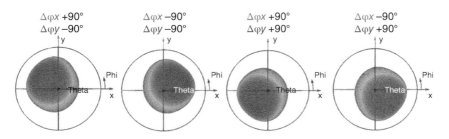

Figure 9.29 Typical radiation patterns with microstrip patches supplied by phase-shifted signals.

the same way of steering as in 4×1 patch antenna array, but in this case, phase shifts are introduced between the array elements placed in both planes of the array. Typical radiation patterns for the simulated 2×2 patch antenna array, including applied phase shifts in both planes of the array, can be seen in Figure 9.29.

To realize the beam switching, the Butler matrix could potentially be used, as in the case of the 2D WPT system. As in the Butler matrix, a rather higher number of

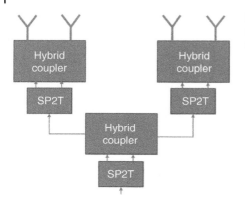

Figure 9.30 Block diagram of beam switching circuit for 2 × 2 patch antenna array.

outputs is used, in the case of 2 × 2 antenna array, half of the outputs would have to be terminated. A simple solution to generate equal phase shifts in both planes is to use a 3 dB hybrid coupler, which provides output signals characterized by mutual ±90° phase shifts, what is presented in the individual Figure 9.29a–d. The biggest advantage of the hybrid coupler approach in comparison with the Butler matrix is the size. Due to the fact that the used transmitter antenna array includes two radiators in each plane, it is necessary to use three SP2T microwave switches (Figure 9.30) instead of one SP4T, as used for the 2D case, to allow for switching of the radiation direction in two planes. As the switches and hybrid couplers can be obtained in highly integrated forms, the area used by the feeding circuit could potentially be significantly decreased. One important thing which has to be kept in mind is the fact that all the SP2T switches used in the circuit need to be switched at the same time in order to send the RF power in the desired direction.

An exemplary board including three SP2T switches and three hybrid couplers can be seen in Figure 9.31. The switches can be connected using steering pins with a microcontroller switching the radiation direction of the transmitter between the beams as shown in Figure 9.29, in order to power up network nodes distributed in the 3D space.

Using the components as shown previously and standard receivers as described in Chapter 3, wireless charging of multiple nodes distributed in 3D space is made possible by switching the radiation pattern of the transmitter in two planes at the same time (electric and magnetic planes of a single planar radiator). Using the proposed switching algorithm, full automation of the charging process and also equalization of voltage levels for all the WSN nodes should be possible. With additional node identification procedure on a 3D sphere, the moving WSN nodes could be easily charged.

In order to achieve higher resolution of the WPT transmitter, a higher number of radiators in each plane should be utilized. This entails the necessity to prepare a more complicated power beam switching network based on multiple Butler

Figure 9.31 Exemplary board including SP2T switches and hybrid couplers required to perform beam switching.

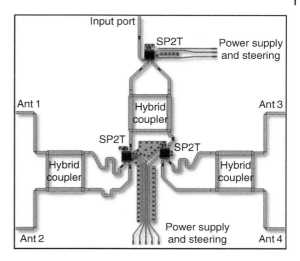

matrices or Rotman lenses. An alternative solution would be to use an integrated phase shifter and an attenuator at each radiator, thus allowing for changing the beam direction with high resolution or even continuously. The third option would be to use a mechanical solution, which would be used to mechanically turn the antenna array toward the target. This solution will be presented in the following chapter.

9.5 Locating System and Electromagnetic Power Supply for WSN in 3D Space

The two WPT systems described previously are designed for sending microwave energy in the predefined directions only. This chapter proposes a solution which can be used to send the microwave energy in any direction in 3D space. Moreover, it can autonomously detect where WSN nodes are located and can even track the node in case of its movement.

To realize this concept, we propose to use a system initially designed for tracking a research rocket and data exchange between the rocket and the ground station. In this chapter, we present the system in its initial version [4] indicating at the end of the chapter which elements of the system need to be replaced to meet the requirements of the WPT application for powering WSN nodes, either static or moving.

In the concept [4], the data on board the rocket were generated using WSN configured with the help of the routing algorithm presented previously. The system consisted of a tracking subsystem operating in 869 MHz ISM band, and a data exchange subsystem operating in 2.4 GHz ISM band. The first system [18] utilized

Figure 9.32 Tracking and data exchange system with indication of rotation planes.

one-way transmission, whereas the second system allowed for bidirectional data exchange between the rocket and the ground station. The tracking subsystem included a signal generator (beacon) placed on board the rocket and a movable ground platform equipped with four circularly polarized antennas (Figure 9.32), which were used to direct the maximum gain of the 2.4 GHz data transmission antenna toward the rocket, thus maximizing the transmission range.

Figure 9.32 shows the main part of the tracking and data exchange system, which is a rotatable platform including five antennas. The platform can rotate in two axes – vertical and horizontal, to allow for half-sphere coverage of the system, i.e. to allow for tracking within 360° in a horizontal plane and 180° in a vertical plane.

9.5.1 Tracking Subsystem

The tracking subsystem located on board the rocket consisted of an 869 MHz high-power signal generator, working as a beacon, equipped with an isotropic antenna. The generator sent a continuous wave signal at 869 MHz, which was received by the tracking system (Figure 9.32). The four antennas were high directivity helical antennas [19], each tilted in a different direction, allowing for reception of different power levels. The different received power levels were caused by antenna gain variation, with the azimuth and elevation angles of the antenna determined by the line of sight (LOS) for all the four antennas.

In perfect conditions, i.e. perfectly symmetric antennas and absolute values of elevation angles equal for all the antennas, the received power levels should be equal. Each antenna was connected to a receiver, allowing for amplification,

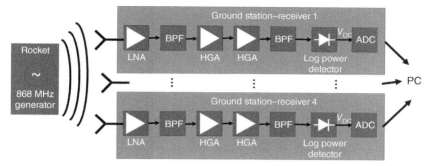

Figure 9.33 Block diagram of the tracking system. The total of four receivers was used in this version of the system. LNA – low noise amplifier, HGA – high gain amplifier.

Figure 9.34 Back of rotatory platform including receivers and stepper motors.

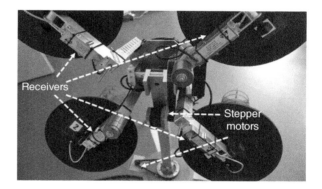

filtering, and detection of the received continuous wave 869 MHz signals. A simplified block diagram of the tracking subsystem can be found in Figure 9.33.

The RF receiver path included a medium gain, low noise amplifier (around 0.7 dB of Noise Figure), which mainly determines the noise figure of the whole receiver path. The two additional amplification stages (high gain amplifiers) used in each receiver allow for obtainment of the total gain of the receiver path, equal to approximately 50 dB, which, together with NF at the level below 1 dB, significantly increased the range of the system. The receivers can be seen in Figures 9.34 and 9.35.

Because of high bandwidth and gain of the receiver system based on amplifiers only, narrowband surface acoustic wave (SAW) filters were used to condition the signal before putting it on a logarithmic detector. The logarithmic detector allows for obtainment of DC voltages proportional to the input power level expressed in (dBm). The voltages generated in such a way were digitized using 8 bit analog-to-digital converter (ADC) and sent – using an USB interface – to the processing unit – a laptop. With the digitized voltage level change rate of

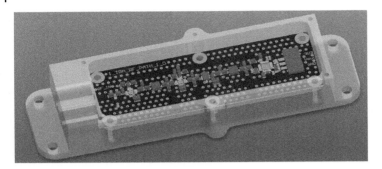

Figure 9.35 View of PCB receiver including pads for LNA, two SAW filters, two high-gain amplifiers, and a logarithmic power detector.

20 Hz, tracking of a fast moving object (start of the rocket) was made possible. Between digitization of the subsequent values of voltages, rotation of the whole platform was performed in two planes with the use of dedicated stepper motors, each assigned to rotation in one plane. The platform was rotated in the direction indicated by the antenna/receiving path with the highest RF power/voltage level until steady state (equal voltages on all antennas) was achieved. The rotation is necessary to direct the data exchange subsystem antenna toward the rocket, thus maximizing the range of the subsystem.

9.5.2 Data Exchange System

The data exchange system consisted of two 0.5 W MIMO Wi-Fi cards, one placed on board the rocket and the second located on the ground. The MIMO ports of the on-board Wi-Fi card were connected to two circularly polarized quadrifilar helix antennas [20], each with inverted polarization, i.e. one left-hand circular polarization (LHCP) antenna and one right-hand circular polarization (RHCP) antenna. The quadrifilar helix antennas possess a half-sphere radiation characteristic, allowing for illumination of the ground station during the movement of the rocket. The Wi-Fi card, including the antennas and the radiation pattern, is shown in Figure 9.36.

The Wi-Fi card located on the ground was connected to a highly directive 2.4 GHz helical antenna, located in the middle of the rotating platform (rocket tracking subsystem). With correct alignment of the rocket tracking subsystem, the highest possible gain of the 2.4 GHz antenna can be obtained in the LOS between the antenna and the sounding rocket's antenna. In this way, the highest possible data transmission range can be achieved. The second isotropic circularly polarized antenna (with inverse polarization) is used as an emergency solution in case of any mechanical problems with the tracking system.

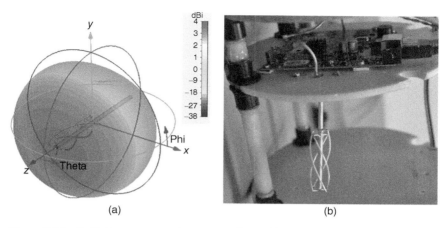

Figure 9.36 Radiation pattern and structure of quadrifilar helix antenna (a) and Wi-Fi card including two quadrifilar helix antennas mounted on rocket frame (b).

9.5.3 Angular Position Estimation of Moving WSN Node

Steering of the tracking and data exchange system was carried out with the use of two stepper motors, each controlled by state estimate obtained with a Kalman filter [21]. A state vector of the modeled system (9.2) consists of the angular positions, the distance, and the velocities,

$$\mathbf{x}(t) = \begin{bmatrix} \varphi_p(t) & \theta_p(t) & r_p(t) & \varphi_v(t) & \theta_v(t) & r_v(t) \end{bmatrix}^T \tag{9.2}$$

where $\varphi_p(t)$, $\theta_p(t)$ denote the horizontal and the vertical tracking error angles ($°$), $r_p(t)$ denotes the distance to the tracking object (m). The next $\varphi_v(t)$, $\theta_v(t)$ denote the angular velocity of the tracking error angles (deg/s), r_v denotes the speed of tracking object moving away (m/s). Introducing a sampling period as $T_s[s]$ state transition can be represented in state space Eq. (9.3),

$$x(t+1) = Ax(t) + v(t) \tag{9.3}$$

where

$$A = \begin{bmatrix} 1 & 0 & 0 & T_s & 0 & 0 \\ 0 & 1 & 0 & 0 & T_s & 0 \\ 0 & 0 & 1 & 0 & 0 & T_s \\ 0 & 0 & 0 & 1 & 0 & 0 \\ 0 & 0 & 0 & 0 & 1 & 0 \\ 0 & 0 & 0 & 0 & 0 & 1 \end{bmatrix} \tag{9.4}$$

$v(t)$ denotes the process noise vector. The system output consists of the voltages read from the power detectors. The system output equations can be represented as a set of a nonlinear equations,

$$y(t) = h[x(t)] + w(t) = \begin{bmatrix} V_1(t) \\ V_2(t) \\ V_3(t) \\ V_4(t) \end{bmatrix} + w(t) \tag{9.5}$$

where

$$V_i(t) = S_{\mathrm{pd}_i} \left(\begin{array}{c} I_{\mathrm{pd}_i} + P_{Ti} + G_{Ti} + G_{Ri} \left(\varphi_p - \psi, \theta_p - \psi \right) + \\ 20 \log_{10} \frac{\lambda}{4\pi r_p} + G_{Ai} \end{array} \right) \qquad i = 1, \ldots, 4 \tag{9.6}$$

$w(t)$ represents an input noise vector, $V_n(t)$ represents a voltage read from nth power detector. S_{pd} denotes a power detector logarithmic slope $\left(\frac{\mathrm{mV}}{\mathrm{dBm}} \right)$, I_{pd} denotes a power detector logarithmic intercept (dBm). P_T represents the transmitter output power (dBm). G_T, $G_R(\alpha, \beta)$ represent the transmitter and the receiver antenna gains (dBi). In the presented system values, S_{pd_i}, $I_{\mathrm{PD}i}$, G_{Ti}, G_{Ri}, G_{Ai} are the same for $i = 1, \ldots 4$. Value ψ denotes the receiver antennas deflection angle (°), λ stands for wavelength (m). Finally, G_A stands for the total amplifier gain (low noise and medium power amplifiers) (dB). The $G_R(\alpha, \beta)$ was approximated in the valid region shown in Figure 9.37 and is expressed as:

$$G_R(\alpha, \beta) = -0.1050 \cdot \left(\alpha^2 + \beta^2 \right) + 16 \tag{9.7}$$

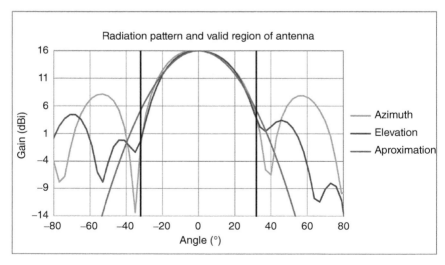

Figure 9.37 Antenna directivity with marked valid region of radiation pattern used for gain curve approximation and Kalman filter setup.

The estimate of system state variables $\hat{\varphi}_p(t)$ and $\hat{\theta}_p(t)$, obtained from Kalman filter, was used to control both motors of the tracking device in accordance with the following formulas:

$$M_H(t) = P\hat{\varphi}_p(t) \tag{9.8}$$

$$M_V(t) = P\hat{\theta}_p(t)$$

where $M_H(t)$, $M_V(t)$ denote the set speed for the horizontal motor and the vertical motor $\left(\frac{RPM}{s}\right)$, and P represents the proportional gain.

9.5.4 Experimental Verification

The operation of the whole system was verified during rocket flight campaign. The system was able to continuously track the rocket during take-off, the main flight phase, and the landing of the rocket components on parachutes. The tracking system was located approximately 600 m away from the launch pad in order not to destroy low noise receivers and to decrease the angular velocity of the tracking system. During the test, information about the arm movement was stored. The data recalculated to angular position in a three-dimensional coordinate system in relation to the starting point can be found in Figure 9.38.

It can be seen that the tracking was uninterrupted and lasted from launch to landing. The maximum measured range of the system during the experiment was 10 km (the maximum recorded distance between the rocket and the tracking system). In this range, the Wi-Fi connectivity was continuously assured, which was later confirmed by comparing the received data with the data stored in the

Figure 9.38 3D chart showing angles of tracking system during sounding rocket flight.

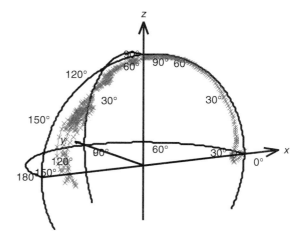

on-board memory. It was confirmed that 100% of data generated by the WSN was successfully transmitted to the ground station, thus verifying the correctness of system operation. The theoretical ranges (using Friis' equation) are more than 100 km for 869 MHz tracking system and approximately 30 km for Wi-Fi transmission, which should be useful even for big rockets that are planned to be used with the system in the near future. Even for bigger rockets either higher power transmitters and more sensitive receivers for both systems or lower frequency bands can be chosen to further increase the maximum ranges for both systems.

9.5.5 Adaptation of the System to WPT for WSN

The primary purpose of the system presented previously was to track the research rocket and to allow for data exchange between the rocket and the ground station. The presented system is a good base to be implemented in wireless powering of WSN nodes, i.e. to build a WPT system for powering WSN nodes distributed in 3D space, which in general could be static or moving.

Below there is a list of subsystems which need to be replaced to adopt the system so that it can be used for WPT for WSN nodes distributed in 3D space.

9.5.5.1 Tracking System
- Four main antennas of the system need to be changed from 869 MHz to 2.4 GHz ones.
- One extra antenna working at 869 MHz for WPT needs to be added and located in the center of the system next to the 2.4 GHz antenna, used for data exchange between the tracking WPT system and WSN nodes.
- Filters, amplifiers, and detectors need to be changed from 869 MHz to 2.4 GHz ones.

9.5.5.2 WSN Node
- Rectenna (antenna and rectifier) working at 869 MHz needs to be added to each single WSN node for WPT purposes.

Figure 9.39 shows the original solution [4] (Figure 9.39a), and its adaptation after changes is described in Section 9.5.5 (Figure 9.39b), making it possible to be used for WTP and tracking of WSN nodes. Smaller helical antennas are designed to work at 2.4 GHz, and larger ones – to work at 869 MHz.

The concept presented here is a fresh one and has not been realized yet. It is a subject for further development of the system, and new results will be published as soon as they have been obtained.

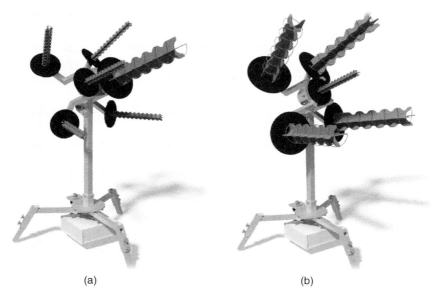

(a)   (b)

Figure 9.39 3D visualizations of (a) tracking and data exchange system for sounding rocket and (b) modified version proposed for WPT for WSN.

9.6 Summary

In this publication, we proposed solutions intended for WPT used for powering WSN nodes distributed, in general, in 2D or 3D space.

In Section 9.2, we proposed a novel algorithm for data routing between nodes. The most important feature of the algorithm is very fast switching of data route, even in case of multiple node failure, minimizing the total loss of data. The drawback which needs to be mentioned here is that routing paths need to be found when some changes in net configuration are made, e.g. when new nodes are added or some of the existing ones are removed.

The 2D WPT system presented in Section 9.3 was designed and manufactured for wireless charging of supercapacitors powering WSN nodes located at the spatial angle of approximately 120°. The presented system allowed for charging multiple WSN nodes by switching the transmitter radiation pattern. The implemented algorithm allowed for full automation of the charging process and also for equalization of voltage levels for all WSN nodes, independent of the node's measurement frequency and the initial voltage level. The obtained performance made it possible to achieve minimum five measurements per minute with standard ZigBee WSN nodes, placed in the maximum radiation directions of the obtained radiation patterns, at the distance of 65 cm from the transmitter.

Section 9.4 showed the concept of evolution of the system for WSN nodes distributed in 3D space. The proposed concept of switched beam antenna arrays placed on walls of a tetrahedral structure should make it possible to power up the WSN network nodes distributed randomly around the transmitter. By using less lossy laminates with lower permittivity, the range of the presented systems should be significantly increased, and the improvement of system efficiency could be achieved due to obtaining higher radiation efficiencies and lower dielectric losses. However, lower permittivity in the antennas would result in the increased size of the radiators.

Finally, Section 9.5 described the system used for tracking a research rocket and data transmission between the rocket and the ground station. The presented system was designed, manufactured, and successfully tested during the Candle2 rocket flight campaign. The system showed perfect data transmission efficiency using standard, off-the-shelf components at distances of around 10 km. A significant part of this chapter was devoted to Kalman filtering, which allows for smooth movement of the antenna platform in case of lack of the received signals or in case such signals are noisy.

In our concept, we proposed to use this system as the base to build a WPT system able to transmit electromagnetic energy in any direction in 3D space. If the system was equipped with high power RF amplifier with highly directional transmitter antenna, it could be used as a 3D WPT system, meeting the requirements of WSN implemented by us. Bearing in mind the system's capability of tracking objects (WSN nodes) which are in motion, we could use it to search for WSN nodes in 3D space, and to make a map showing their location. Also, the placement of each sensor can be predefined. All in all, the idea looks promising and brings more advantages, compared to the systems shown in Sections 9.3 and 9.4, because it is not necessary to use rather big switching circuits (Butler matrix, or hybrid couplers with RF switches), with antennas realized as patch arrays. Additionally, the function of variable power of the transmitted signal would be profitable, because some nodes could be placed close to the device and some of them quite far away from it. At this moment, this is just an idea but it should be realized soon to verify the concept.

References

1 Cost Action IC1301 Team (2017). Europe and the future for WPT: European contributions to wireless power transfer technology. *Institute of Electrical and Electronics Engineers, IEEE Microwave Magazine* 18 (4): 56–87.

2 Riviere, S. (2009). An integrated model of a wireless power transportation for RFID and WSN applications. *16th IEEE International Conference on Electronics, Circuits, and Systems*: 235–238.

3 Riviere, S. (2010). Study of complete WPT system for WSN applications at low power level. *Electronics Letters* 46 (8): 597–598.

4 Kant, P. and Michalski, J.J. (2017). Wireless sensor network for rocket applications. In: *Microwave Technology and Techniques Workshop*, (3–5 April 2017), The Netherlands: ESA-ESTEC.

5 Kant, P., Chelstowski, T., Dobrzyniewicz, K., and Michalski, J. J. (2016). Wireless sensor network analysis and optimization by 3D electromagnetic simulations for research rocket application. *21st International Conference on Microwave, Radar and Wireless Communications (MIKON)*, Cracow, Poland (9–11 May 2016).

6 Al-Karaki, J.N. and Kamal, A.E. (2004). Routing techniques in wireless sensor networks: a survey. *Wireless Communications, IEEE* 11 (6): 6–28.

7 Avresky, D.R. (1999). Embedding and reconfiguration of spanning trees in faulty hyper cubes. *IEEE Transactions on Parallel and Distributed Systems* 10: 211–222.

8 Ji, P., Wu, C., Zhang, Y., and Jia, Z. (2007). Research of directed spanning tree routing protocol for wireless sensor networks. *International Conference on Mechatronics and Automation, ICMA*: 1406–1410.

9 Das, D. and Misra, R. (2014). Improvised tree selection algorithm in greedy distributed spanning tree routing. *Engineering and Computational Sciences (RAECS), 2014 Recent Advances*: 1–6.

10 Gomes, T., Tipper, D., and Alashaikh, A. (2014). A novel approach for ensuring high end-to end availability: the Spine concept. In: *Design of Reliable Communication Networks (DRCN), 2014 10th International Conference*, 1–8. IEEE.

11 Kruskal, J.B. (1956). On the shortest spanning subtree of a graph and the traveling salesman problem. *Proceedings of the American Mathematical Society* 7 (1): 48–50.

12 Kant, P. and Michalski, J. J. (2015). Concept of wireless power distribution system for wireless sensor networks. *IRMAST 2015, 1st International Conference on Innovative Research and Maritime Applications of Space Technology*, Gdansk, Poland (23–24 April 2015).

13 Michalski, J. J. and Kant, P. (2017). Reliable and fast switching wireless sensor network for space application. *International Microwave Symposium, Technology for Space Low Cost Satellites Session*, Honolulu, Hawaii (4–7 June 2017).

14 He, J., Wang, B.-Z., He, Q.-Q. et al. (2007). Wideband X-band microstrip butler matrix. *PIER* 74: 131–140.

15 Khan, O.U. (2006). Design of X-band 4×4 Butler matrix for microstrip patch antenna array. *Electronic Engineering Department, NED University of Engineering & Technology.*

16 Oualkadi, E.A. and Zbitou, J. (2016). *Handbook of Research on Advanced Trends in Microwave and Communication Engineering.* IGI Global.

17 Congedo, F. (2013). A 2.45-GHz Vivaldi Rectenna for the remote activation of an end device radio node. *IEEE Sensors Journal* 13 (9).

18 Kant, P., Szwaba, A., and Michalski, J. J. (2017). Fully autonomous tracking and data transmission system for sounding rocket. *Proceedings of the 47th European Microwave Conference, Special Session: Radar Technology and System Applications in Central Europe*, Nuremberg.

19 Balanis, C.A. (2005). *Antenna Theory*, 3e. Wiley.

20 Slade, B. (2015). *The Basics of Quadrifilar Helix Antennas.* Orban Microwave.

21 Haykin, S. (2010). *Kalman Filtering and Neural Networks.* Wiley.

10

Smartphone Reception of Microwatt, Meter to Kilometer Range Backscatter Resistive/Capacitive Sensors with Ambient FM Remodulation and Selection Diversity

Georgios Vougioukas and Aggelos Bletsas

School of Electrical and Computer Engineering, Technical University of Crete, Chania, Crete, Greece

10.1 Introduction

Communication by means of reflection, first introduced in [1], has recently attracted a lot of research interest due to its simplicity and ultra-low power characteristics. The most notable application of this type of communication technique, which is also known as backscatter radio, is found in common radio frequency identification (RFID) tags. Except from RFID tags, exactly due to its aforementioned attractive characteristics, backscatter radio has been employed in the development of ultra-low power wireless sensor networks [2–5]. Such deployments demonstrate that backscatter communication is a key enabler for implementing the low-level communication fabric of the upcoming internet-of-things (IoT) revolution, in an *ultra*-low power, *ultra*-low cost manner.

The operating principle of backscatter-based communication is the reflection occurring when a radio frequency (RF) signal impinges on an antenna that is not appropriately terminated. An RF signal is assumed to illuminate the tag's antenna. By carefully varying the termination of the antenna between different loads, a version of the impinged signal is reflected back with, however, altered phase, amplitude, frequency, or combinations of them. That way, backscattered signal can attain different modulation schemes, utilizing just an RF switch (to choose between the appropriate termination loads) and proper illumination. Contrary to conventional Marconi radios, in this scheme, power-consuming signal conditioning components (such as amplifiers, high frequency oscillators, filters, mixers) are not necessary at the tag's side and ultra-low power communication can be achieved. For backscatter communication, three units are necessary. First,

Wireless Power Transmission for Sustainable Electronics: COST WiPE - IC1301,
First Edition. Edited by Nuno Borges Carvalho and Apostolos Georgiadis.
© 2020 John Wiley & Sons, Inc. Published 2020 by John Wiley & Sons, Inc.

an active RF transmitter to provide tag(s) with the necessary illumination. Second, the backscattering tag and third, a receiver unit to recover tags' information from the backscattered signal. The setup where the illuminator (active RF transmitter) and the receiver are co-located (common RFID tag setup) is called monostatic. When the receiver of the backscattered signal and the illuminator are different devices, the setup is called bistatic. A multistatic setup is considered when multiple emitters are used to illuminate the tag(s) [6].

As mentioned in the previous paragraph, in a bistatic setup, the unit providing tag(s) with illumination is a separate unit from the receiver. Examples of digital bistatic backscatter radio setups can be found in [2, 5, 7, 8]. In [8], backscatter frequency shift keying (FSK) and on-off keying (OOK) were considered, and software defined radio (SDR)–based, noncoherent, and symbol-by-symbol receivers were designed. In [7], coherent, SDR-based receivers for backscatter FSK were proposed along with utilization of small block length channel coding. Employing backscatter FSK, work in [2] offered novel noncoherent, both symbol-by-symbol and sequence receivers, with or without utilizing channel coding, suitable for short packet communication. In all the described setups, the achieved communication range (tag-2-reader distance) was in the order of 150 m, for an illuminator-2-tag distance in the order of 10 m, illuminator transmission power at 13 dBm and bit rate at the kHz regime. These results demonstrate an increase (in communication range), of an order of magnitude, compared to conventional RFID tags.

Examples of analog bistatic backscatter networks can be found in [3, 4, 9]. Work in [3] was based on capacitive sensing and analog timer principles, offering environmental humidity measurements with a root mean square (RMS) error of 2%, conveyed through backscatter frequency modulation (FM). The ultra-low cost tags (in the order of 3 euro/tag) consumed $\approx$500 μW. A custom soil moisture sensor was designed in [9] demonstrating RMS error of 1.9%. Using the same analog backscatter principles as in [3] and utilizing the aforementioned custom sensor, backscatter tags consuming $\approx$200 μW were designed and implemented. Illumination came from a dedicated RF source, and the receiver used was a low-cost SDR (Realtek [RTL] dongle). Analog FM backscatter was utilized in [4] to convey a plant's electric potential (EP), measured across two electrodes placed in the stem of the plant, toward an SDR reader, under illumination from a dedicated carrier. The implemented tags, due to their ultra-low power consumption ($\approx$20 μW), were powered by the plant itself using a second pair of electrodes. An interesting correlation between the EP signal, solar irradiance, and the moments when the plant was watered was observed.

The capability of backscatter radio to generate signals attaining the specifications of commercial protocols was recently demonstrated [10–13]. In [10], backscatter FSK was utilized to mimic the signals of bluetooth low energy

(BLE); reception of backscatter-generated bluetooth signals was demonstrated using a smartphone, achieving communication ranges in the order of 6.5 m while utilizing an emitter at 17 dBm, located at a distance of 2.9 m from the tag. In [13], a microcontroller-based tag was designed, and utilizing backscatter FSK, the signal characteristics employed by commercial embedded radio receivers were attained. The tag, while being 3 m away from an illuminating carrier at 16 dBm was able to achieve communication ranges up to 270 m, utilizing a Silicon Laboratories SI1064 embedded receiver. Illumination from a bluetooth (with appropriate software modifications) transmitter, with reception from a WiFi receiver was demonstrated in [11]. Additionally, a way for attaining single sideband backscatter signaling was proposed and demonstrated. Backscatter based, LoRa transmission was demonstrated in [12]. While the illuminator transmitted signals @ 36 dBm, tag-2-reader ranges in the order of 2.8 km were observed (for illuminator-2-tag distance of 5 m).

Efforts described in the previous paragraphs assumed dedicated illumination i.e. a dedicated, active RF transmitter is necessary to emit an *unmodulated* carrier, providing illumination to the tag(s). Utilizing a dedicated illuminator increases communication range, while employing multiple illuminators can increase coverage [6], at the expense of increased deployment cost. A special case of bistatic backscatter, ambient backscatter, omits the need for a dedicated *unmodulated* illuminator at the expense of reduced communication range. In ambient backscatter, ambient *modulated* signals, already present "in the air," are exploited for illuminating the tag(s). Such signals may come from FM radio and digital television (DTV) broadcasting stations or other nondedicated ambient sources. It is immediately evident that recovering tag's information from the backscattered signal becomes a challenging task, given that the illuminating signal is no longer constant (it is modulated) and its parameters are, in the general case, unknown. Ambient backscatter was first introduced in [14], where modulated DTV signals were exploited to achieve tag-2-tag communication. Using envelope detection, the communication range achieved was in the order of 60 cm. Interference cancellation based on multiantenna design, as well as spread spectrum techniques was implemented in [15], using analog circuitry, thus attaining low-power consumption. Multiantenna, analog design offered bit rates in the order of 1 Mbps (exploiting an impinged DTV signal power of −10 dBm) for a tag-2-tag communication range in the order of 2 m, while analog spread spectrum techniques offered communication ranges in the order of 6 m (for an impinged DTV signal power of −15 dBm) at the expense of reduced bit rate, namely 3.3 bps. Utilizing a computer connected to a function generator which drove an RF switch, signals from FM radio stations were exploited in [16]. Using the experimental setup, ranges in the order of 18 m (for analog audio transmission) were achieved. Besides analog audio transmission, 2 and 4 audio FSK was utilized to achieve digital communication. With the proposed switching method, *any* FM radio receiver can be exploited as a reader to recover tag's information. Additionally, an integrated circuit performing the proposed switching was designed and simulated. The same switching methodology presented in [16] was independently given in [17][1] while additionally providing a full tag circuit prototype implementation. Illumination from FM radio stations was also exploited in [18]. The implemented tag achieved communication ranges in the order of 5 m while consuming 1.78 mW (for a bit rate of 1 kbps) in duty-cycled operation, using the RTL-SDR dongle, as receiver. Ambient backscatter communication was recently studied from a theoretical perspective. Related works can be found in [19–21] and references therein. Current overview of the ambient backscatter literature can be found in [22].

Contrary to prior art discussed in the previous paragraphs, this chapter presents:

A backscatter modulation method, *FM remodulation.* Assuming illumination from FM radio stations and utilizing FM remodulation, the backscattered signal can be recovered by *any* FM radio receiver, including those found in modern smartphones, without the need for providing dedicated illumination.

Detailed analysis for the impact the (input) noise has at the output of a conventional FM receiver, both in high and (through approximations) low signal-to-noise-ratio (SNR) regimes. Closed form expressions are given for the signal to interference plus noise ratio (SINR) (with respect to tag's signal) at the output of the receiver. Simulation results are offered to validate the derived expressions.

The design and implementation of an *ultra*-low power tag, exploiting the aforementioned FM remodulation scheme and its advantages. The tag was implemented using commercial-off-the-shelf (COTS) low-cost components. It was able to facilitate any capacitive or resistive sensor conveying the measured value toward an FM radio equipped smartphone 26 m away, while the illuminating FM radio station was located 6.5 km away from the tag. The tag consumed 24 μW in *continuous* (non duty-cycled) operation (Figure 10.1).

Analysis of selection diversity, through the exploitation of which, 26 m of communication range were achieved.

The rest of the document is organized as follows. Section 10.2 offers the basic principles of scatter radio FM remodulation. Section 10.3 provides the detailed analysis of noise on the output of the FM receiver, given the backscattered signal as input. In Section 10.4, an approximation for the occupied bandwidth is given. Benefits of selection diversity are quantified in Section 10.5. Details about the implementation of the proposed tag are given in Section 10.6. Simulation and experimental results are given in Section 10.7, and finally, conclusions are drawn in Section 10.8.

---

1 Parts of this work were originally presented in [17].

Figure 10.1 Remodulation with backscatter and selection diversity.

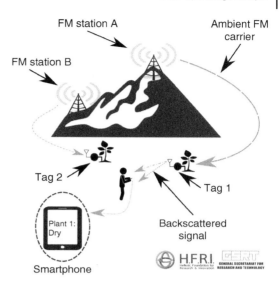

10.2 Operating Principle

10.2.1 Backscatter Communication

A radio frequency wave is produced by a source with impedance Z_{source}, which is connected to load Z_{load}. In the case that the load is not matched with the source, i.e $Z_{source} \neq Z_{load}^*$, a portion of the power destined for the load will be reflected back to the source. That portion depends on the *reflection coefficient*, defined as follows:

$$\Gamma = \frac{Z_{load} - Z_{source}^*}{Z_{load} + Z_{source}} \tag{10.1}$$

For example, if the RF source is an antenna and an open circuit load is chosen, the antenna is forced to reflect. If in another case the antenna is matched with the load, the load absorbs (ideally) all the power and no reflection occurs. By carefully choosing the loads and the alternation between them, the reflected signal can be manipulated toward different modulation schemes and characteristics.

A tag is considered with an antenna terminated, alternatively, at two different loads Z_0, Z_1 (resulting to Γ_0, Γ_1, respectively), using an RF switch. A square wave $v(t)$ drives the switch with 50% duty cycle and frequency F_{sw}. By keeping only the fundamental frequency component of $v(t)$,[2] the signal driving the switch can be expressed as:[3]

$$x_{sw}(t) = A_{sw} \cos(2\pi F_{sw} t) \tag{10.2}$$

---

2 It holds $\approx 80\%$ of $v(t)$'s power.
3 For simplifying the explanation, microwave antenna-related parameters are omitted [8].

It can be shown [8] that if the antenna of the aforementioned setup is illuminated by a high frequency signal $c(t)$, the backscattered signal from the antenna is given by:

$$y(t) = \sqrt{\eta} c(t) x_{\text{sw}}(t) = \sqrt{\eta} A_{\text{sw}} c(t) \cos(2\pi F_{\text{sw}} t) \tag{10.3}$$

where η denotes the *scattering efficiency*, depending on the chosen loads and the antenna characteristics. Equation (10.3) shows that by alternating the termination of an antenna illuminated by $c(t)$, a modulation operation is attained.

10.2.2 FM Remodulation

The signal of a high frequency carrier that is frequency-modulated (FM) by $\phi_s(t)$, is given by [23]:

$$c_s(t) = A_s \cos\left(2\pi F_s t + 2\pi k_s \int_0^t \phi_s(\tau) d\tau\right) \tag{10.4}$$

where A_s is the carrier's amplitude, F_s its center frequency, while k_s is the modulator's frequency sensitivity, measured in Hz/V. FM modulation index is given by $\beta_s = \Delta f_{\text{max}}/W = k_s \max |\phi_s(t)|/W$, where W is the (baseband) bandwidth of $\phi_s(t)$. The model in Eq. (10.4) applies to any FM radio station and thus, $\phi_s(t)$ is comprised of a processed combination of Left/Right (L/R) audio channels, along with (optional) digital information about the station (RDS).

Assume that $x_{\text{sw}}(t)$ of Eq. (10.3) admits the following form:

$$x_{\text{sw}}(t) \equiv x_{\text{FM}}(t) = A_{\text{sw}} \cos\left(2\pi F_{\text{sw}} t + 2\pi k_{\text{sw}} \int_0^t \mu(\tau) d\tau\right) \tag{10.5}$$

which means that a square wave with fundamental frequency modulated by $\mu(t)$ drives the RF switch. If at the same time, the carrier illuminating tag's antenna is an FM-modulated signal (e.g. from an FM broadcasting radio station), then the tag-backscattered signal attains the following form:

$$y_{\text{bs}}(t) = \sqrt{\eta} c_s(t) x_{\text{FM}}(t) = \frac{\gamma_s}{2} \cos(2\pi(F_s + F_{\text{sw}})t + \phi_d^t + \mu_d^t)$$

$$+ \frac{\gamma_s}{2} \cos(2\pi(F_s - F_{\text{sw}})t + \phi_d^t - \mu_d^t) \tag{10.6}$$

where $\gamma_s = \sqrt{\eta} A_s A_{\text{sw}}$, $\phi_d^t = 2\pi k_s \int_0^t \phi_s(\tau) d\tau$, and $\mu_d^t = 2\pi k_{\text{sw}} \int_0^t \mu(\tau) d\tau$.

Signals appearing in Eq. (10.6) are considered FM signals, since their instantaneous frequency depends on $k_s \phi_s(t) \pm k_{\text{sw}} \mu(t)$. Thus, backscattering results to FM signaling when the illuminating signal is FM and the switching signal is also FM. Such operation may be coined as FM remodulation [17]. It can be observed that if $\phi_s(t)$ does not change (i.e. $\phi_s(t) = \phi_s$), the model simplifies to the bistatic backscatter radio model with unmodulated illumination (see e.g. [24]).

FM remodulation principle demonstrates that any conventional FM radio receiver, tuned at either of $F_s \pm F_{\text{sw}}$, can demodulate the backscattered signal, as long as the following hold:

1. Tag's $\mu(t)$ is band-limited to the audible spectrum or up to the maximum frequency of 53 kHz (assuming receiver with stereo capabilities) or slightly above (assuming RDS capability).
2. One of $F_s \pm F_{sw}$ falls within the tunable range of the receiver. For an FM broadcast receiver in Greece that range would be 88–108 MHz.
3. Audio level of the backscattered demodulated tag signal (given that it's limited within the audible spectrum) is above a required threshold for successful FM reception.

In the proposed system, sensor's signal $\mu(t)$ is limited in the audible spectrum, potentially amenable to FM station's interference, while $k_{sw} \neq k_s$. To reduce interference from the FM station signal, $\mu(t)$ may be designed such that its frequency components fall on (audio) bands not occupied by $\phi_s(t)$. Additionally, as shown in the next Section 10.3 and verified through numerical simulations and laboratory experiments with the implemented system, increasing the frequency deviation of the switching signal (up to a certain value, so that the FM threshold phenomenon does not kick in), higher audio levels of $\mu(t)$ are offered compared to interference.

10.3 Impact of Noise

The analysis performed in order to examine the effect of noise at the output of the FM receiver (when the input is the remodulated backscattered signal) is based on similar analysis performed in [23, 25]. Contrary to the methods followed in [23, 25], this work studies the performance of an FM receiver when excited by an FM signal resulted from remodulation due to backscatter from a tag, while the information-bearing, deterministic, tag signal is affected by stochastic FM interference, as well as thermal noise.

In order to receive the backscattered signal, an FM receiver tuned at $F_t \triangleq F_s + F_{sw}$ is considered. Based on Eq. (10.6), the backscattered signal appearing at that frequency is given by:

$$y_b(t) = \gamma_b \cos(2\pi F_t t + \phi_d^t + \mu_d^t) + n(t) \tag{10.7}$$

where $\gamma_b = \frac{\gamma_s}{2}$ and $n(t)$ a zero-mean, wide-sense stationary (WSS) Gaussian process. The noise in the passband model of Eq. (10.7) can be equivalently expressed as the following:

$$
\begin{aligned}
n(t) &= n_I(t) \cos(2\pi F_t t) - n_Q(t) \sin(2\pi F_t t) \\
&= \sqrt{n_I^2(t) + n_Q^2(t)} \cos\left(2\pi F_t t + \tan^{-1}\left(\frac{n_Q(t)}{n_I(t)}\right)\right) \\
&= V_n(t) \cos(2\pi F_t t + \Phi_n(t))
\end{aligned}
\tag{10.8}
$$

where $n_Q(t)$, $n_I(t)$ are independent, baseband, band limited, zero mean WSS Gaussian processes with $R_{n_I}(\tau) = R_{n_Q}(\tau) = R_{nb}(\tau)$. The power spectral density of $n_I(t)$, is given by $S_{nb}(f) = \mathcal{F}\{R_{nb}(\tau)\}$:[4]

$$S_{nb}(f) = \begin{cases} N_0, & |f| \le \dfrac{B}{2} \\ 0, & \text{otherwise.} \end{cases} \tag{10.9}$$

The SNR at the input of the receiver can be defined as:

$$\text{SNR} = \frac{P_c}{P_n} = \frac{\gamma_b^2/2}{N_0 B} \tag{10.10}$$

with $P_c = \gamma_b^2/2$ and $P_n = N_0 B$, where N_0 the noise spectral density. By combining Eqs. (10.7) and (10.8) and then performing vector addition, the received signal can be expressed as:

$$y_b(t) = \sqrt{\gamma_b^2 + V_n^2(t) + 2\gamma_b V_n(t)\cos(\Phi_n(t) - \phi_d^t - \mu_d^t)}$$
$$\times \cos\left(2\pi F_t t + \phi_d^t + \mu_d^t + \tan^{-1}\left(\frac{V_n(t)\sin(\Phi_n(t) - \phi_d^t - \mu_d^t)}{\gamma_b + V_n(t)\cos(\Phi_n(t) - \phi_d^t - \mu_d^t)}\right)\right) \tag{10.11}$$

The transformation of Eqs (10.7) into (10.11) was made in order for the noise terms to appear within the signal's phase which holds the information. That allows to analyze the effect that the input noise has at the output of the receiver.

10.3.1 High SNR Case

For simplifying the analysis, in this section, $\gamma_b \gg V_n(t)$ is assumed, which translates to SNR > 10 dB. Below that 10 dB threshold, the output SINR[5] drops rapidly, as it will be later shown, with respect to input SNR, dictating the need for ambient selection diversity, explained later in the text.

Under the assumption mentioned previously, the following approximation can be used:

$$\sqrt{\gamma_b^2 + V_n^2(t) + 2\gamma_b V_n(t)\cos(\Phi_n(t) - s(t))} \approx \gamma_b + V_n(t)\cos(\Phi_n(t) - s(t)) \approx \gamma_b \tag{10.12}$$

where $s(t) = \phi_d^t + \mu_d^t$. Using the aforementioned approximation and the fact that $\tan(\phi) \approx \phi \Leftrightarrow \phi \approx \tan^{-1}(\phi)$, for small ϕ, Eq. (10.11) can be expressed, in the high

---

4 Where $\mathcal{F}$ the Fourier transform.
5 Considering the station's demodulated signal as interference and the tag's signal as useful.

SNR-regime, as follows:

$$y_b(t) = a(t)\cos(2\pi F_t t + s(t) + n_{\text{out}}(t)) = \Re\left\{a(t)e^{+j(s(t)+n_{\text{out}}(t))}e^{+j(2\pi F_t t)}\right\}$$

(10.13)

with $a(t) = \gamma_b + V_n(t)\cos(\Phi_n(t) - s(t)) \approx \gamma_b$ and $n_{\text{out}}(t) = \frac{V_n(t)}{\gamma_b}\sin(\Phi_n(t) - s(t))$.

The receiver has to obtain the tag's signal $\mu(t)$, which is superimposed to the station's signal $\phi_s(t)$. To achieve this, the signal of Eq. (10.13) is first downconverted to baseband and then the complex envelope is obtained:

$$y_b^{\text{dc}}(t) = a(t)e^{+j(s(t)+n_{\text{out}}(t))}$$

(10.14)

The phase of the complex envelope is then extracted:

$$y_b^{\text{arg}}(t) = \angle y_b^{\text{dc}}(t) = s(t) + n_{\text{out}}(t)$$

$$= 2\pi k_s \int_0^t \phi_s(\tau)d\tau + 2\pi k_{\text{sw}} \int_0^t \mu(\tau)d\tau + n_{\text{out}}(t)$$

(10.15)

As a final step, the receiver derivates and divides by 2π, $y_b^{\text{arg}}(t)$ to obtain:

$$r(t) = \frac{1}{2\pi}\frac{d}{dt}y_b^{\text{arg}}(t) = k_{\text{sw}}\mu(t) + k_s\phi_s(t) + \overbrace{\frac{1}{2\pi}\frac{d}{dt}\left[\frac{V_n(t)}{\gamma_b}\sin(\Phi_n(t) - s(t))\right]}^{w_n(t)}$$

(10.16)

In Eq. (10.16), term $k_{\text{sw}}\mu(t)$ is considered as useful signal (tag's signal). Term $k_s\phi_s(t)$ is interference to tag's signal and the last term is noise. In order to obtain a closed form expression for the SINR at the output of the FM receiver, the characteristics of the noise term, $\frac{1}{2\pi}\frac{d}{dt}\left[\frac{V_n(t)}{\gamma_b}\sin(\Phi_n(t) - s(t))\right]$ must first be derived. $w_n(t)$ can be rewritten as:

$$w_n(t) = \frac{V_n(t)}{\gamma_b}\sin(\Phi_n(t) - s(t)) \overset{*}{=} \frac{V_n(t)}{\gamma_b}$$

$$\times \left[\sin(\Phi_n(t))\cos\left(\phi_d^t + \mu_d^t\right) - \cos\left(\Phi_n(t)\right)\sin\left(\phi_d^t + \mu_d^t\right)\right]$$

(10.17)

where in $*$ the identity $\sin(a - b) = \sin(a)\cos(b) - \cos(a)\sin(b)$ was used. Using Eq. (10.8), $w_n(t)$ becomes:

$$w_n(t) = \frac{1}{\gamma_b}\left[n_Q(t)\cos\left(\phi_d^t + \mu_d^t\right) - n_I(t)\sin\left(\phi_d^t + \mu_d^t\right)\right]$$

(10.18)

That way, the mean value of $w_n(t)$ can be calculated as follows:

$$\mathbb{E}[w_n(t)] = \frac{1}{\gamma_b}\left(\mathbb{E}[n_Q(t)]\cos\left(\phi_d^t + \mu_d^t\right) - \mathbb{E}[n_I(t)]\sin\left(\phi_d^t + \mu_d^t\right)\right) = 0$$ (10.19)

In Eq. (10.19), the facts that the thermal noise components $n_I(t)$, $n_Q(t)$ are independent of the signal as well as $\mathbb{E}[n_I(t)] = \mathbb{E}[n_Q(t)] = 0$ were used. With the mean value known, the autocorrelation function of $w_n(t)$ can be calculated:

$$R_{w_n}(t + \tau, t) = \mathbb{E}[w_n(t + \tau)w_n(t)]$$

$$= \frac{1}{\gamma_b^2}\mathbb{E}\left[\left(n_Q(t + \tau)\cos(\phi_d^{t+\tau} + \mu_d^{t+\tau}) - n_I(t + \tau)\sin(\phi_d^{t+\tau} + \mu_d^{t+\tau})\right)\right.$$

$$\left. \cdot (n_Q(t)\cos(\phi_d^t + \mu_d^t) - n_I(t)\sin(\phi_d^t + \mu_d^t))\right] \tag{10.20}$$

with $\mathbb{E}[n_Q(t + \tau)n_Q(t)] = \mathbb{E}[n_I(t + \tau)n_I(t)] = R_{nb}(\tau)$ and using the fact that $R_{n_I n_Q}(\tau) = 0$, Eq. (10.20) becomes:

$$R_{w_n}(t + \tau, t) = \frac{1}{\gamma_b^2}R_{nb}(\tau)\mathbb{E}\left[\cos\left(\phi_d^{t+\tau} + \mu_d^{t+\tau}\right)\cos\left(\phi_d^t + \mu_d^t\right)\right.$$

$$\left. + \sin\left(\phi_d^{t+\tau} + \mu_d^{t+\tau}\right)\sin\left(\phi_d^t + \mu_d^t\right)\right] \tag{10.21}$$

Taking advantage of the identity $\cos(a - b) = \cos(a)\cos(b) + \sin(a)\sin(b)$, Eq. (10.21) becomes:

$$R_{w_n}(t + \tau, t) = \frac{1}{\gamma_b^2}R_{nb}(\tau)\mathbb{E}\left[\cos\left(\phi_d^{t+\tau} - \phi_d^t + \mu_d^{t+\tau} - \mu_d^t\right)\right] \tag{10.22}$$

The signal ϕ_d^t, is assumed to be a zero mean, WSS Gaussian process. That way, at any time t, $\phi_d^{t+\tau}, \phi_d^t$, are zero mean, jointly Gaussian random variables [26, Def. 2.1] and the variance of $q^{t,\tau} = \phi_d^{t+\tau} - \phi_d^t$ can be calculated as:

$$\sigma_q^2 = \mathbb{E}[(q^{t,\tau})^2] = \mathbb{E}[(\phi_d^{t+\tau} - \phi_d^t)^2] = \mathbb{E}[(\phi_d^{t+\tau})^2] - 2R_{\phi_d}(\tau) + \mathbb{E}[(\phi_d^t)^2]$$

$$= 2R_{\phi_d}(0) - 2R_{\phi_d}(\tau) \tag{10.23}$$

That way $R_{w_n}(t + \tau, t)$ becomes:

$$R_{w_n}(t + \tau, t) = \frac{1}{\gamma_b^2}R_{nb}(\tau)\mathbb{E}\left[\cos\left(q^{t,\tau} + \mu_d^{t+\tau} - \mu_d^t\right)\right]$$

$$= \frac{1}{\gamma_b^2}R_{nb}(\tau)\mathbb{E}\left[\Re\left\{e^{+jq^{t,\tau}}e^{+j(\mu_d^{t+\tau} - \mu_d^t)}\right\}\right] \tag{10.24}$$

It is assumed that $\mu(t)$ is a deterministic signal (audio tone), more specifically $\mu(t) = -\sin(2\pi F_{\text{sens}}t)$. That way $\mu_d^t = 2\pi k_{\text{sw}}\int_0^t \mu(\tau)d\tau = \frac{k_{\text{sw}}}{F_{\text{sens}}}\cos(2\pi F_{\text{sens}}t) + c$ and Eq. (10.24) becomes:

$$R_{w_n}(t + \tau, t) = \frac{1}{\gamma_b^2}R_{nb}(\tau)\Re\left\{\mathbb{E}\left[e^{+jq^{t,\tau}}\right]e^{+j(\mu_d^{t+\tau} - \mu_d^t)}\right\}$$

$$\overset{*}{=} \frac{1}{\gamma_b^2}R_{nb}(\tau)\Re\left\{e^{-\frac{\sigma_q^2}{2}}e^{+j(\mu_d^{t+\tau} - \mu_d^t)}\right\}$$

$$= \frac{1}{\gamma_b^2}R_{nb}(\tau)e^{-\frac{\sigma_q^2}{2}}\cos(\mu_d^{t+\tau} - \mu_d^t) \tag{10.25}$$

where in point * the characteristic function [27, pp. 25] of the Gaussian random variable $q^{t,\tau}$ was used. In Eq. (10.25), the fact that $\mathbb{E}[\mathfrak{R}\{z\}] = \mathfrak{R}\{\mathbb{E}[z]\}$ with $z \in \mathbb{C}$, was also used. Finally, the autocorrelation function of $w_n(t)$ is given by:

$$R_{w_n}(t + \tau, t) = \mathbb{E}[w_n(t + \tau)w_n(t)] = \frac{1}{\gamma_b^2} R_{nb}(\tau) e^{R_{\phi_d}(\tau) - R_{\phi_d}(0)} \cos(\mu_d^{t+\tau} - \mu_d^t) \tag{10.26}$$

Due to dependence of $R_{w_n}(t + \tau, t)$ on t, $w_n(t)$ cannot be characterized as a WSS process. However, μ_d^t is a periodic function with period $T_{sens} = \frac{1}{F_{sens}}$. That way $\mu_d^{t + T_{sens}} = \mu_d^t$ and

$$
\begin{aligned}
R_{w_n}(t + T_{sens} + \tau, t + T_{sens}) &= \frac{1}{\gamma_b^2} R_{nb}(\tau) e^{R_{\phi_d}(\tau) - R_{\phi_d}(0)} \cos\left(\mu_d^{t + T_{sens} + \tau} - \mu_d^{t + T_{sens}}\right) \\
&= \frac{1}{\gamma_b^2} R_{nb}(\tau) e^{R_{\phi_d}(\tau) - R_{\phi_d}(0)} \cos\left(\mu_d^{t+\tau} - \mu_d^t\right) \\
&= R_{w_n}(t + \tau, t) \tag{10.27}
\end{aligned}
$$

Equations (10.27) and (10.19) show that process $w_n(t)$ is cyclostationary with period T_{sens}. The t-averaged, autocorrelation function is given by:

$$\overline{R}_{w_n}(\tau) = \frac{1}{\gamma_b^2} R_{nb}(\tau) e^{R_{\phi_d}(\tau) - R_{\phi_d}(0)} \frac{1}{T_{sens}} \int_{T_{sens}} \cos\left(\mu_d^{t+\tau} - \mu_d^t\right) dt \tag{10.28}$$

Next, the integral involved in Eq. (10.28) will be calculated:

$$\int_{T_{sens}} \cos(\mu_d^{t+\tau} - \mu_d^t) dt = \int_{T_{sens}} \cos\left[\frac{k_{sw}}{F_{sens}} \left[\cos(2\pi F_{sens} t + 2\pi F_{sens} \tau) - \cos(2\pi F_{sens} t)\right]\right] dt \tag{10.29}$$

Using the trigonometric identity $\cos(a) - \cos(b) = -2\sin\left(\frac{a-b}{2}\right)\sin\left(\frac{a+b}{2}\right)$, Eq. (10.29) becomes:

$$
\begin{aligned}
\int_{T_{sens}} \cos(\mu_d^{t+\tau} - \mu_d^t) dt &= \int_{T_{sens}} \cos\left(-2\frac{k_{sw}}{F_{sens}} \sin(\pi F_{sens}\tau) \sin(2\pi F_{sens}t + \pi F_{sens}\tau)\right) dt \\
&= \int_{T_{sens}} \cos(\tau_0 \sin(2\pi F_{sens}t + \tau_1)) dt \tag{10.30}
\end{aligned}
$$

where

$$\tau_0 \stackrel{*}{=} 2\rho\sin(\pi F_{sens}\tau), \quad \rho = \frac{k_{sw}}{F_{sens}} \tag{10.31}$$

$$\tau_1 = \pi F_{sens}\tau \tag{10.32}$$

where in *, the symmetry of cosine function was exploited. Applying Equation [28, 9.1.42]:

$$\cos(z\sin(\theta)) = J_0(z) + 2\sum_{k=1}^{\infty} J_{2k}(z)\cos(2k\theta), \quad z \in \mathbb{C} \tag{10.33}$$

where $J_k(z)$ is the Bessel function of the first kind of order k, to Eq. (10.30):

$$\int_{T_{\text{sens}}} \cos(\tau_0 \sin(2\pi F_{\text{sens}}t + \tau_1))dt$$

$$= \int_{T_{\text{sens}}} J_0(\tau_0) + 2\sum_{n=1}^{\infty} J_{2n}(\tau_0)\cos(2n(2\pi F_{\text{sens}}t + \tau_1))dt$$

$$= \int_{T_{\text{sens}}} J_0(\tau_0)dt + \int_{T_{\text{sens}}} 2\sum_{n=1}^{\infty} J_{2n}(\tau_0)\cos(2n(2\pi F_{\text{sens}}t + \tau_1))dt$$

$$= J_0(\tau_0)\int_{T_{\text{sens}}} dt + 2\sum_{n=1}^{\infty} J_{2n}(\tau_0)\overbrace{\int_{T_{\text{sens}}} \cos(2n(2\pi F_{\text{sens}}t + \tau_1))dt}^{=0}$$

$$\overset{*}{=} J_0(\tau_0)T_{\text{sens}} \tag{10.34}$$

$\overline{R}_{w_n}(\tau)$ is thus given by:

$$\overline{R}_{w_n}(\tau) = \frac{1}{\gamma_b^2}R_{nb}(\tau)e^{R_{\phi_d}(\tau)-R_{\phi_d}(0)}J_0(\tau_0)$$

$$= \frac{1}{\gamma_b^2}R_{nb}(\tau)e^{R_{\phi_d}(\tau)-R_{\phi_d}(0)}J_0(2\rho\sin(\pi F_{\text{sens}}\tau)) \tag{10.35}$$

The spectral density of $w_n(t)$ is given by:

$$S_w(f) = \mathcal{F}\left\{\overline{R}_{w_n}(\tau)\right\} = \frac{1}{\gamma_b^2}e^{-R_{\phi_d}(0)}\mathcal{F}\left\{R_{nb}(\tau)e^{R_{\phi_d}(\tau)}J_0(2\rho\sin(\pi F_{\text{sens}}\tau))\right\} \tag{10.36}$$

The product in the time domain can be expressed as convolution in the frequency domain. By denoting $p(\tau) = e^{R_{\phi_d}(\tau)}$, $P(f) = \mathcal{F}\{p(\tau)\}$ and $M_J(f) = \mathcal{F}\{\mu_J(\tau)\}$ with $\mu_J(\tau) = J_0(2\rho\sin(\pi F_{\text{sens}}\tau))$, Eq. (10.36) becomes

$$S_w(f) = \frac{1}{\gamma_b^2}e^{-R_{\phi_d}(0)}[S_{nb}(f) * P(f) * M_J(f)] \tag{10.37}$$

$S_{nb}(f)$ is defined in Eq. (10.9) as a rectangular window, band limited in $\left[-\frac{B}{2}, \frac{B}{2}\right]$. $P(f)$ is considered a signal with most of its power concentrated around zero, with its tails near $-\frac{B}{2}, \frac{B}{2}$, attaining negligible values. That way, the convolution between them can be evaluated as follows:

$$S_{nb}(f) * P(f) = \int_{-\infty}^{+\infty} S_{nb}(\xi)P(f - \xi)d\xi = N_0 \int_{-\frac{B}{2}}^{\frac{B}{2}} P(f - \xi)d\xi$$

$$\overset{u=f-\xi}{=} N_0 \int_{f-\frac{B}{2}}^{f+\frac{B}{2}} P(u)du \tag{10.38}$$

Let W_m denote the (baseband) bandwidth of the signal at the output of the receiver (or the bandwidth of the filter used at the output of the receiver), i.e the bandwidth of $r(t)$ (see Eq. (10.16)). Both interference $\phi_s(t)$ and tag's $\mu(t)$ signals are considered band limited within $[-W_m, W_m]$. Thus, frequencies of interest (for SINR calculation) are the ones satisfying $|f| \leq W_m$. Additionally, it is assumed that $W_m \ll \frac{B}{2}$, which holds in case of large modulation index β_s (wideband FM).[6] Under those assumptions, it can be seen that the area calculated for frequency shifts $|f| \leq W_m$ (see Figure 10.2) is approximately constant. Because $P(f)$ takes negligible values at its tails, it follows that for $|f| \leq W_m$:

$$S_{nb}(f) * P(f) = N_0 \int_{f-\frac{B}{2}}^{f+\frac{B}{2}} P(u)du \approx N_0 \int_{-\infty}^{+\infty} P(u)du = p(0)$$

$$= N_0 e^{R_{\phi_d}(0)}, \quad \text{for } |f| \leq W_m \tag{10.39}$$

Using the above result, Eq. (10.37) becomes:

$$S_w(f) = \left(\frac{1}{\gamma_b^2} e^{-R_{\phi_d}(0)} N_0 e^{R_{\phi_d}(0)} \right) * M_J(f) = \left(\frac{N_0}{\gamma_b^2} \right) * M_J(f), \quad \text{for } |f| \leq W_m \tag{10.40}$$

As it is shown, both analytically and via simulations in Appendix 10.9, $M_J(f)$ is band-limited within $[-W_m, W_m]$.[7] Because of the large value of B compared to W_m, function $\alpha(f) = (S_{nb} * P)(f)$ can be safely assumed to attain value $N_0 e^{R_{\phi_d}(0)}$ for $|f| \leq W_m$. That way the convolution in Eq. (10.40) can be derived as follows:

$$S_w(f) = \frac{1}{\gamma_b^2} e^{-R_{\phi_d}(0)} \int_{-\infty}^{\infty} M_J(v)\alpha(f - v)dv$$

$$\overset{(a)}{=} \frac{1}{\gamma_b^2} e^{-R_{\phi_d}(0)} \int_{-W_m}^{W_m} M_J(v)\alpha(f - v)dv$$

$$\overset{(b)}{=} \frac{N_0}{\gamma_b^2} \int_{-W_m}^{W_m} M_J(v)dv \overset{(a)}{=} \frac{N_0}{\gamma_b^2} \int_{-\infty}^{+\infty} M_J(v)dv$$

$$= \frac{N_0}{\gamma_b^2} \mu_J(0) = \frac{N_0}{\gamma_b^2} J_0(0) = \frac{N_0}{\gamma_b^2}, \quad \text{for } |f| \leq W_m \tag{10.41}$$

6 Remodulated signal attains such an index due to both large modulation index utilized at the tag and illumination from FM stations (being wideband FM).

7 For the values of ρ utilized in this work.

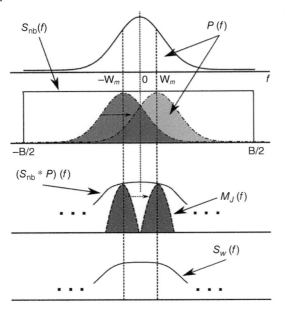

Figure 10.2 Qualitative illustration of the convolutions involved in calculating the spectral density of noise at the output of the receiver.

where in points (a) the band-limited nature of $M_J(\cdot)$ was exploited, while in point (b) the aforementioned fact regarding $\alpha(\cdot)$. Process $w_n(t)$ is filtered by a derivator, with frequency response $H(j\omega) = j\frac{\omega}{2\pi}$, to result $w_d(t) = \frac{1}{2\pi}\frac{d}{dt}w_n(t)$. That way, the spectral density of $w_d(t) = \frac{1}{2\pi}\frac{d}{dt}w_n(t)$ can be expressed as:

$$S_{w_d}(f) = S_w(f)|H(j\omega)|^2 = \frac{N_0}{\gamma_b^2}f^2, \quad \text{for } |f| \le W_m \tag{10.42}$$

Finally, the noise power at the output of the receiver can be evaluated as follows:

$$P_{w_d} = \int_{-W_m}^{W_m} S_{w_d}(f)df = \frac{N_0}{\gamma_b^2}\int_{-W_m}^{W_m} f^2 df = \frac{N_0}{\gamma_b^2}\frac{f^3}{3}\Big|_{-W_m}^{W_m} = \frac{2N_0 W_m^3}{3\gamma_b^2} \tag{10.43}$$

Notice that an illuminating FM station offering a stronger carrier at the high SNR regime, results to smaller P_{w_d}. Having derived the noise power at the output of the receiver and using Eq. (10.16), output SINR (for tag's signal, when operating at the high SNR regime), can be defined as:

$$\text{SINR} = \frac{k_{sw}^2 P_{tag}}{k_s^2 P_{st} + P_{w_d}} \tag{10.44}$$

where $P_{tag} = \mathbb{E}[\mu^2(t)]$ the power of the tag's signal $\mu(t)$ and $P_{st} = \mathbb{E}[\phi_s^2(t)]$ the power of the station's signal $\phi_s(t)$. The denominator in Eq. (10.44) can be written in that form given that in the high SNR regime, the output noise in $[-W_m, W_m]$ band of interest can be considered independent of both the interference and the tag's information signal. This stems from the fact that spectral density of noise at the output of the receiver (Eq. (10.42)) is only a function of frequency f, noise's spectral density N_0, and impinged signal's amplitude γ_b. What this relationship (and Eq. (10.43)) effectively shows is that in FM, increasing the transmission power (and by extension the impinged power) lowers the noise level at the FM receiver's output.

10.3.2 Low SNR Case

It was assumed in Section 10.3.1 that the noise level at the input of the receiver is at least 10 dB lower than that of the signal of interest. It can be shown that when the power of noise at the input of the receiver is comparable to the signal's power, the noise at the output of the receiver has the form of shots, superimposed to the noise described in the previous Section 10.3.1 [29]. Following the analysis presented in [29], the power of shot noise is given by the following equation:

$$N_s = 8\pi^2 W_m \overline{|\,\delta f\,|} e^{-\text{SNR}} \tag{10.45}$$

where $\overline{|\,\delta f\,|}$ is the average, absolute frequency deviation which needs to be calculated. In this work, $\delta f = k_s \phi_s(t) + k_{sw}\mu(t)$. Using the triangle inequality, $|\delta f|$ can be bounded as follows:

$$| \delta f | = | k_s\phi_s(t) + k_{sw}\mu(t) | \leq | k_s\phi_s(t) | + | k_{sw}\mu(t) |$$
$$\stackrel{k_s, k_{sw} \geq 0}{=} k_s \, | \, \phi_s(t) \, | + k_{sw} \, | \, \mu(t) \, | \tag{10.46}$$

The (time) average value of $k_{sw} | \mu(t)|$ can be calculated as:

$$
\begin{aligned}
\overline{k_{sw}|\,\mu(t)\,|} &= \frac{k_{sw}}{T_{\text{sens}}} \int_{T_{\text{sens}}} |\,\mu(t)\,| \, dt \\
&= \frac{k_{sw}}{T_{\text{sens}}} \int_0^{T_{\text{sens}}} |\sin(2\pi F_{\text{sens}}t)| \, dt \\
&= \frac{2k_{sw}}{T_{\text{sens}}} \int_0^{T_{\text{sens}}/2} \sin(2\pi F_{\text{sens}}t) dt \\
&= \frac{2k_{sw}}{T_{\text{sens}}} \left[\frac{-1}{2\pi F_{\text{sens}}} \cos(2\pi F_{\text{sens}}t) \Big|_0^{T_{\text{sens}}/2} \right] = \frac{2k_{sw}}{\pi} \tag{10.47}
\end{aligned}
$$

For any t, $\phi_s(t)$ is assumed a zero mean, normally distributed random variable. It can be shown (see Appendix 10.10) that for any t, the associated random variable $k_s | \phi_s(t)|$ has mean value:

$$k_s\overline{|\,\phi_s(t)\,|} = k_s\mathbb{E}[|\,\phi_s(t)\,|] = \sqrt{\frac{2}{\pi}}k_s\sqrt{P_{\text{st}}} \tag{10.48}$$

Using Eqs. (10.47) and (10.48) on (10.46),

$$\overline{\mid \delta f \mid} \le k_{sw}\overline{\mid \mu(t) \mid} + k_s\overline{\mid \phi_s(t) \mid} = \frac{2k_{sw}}{\pi} + \sqrt{\frac{2}{\pi}}k_s\sqrt{P_{st}} \triangleq \hat{\delta f} \tag{10.49}$$

The upper bound $\hat{\delta f}$ will be used in place of $\overline{\mid \delta f \mid}$ in Eq. (10.45). An approximation for the power of the noise at the output of the FM receiver, considering an input SNR attaining values less than 10 dB, can be then given by:

$$\tilde{N}_s = 8\pi^2 W_m \hat{\delta f} e^{-SNR} \tag{10.50}$$

Considering the results from Sections 10.3.1 and 10.3.2, the SINR at the output of the receiver can be written as:

$$SINR_{low} \approx \frac{k_{sw}^2 P_{tag}}{k_s^2 P_{st} + P_{w_d} + \tilde{N}_s} = \frac{k_{sw}^2 P_{tag}}{k_s^2 P_{st} + P_{w_d} + 8\pi^2 W_m \hat{\delta f} e^{-SNR}} \tag{10.51}$$

Equation (10.51) shows that at the high SNR regime, i.e. when $SNR \ge 10$ dB, term e^{-SNR} can be assumed zero. However, when $SNR < 10$ dB, $\tilde{N}_s$ cannot be neglected, and $SINR_{low}$ gives an approximation for the attained SINR at the FM receiver's output.

10.4 Occupied Bandwidth

An upper limit for the occupied bandwidth can be obtained using Carson's rule [30]. The rule states that 98% of the modulated signal's power is held within bandwidth $B_o = 2W_m(\beta_s + 1) = 2(\Delta f_{max} + W_m)$,[8] where $\Delta f_{max} = \max \mid \delta f \mid$.

As with Section 10.3.2, triangle inequality will be used:

$$\Delta f_{max} = \max \mid \delta f \mid \le \max(k_{sw} \mid \mu(t) \mid + k_s \mid \phi_s(t) \mid)$$
$$\overset{*}{=} k_{sw}\max \mid \mu(t) \mid + k_s\max \mid \phi_s(t) \mid$$
$$= \Delta f_{tag,max} + \Delta f_{s,max} \tag{10.52}$$

where in point $*$, the fact that all the quantities involved are positive was used. B_o is then upper bounded by B_u as:

$$B_o \le B_u = 2(\Delta f_{tag,max} + \Delta f_{s,max} + W_m) \tag{10.53}$$

B_u can be used for the calculation of the passband area of the band selection filter in the receiver.

---

8 Rule holds for both wideband and narrowband FM. For narrowband FM transmission $\beta \ll 1$, that way $B_o \approx 2W_m$.

10.5   Ambient Selection Diversity

As described in Section 10.2.1, backscattering can be seen as a modulation operation performed in passband. Tag's switching according to Eq. (10.5) offers remodulated backscattered signal components at $\{F_s \pm F_{sw}\}$, for *all* FM stations $s \in \{1, 2, \ldots, L\}$. Thus, a question immediately arises: from the available FM stations/illuminators $s \in \{1, 2, \ldots, L\}$, which should the FM receiver select to tune at $F_s \pm F_{sw}$? As it can be observed, there are $2L$ possible (passband) frequencies for the utilized FM receiver to tune at, when there are L FM stations/illuminators available. Measurements regarding the RF power offered by two nonline of sight (NLOS) FM radio stations, located 6.5 km away were conducted. It was found that the expected power varied significantly between the two stations, namely station No. 1 offered -56 dBm, while station No. 2 offered -69 dBm.[9]

As shown in Section 10.3, higher carrier amplitude γ_s results to lower impact of thermal noise at the output of the receiver. Backscatter communication is by nature link-budget limited. Thus, selecting the FM station offering the strongest received power (i.e. highest γ_s) could offer better performance. The impinged power at tag's antenna from station $s \in \{1, 2, \ldots, L\}$, γ_s, is assumed to follow a Gamma distribution with shape and scale parameters k_s, θ_s, respectively. Similarly, the power received at the smartphone is also a Gamma-distributed random variable, γ_0, with shape and scale parameters k_0, θ_0. The latter two parameters incorporate the tag-fixed scattering efficiency η, as well as link-budget average loss due to tag-to-smartphone distance d_0. The following proposition holds for the end-2-end received power $\gamma_s \gamma_0$ of the backscattered signal at the smartphone, under selection of the stronger FM illuminator, performed among L potential FM stations:

Proposition 10.1   *For $\gamma_i : \sim$ Gamma $(\cdot; k_i; \theta_i)$, $i \in \{0, 1, \ldots, L\}$, the benefits of selection diversity can be assessed by the following outage probability:*

$$\Pr\left(\max_{i \in \{1,2,\ldots,L\}} \gamma_i \, \gamma_0 < \Theta_{RF}\right) = \frac{1}{\theta_0^{k_0}} \frac{1}{\prod\limits_{j=0}^{L} \Gamma(k_j)} \int_0^{+\infty} x^{k_0-1} e^{-\frac{x}{\theta_0}} \prod_{i=1}^{L} \gamma\left(k_i, \frac{\Theta_{RF}}{\theta_i x}\right) dx$$

$$(10.54)$$

where $\gamma(s, x) = \int_0^x t^{s-1} e^{-t} dt$ is the lower incomplete gamma function, $\Gamma(s) = \int_0^{+\infty} t^{s-1} e^{-t} dt$ is the Gamma function, and Θ is a test (fixed) threshold value. As can be seen in Figure 10.3, the aforementioned probability decreases with increasing L. The proof is given in Appendix 10.11.

---

9 Measurements were performed for the duration of one hour, every two seconds for each station, using a portable spectrum analyzer.

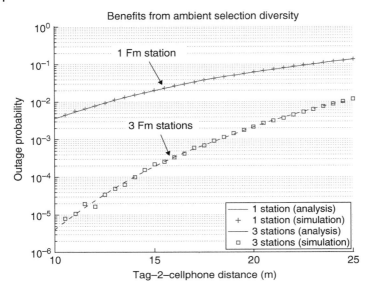

Figure 10.3 Probability of outage according to Eq. (10.54) when Θ_{RF} corresponds to input SNR = 10 dB, value chosen according to observations in Figure 10.8.

Nevertheless, the previously mentioned selection does not necessarily minimize the interference from each FM station's own $\phi_s(t)$ signal on the tag's signal $\mu(t)$. Additionally, the aforementioned selection process assumes that there is no *other* interfering signal (e.g. from another FM station) at $F_s \pm F_{sw}$ as well as at frequencies around them.

An alternative to the aforementioned selection process is to tune at the frequency where the demodulated sensor's audio signal is maximized. That can be easily implemented with a single smartphone FM receiver, tuning sequentially at $2L$ candidate frequencies $F_s \pm F_{sw}$, $s \in \{1, 2, ..., L\}$ (for fixed F_{sw}), and selecting the one where the sensor's audio tone level is above a predefined (by the user) threshold Θ. In the experiments, Θ was selected 10 dB above (audio) noise (thermal and interference) floor. The described procedure is depicted in Figure 10.4.

10.6 Analog Tag Implementation

The main components of the tag are the two oscillators. The first produces sensor's modulating signal $\mu(t)$, and the second is driven by the first. That way, the signal (to be backscattered) of Eq. (10.5) can be produced. As can be seen in Figure 10.5-left, a capacitive or resistive sensing element is assumed, connected to Oscillator A

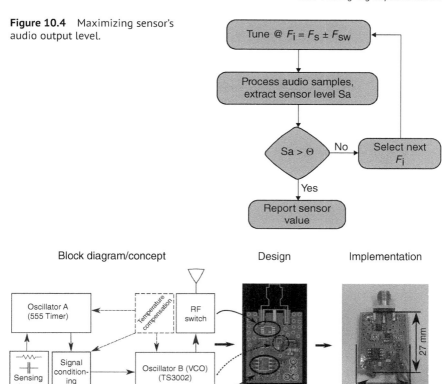

Figure 10.4 Maximizing sensor's audio output level.

Figure 10.5 Tag implementation.

(first oscillator). As will be explained below, any capacitive or resistive element can be used.

10.6.1 Sensing Capacitor and Control Circuit

10.6.1.1 Generating $\mu(t)$ – First Modulation Level

To generate sensor-value dependent $\mu(t)$, an ultra-low power variant of the 555 timer, shown as Oscillator A in Figure 10.5-left, is used. The timer produces a square wave with its fundamental frequency determined by values of external, passive components (resistors and capacitors). The sensing element is part of those components. Thus, a change in the measured quantity results to a change in $\mu(t)$'s fundamental frequency. The circuit, as discussed in Section 10.2.2, must be designed in a way that $\mu(t)$ will remain audible throughout the sensor's range. Thus, provided an appropriate design, any sensing capacitive/resistive element can be used. In this work, a soil moisture capacitive sensing element, part of the sensor in [9, 31] was used. An environmental humidity sensing capacitor (HCH-1000), part of the sensor in [3, 32] was tested as well.

10.6.1.2   Generating $x_{FM}(t)$ – Second Modulation Level

To generate $x_{FM}(t)$, Oscillator B is implemented with a Silicon Laboratories TS3002 timer, configured as a voltage-controlled oscillator (VCO) (Figure 10.5-left). Signal $\mu(t)$ is set as the VCO's control voltage. That way, $\mu(t)$ controls the instantaneous frequency of the VCO's output signal. By extension, maximum and minimum values of $\mu(t)$ define the maximum and minimum frequency values, produced by the VCO. To produce an FM signal with specific frequency limits, before driving the VCO's input, $\mu(t)$ is scaled using a signal conditioning block (Figure 10.5-left) comprised of variable resistors.

10.6.2   RF-Switch

Switching between loads terminating the tag antenna was conducted with the Analog Devices ADG919 switch. Due to the ultra-low power requirements of the overall tag, the switch was tested outside its typical operating conditions, regarding supply voltage (the minimum operating voltage is 1.65 V, according to device's datasheet). To measure the reflection coefficients, a vector network analyzer (VNA) was used to stimulate the switch at 90 MHz. For a supply voltage of 1.2 V, $|\Delta\Gamma|_{1.2V} \overset{\Delta}{=} |\Gamma_1 - \Gamma_0| = 1.70$ and for 1.7 V, $|\Delta\Gamma|_{1.7V} = 1.78$. Ignoring expected degradation in backscattering performance, the switch could operate even at 1.2 V.

10.6.3   Power Consumption and Supply

For fixed sensor value (sensor dry), second row of Table 10.1 offers the total current consumption of the overall tag, as a function of supply voltage; it is clearly shown that the system is capable of $<20\,\mu$A @ 1.2 V, resulting in 24 μW of power consumption, even in *continuous* (nonduty cycled) operation. Using a standard capacitor in place of the sensing element (i.e. fixing Oscillator A's output frequency), Table 10.1 also offers the dependence of Oscillator A's fundamental frequency on supply voltage, directing the utilization of voltage regulation. Three experiments utilizing

Table 10.1   Overall power consumption and 555 frequency offset.

Supply voltage (V)	1.2	1.3	1.4	1.5	1.6	1.7	1.8
Current consumption (μA)	20	21.1	24.1	26.3	28.5	30.9	33.1
Frequency offset (kHz)	2.5	2.73	2.93	3.12	3.21	3.22	3.24

different power sources were conducted to showcase the ultra-low power character of the proposed system.

10.6.3.1 Batteryless Tag with Photodiode
The tag was tested in duty cycled operation, using a Texas Instruments BQ25504 harvesting IC, in conjunction with a BPW34 photodiode (as harvesting element). The flashlight of the smartphone was used to illuminate the photodiode which charged a capacitor through the IC. The harvesting IC enabled the supply line of the tag when the voltage of the storage capacitor reached a certain level. The charging and operating times were approximately six seconds. The resulting operating time is more than enough to get a valid reading on the smartphone. Of course, other energy harvesting methods, as well as duty cycle ratios, can be used.

10.6.3.2 Batteryless Tag with Solar Panel
A small (31 mm × 31 mm) solar panel was used to power the tag. The panel was measured to provide short-circuit current $I_{sc} = 40$ mA and open circuit voltage $V_{oc} = 2.3$ V, under full sunlight conditions. To avoid large deviations on the received sensor value due to supply voltage variations (see Table 10.1), caused by the solar panel, a 1.8 V voltage reference was utilized. The setup was tested outside, as shown in Figure 10.6. In the setup, the soil moisture sensor was placed in a flowerpot. While watering the plant, the frequency of the sensor's tone (as received by the smartphone) dropped as expected.

10.6.3.3 Batteryless Tag with Lemons
For the third experiment, two lemons were used. Each "battery" consisted of a lemon with two inserted electrodes. Electrode 1 was a zinc-plated nail and electrode 2 was a copper wire. Each lemon offered $V_{oc} \approx 0.9$ V and $I_{sc} \approx 600$ μA. To provide enough voltage for the tag to operate, the two lemons were connected in series, and the same setup as with the solar panel experiment was used without, however, utilizing a voltage regulator. In the last two experiments, the tag was

Figure 10.6 Experimental setup using solar panel as power supply.

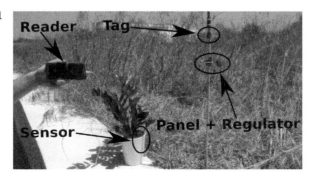

supplied with 1.79 V and consumed 32 µA. The measurements were made with a HP 34401A multimeter.

10.6.4 Receiver

As described in Section 10.2.2, any conventional FM broadcast radio receiver can be used, given that $\mu(t)$ is audible. The following options are proposed:

10.6.4.1 Smartphone

Most modern (smart)phones come equipped with FM radio receivers. After tuning to the frequency offering the strongest tag's audio tone (selection diversity), the sensor value can be acquired by estimating the frequency of that tone. In Figure 10.7, screen captures of an audio spectrum application running on a

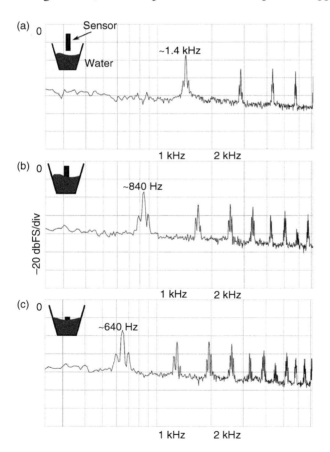

Figure 10.7 Measured spectrum of smartphone audio output; the tag sensing capacitor is being gradually submerged in a glass of water (See graphic in the upper left corner of each plot).

Motorola Moto G3, while the soil moisture sensor is being gradually submerged into a glass of water, can be seen. In the indoor scenario of Figure 10.7, the tag was supplied with 1.2 V while being 1 m away from the smartphone.

10.6.4.2 Computer

Two options are offered in case a computer is needed to read the sensor. First, any conventional FM radio receiver can be connected to the computer's audio input. Alternatively, a SDR receiver, e.g. a low-cost RTL dongle, can be tuned at the appropriate frequency, perform the necessary processing and recover sensor's information, as previously described.

As already noted in Section 10.2.2, the tag is indifferent with respect to the carrier used; thus, a dedicated carrier can be used to illuminate the tag. An SDR-equipped computer can perform the necessary processing and recover the sensor's value. To verify the last statement, experiments have been performed utilizing a dedicated unmodulated carrier generator, tuned at $F_c = 868$ MHz. The receiver used was a low-cost RTL dongle, and the necessary processing for sensor readout was performed in GNU radio. If operation in both ambient and dedicated illumination is necessary, the only additional requirement is that tag's antenna should be designed to operate at both the broadcast FM, as well as the dedicated illuminator's carrier frequencies.

10.7 Performance Characterization

10.7.1 Simulation Results

Equation (10.7) was modeled with $F_t = 2$ MHz and γ_b chosen so that the "transmission power" of the tag was equal to -80 dBm. To obtain different SNR values, the noise level P_n was varied, while maintaining the signal power constant. Noise power spectral density N_0 was calculated as follows:

$$P_n = \int_{-\frac{B}{2}}^{\frac{B}{2}} N_0 df = N_0 \int_{-\frac{B}{2}}^{\frac{B}{2}} df = N_0 B \Leftrightarrow N_0 = \frac{P_n}{B} \qquad (10.55)$$

Signal $\phi_s(t)$ was a six second recorded clip from a local radio station. $\phi_s(t)$ was obtained using GNU-Radio and a RTL-SDR dongle. GNU-Radio performed FM demodulation; the demodulated audio signal was sampled at $F_{s,m} = 100$ kHz and saved to a binary file. The samples contained in the file were then imported in MATLAB and after the necessary processing, included in the model. In a second experiment, $\phi_s(t)$ was modeled as Gaussian noise having variance equal to the power of the recorded clip. $\mu(t)$ was created in MATLAB as defined in Eq. (10.24) with $F_{sens} = 3.2$ kHz. Due to sampling of $\mu(t)$ at the same rate as $\phi_s(t)$ ($F_{s,m}$), the resulting message bandwidth was set to $W_m = 50$ kHz. After interpolating $\phi_s(t)$ and $\mu(t)$, the final signal model was sampled at $F_s = 10$ MHz. No band selection

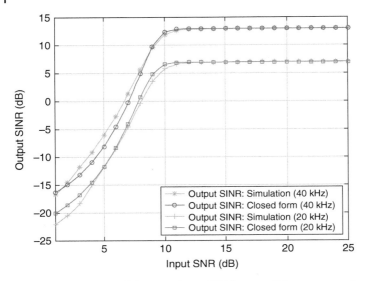

Figure 10.8 Output SINR versus input SNR for two different tag maximum deviation settings. The station's maximum deviation was set to 30 kHz and $\phi_s(t)$ was a recorded sound clip.

filter was used, that way the value used for B was $B/2 = F_s/2$. Then, the receiver was implemented as described in Section 10.3. Closed form SINR was defined as per Eq. (10.51). Simulated SINR was returned by MATLAB's SNR [33] function using as input the receiver's output signal.

Figure 10.8 offers the behavior of the receiver's output SINR when the input SNR was varied and $\phi_s(t)$ was set to be the recorded clip. There are two points worth of attention in this plot. First, it can be seen that around 10 dB, the threshold phenomenon kicks in. Below that 10 dB threshold, the output SINR drops rapidly with respect to input SNR. This observation shows intuitively that the more the illuminating stations are available to choose from, the higher the possibility of attaining drastically better performance, highlighting the importance of selection diversity. Second, it can be observed that for SNR >10 dB, i.e high SNR case, the output SINR is almost constant with respect to input SNR. This comes in agreement with the experimental measurements in Figure 10.11, where for small tag-2-smartphone distances (i.e. higher input SNR), up until 8 m, the output audio level is constant (for the station offering the higher impinged power). It must be noted that this behavior is due to interference from station's signal $\phi_s(t)$. If $\phi_s(t) = 0$, above the 10 dB threshold, an increase in input SNR would result a linear[10] increase in

---

10 In dB scale.

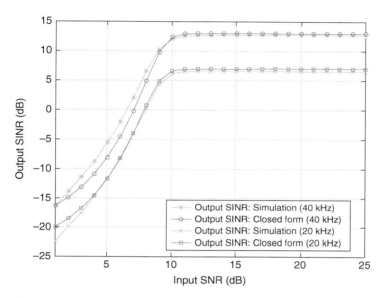

Figure 10.9 Output SINR versus input SNR for two different tag maximum deviation settings. The station's maximum deviation was set to 30 kHz, and $\phi_s(t)$ was Gaussian noise.

output S(I)NR. Figure 10.9, repeats the simulation, with however $\phi_s(t)$ modeled as Gaussian noise with same variance (i.e. equal to the power of the station's recorded clip). The same conclusions as in Figure 10.8 are drawn.

Figure 10.10 shows the dependence of output SINR on the tag's maximum frequency deviation (i.e. on k_{sw}) with fixed station's maximum deviation. The simulation verifies the experimental tests, where by increasing the maximum frequency deviation in the tag's VCO, higher audio levels (for the tag's signal) at the output of the receiver were observed. No threshold effect was observed (in simulations) for the deviation values utilized. That may be explained by the fixed noise bandwidth as well as the small frequency deviation values (for the utilized values of k_{sw}, k_s), attained from Eq. (10.7) compared to that bandwidth. Experimental results were acquired using a function generator acting as the tag and adjusting the deviation setting accordingly. The tag was at a distance of approximately 30 cm from the smartphone. Linear interpolation was applied between measurements, which appeared constant due to lack of decimal digits in the reported measurement.

The apparent gaps between simulated results and closed form expressions are mainly due to the approximations involved in deriving low SNR regime expressions, both regarding Eq. (10.49) and the expressions used [29].

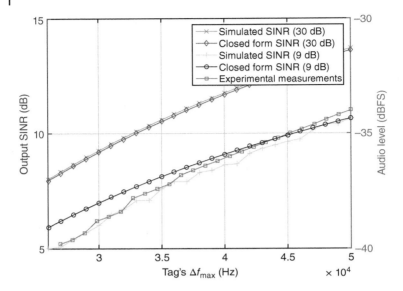

Figure 10.10 Output SINR versus tag maximum frequency deviation $\Delta f_{max} = k_{sw} \max |\mu(t)| = k_{sw}$ for input SNR of 30 and 9 dB. The station's maximum deviation was 30 kHz, and $\phi_s(t)$ was the recorded sound clip. The experimental measurements correspond to the right y-axis.

10.7.2 Tag Indoor and Outdoor Performance

Both indoor and outdoor scenarios were considered to evaluate the achieved range of the proposed system. An audio spectrum application, advanced spectrum analyzer PRO, running on the smartphone was used to report the audio level. Audio level is reported in a scale of dbFS, which represents level with respect to full-scale microphone input. Any tone above −20 dbFS is unbearable when using earphones and anything below −70 dbFS can be considered noise.

A standard value capacitor replaced the sensing element offering a fixed "sensor value" of 3.2 kHz, and a 1.5 V battery was used (as power source) to ease the measurement process. Using the application's markers, the audio level at 3.2 kHz was reported at each location away from the tag. Using a spectrum analyzer, the FM station's RF power was measured at the location of the tag.

Performance results for an outdoor scenario are offered for two FM stations in Figure 10.11; the tag achieves at least 23 m before the audio tone power drops below −60 dbFS, i.e. 10 dB above audio noise floor, resulting to demodulated (backscattered) signal SINR of 10 dB. Two different stations were chosen for the tests, each offering different power level at the tag's location. Due to wireless fading in the end-2-end link (FM station-tag-smartphone), variations in the measurements were observed. Thus, in Figure 10.11, the test was repeated twice

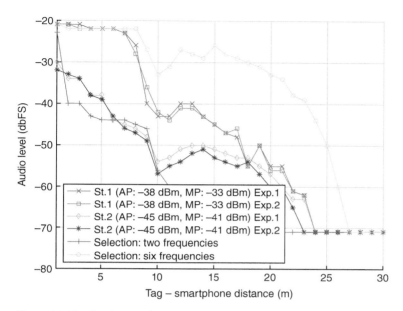

Figure 10.11 Outdoor performance, for two different radio stations. Average power (AP) denotes the average and maximum power (MP) the maximum power impinged at the tag.

(Experiment 1 and 2). In the same figure, results are shown for the case that the smartphone chooses to tune at the frequency offering the maximum audio level at sensor's audio tone. When the selection was performed among $2L = 6$ frequencies (i.e. for a given tag-smartphone distance, smartphone was tuned at each of the six frequencies, and the strongest audio level was reported), the achieved range increased to 26 m. In the case that two frequencies were available to select from, the performance was weaker, highlighting again the importance of selection diversity. Indoor communication range was also studied; 15 m tag-to-receiver range was achieved, with (measured) impinged power at the tag of −55 dBm.

10.8 Conclusions

A backscatter analog modulation scheme, *FM remodulation*, was presented in this chapter, along with detailed analysis and prototype implementation. FM remodulation allowed the exploitation of ambient FM radio signals in conjunction with specific switching at the backscatter tag, to achieve *ultra*-low power communication. At the same time, the method allowed the reception of the backscattered signal by *any* conventional FM radio receiver, including those found in modern smartphones. Besides analyzing the effect the noise has at the output of an FM receiver, when such a scheme is employed, a tag implementing FM remodulation and capable of facilitating any capacitive or resistive sensor was offered. It was able to achieve tag-2-smartphone(receiver) ranges in the order of 26 m while consuming 24 μW in continuous operation, with the illuminating FM station being 6.5 km away. Benefits (i.e., achieving such a communication range) of selection diversity were also quantified in detail. This work is perhaps a concrete, disruptive example on how existing signals can be reused with backscatter radio for applications with important socioeconomic impact.

10.9 Bandwidth of $J_0 \left(2\rho \sin \left(\frac{\omega_{\text{sens}}}{2} t \right) \right)$

In order to obtain the occupied bandwidth of $J_0 \left(2\rho \sin \left(\frac{\omega_{\text{sens}}}{2} t \right) \right)$, where $\omega_{\text{sens}} = 2\pi F_{\text{sens}}$, the Fourier tranform:

$$M(\omega) = \int_{-\infty}^{+\infty} J_0 \left(2\rho \sin \left(\frac{\omega_{\text{sens}}}{2} t \right) \right) e^{-j\omega t} dt \qquad (10.56)$$

needs to be evaluated. First, $J_0 \left(2\rho \sin \left(\frac{\omega_{\text{sens}}}{2} t \right) \right)$ will be expanded using [28, 9.1.10]:

$$J_n(\beta) = \left(\frac{\beta}{2} \right)^n \sum_{k=0}^{+\infty} \frac{(-\beta^2/4)^k}{k!\Gamma(n+k+1)} \overset{n=0}{=} \sum_{k=0}^{+\infty} \frac{(-1)^k (\beta/2)^{2k}}{(k!)^2} \qquad (10.57)$$

and $J_0 \left(2\rho \sin \left(\frac{\omega_{\text{sens}}}{2} t \right) \right)$ becomes

$$J_0 \left(2\rho \sin \left(\frac{\omega_{\text{sens}}}{2} t \right) \right) = \sum_{k=0}^{+\infty} \frac{(-1)^k \left(\rho \sin \left(\frac{\omega_{\text{sens}}}{2} t \right) \right)^{2k}}{(k!)^2}$$

$$= \sum_{k=0}^{+\infty} \frac{(-1)^k \rho^{2k}}{(k!)^2} \sin^{2k} \left(\frac{\omega_{\text{sens}}}{2} t \right) \qquad (10.58)$$

Equation (10.56) then becomes

$$M(\omega) = \int_{-\infty}^{+\infty} \left[\sum_{k=0}^{+\infty} \frac{(-1)^k \rho^{2k}}{(k!)^2} \sin^{2k} \left(\frac{\omega_{\text{sens}}}{2} t \right) \right] e^{-j\omega t} dt$$

$$= \sum_{k=0}^{+\infty} \frac{(-1)^k \rho^{2k}}{(k!)^2} \int_{-\infty}^{+\infty} \sin^{2k} \left(\frac{\omega_{\text{sens}}}{2} t \right) e^{-j\omega t} dt \qquad (10.59)$$

The term $\sin^{2k} \left(\frac{\omega_{\text{sens}}}{2} t \right)$ can be expanded as [34]:

$$\sin^{2k} \left(\frac{\omega_{\text{sens}}}{2} t \right) = \frac{1}{2^{2k}} \binom{2k}{k} + \frac{(-1)^k}{2^{2k-1}} \sum_{i=0}^{k-1} (-1)^i \binom{2k}{i} \cos((k-i)\omega_{\text{sens}} t)$$

$$(10.60)$$

Equation (10.59) then becomes:

$M(\omega)$

$$= \sum_{k=0}^{+\infty} \frac{(-1)^k \rho^{2k}}{(k!)^2} \int_{-\infty}^{+\infty} \left[\frac{1}{2^{2k}}\binom{2k}{k} + \frac{(-1)^k}{2^{2k-1}}\sum_{i=0}^{k-1}(-1)^i\binom{2k}{i}\cos((k-i)\omega_{sens}t)\right]e^{-j\omega t}dt$$

$$= \sum_{k=0}^{+\infty} \frac{(-1)^k \rho^{2k}}{(k!)^2}\left[\frac{1}{2^{2k}}\binom{2k}{k}\int_{-\infty}^{+\infty}e^{-j\omega t}dt + \frac{(-1)^k}{2^{2k-1}}\sum_{i=0}^{k-1}(-1)^i\binom{2k}{i}\int_{-\infty}^{+\infty}\cos((k-i)\omega_{sens}t)e^{-j\omega t}dt\right]$$

$$= \sum_{k=0}^{+\infty} \frac{(-1)^k \rho^{2k}}{(k!)^2}\left[\frac{1}{2^{2k}}\binom{2k}{k}\delta(\omega) + \frac{(-1)^k}{2^{2k}}\sum_{i=0}^{k-1}(-1)^i\binom{2k}{i}\Delta(\omega - (k-i)\omega_{sens})\right] \quad (10.61)$$

where $\delta(x)$ the Dirac delta function and $\Delta(\omega - a) \triangleq \delta(\omega - a) + \delta(\omega + a)$. Eq. (10.61) shows that the spectrum of $J_0(\tau_0)$ is comprised of carriers at integer multiples of F_{sens}. The bandwidth of $J_0(\tau_0)$ can be evaluated by searching for the integer multiple of F_{sens} after which the spectral components of $J_0(\tau_0)$ take negligible values.

Using Matlab and its fft function, the bandwidth of $J_0(\tau_0)$ was evaluated via direct simulation. A sine wave was sampled at $F_s = 400\,\text{kHz}$ with $F_{sens} = 3.2\,\text{kHz}$ and $k_{sw} = 45\,\text{kHz/V}$. Additionally, Eq. (10.61) was numerically evaluated for a given number of harmonics, and the same parameters mentioned earlier. Simulation and numerical evaluation results are given in (Figure 10.12).

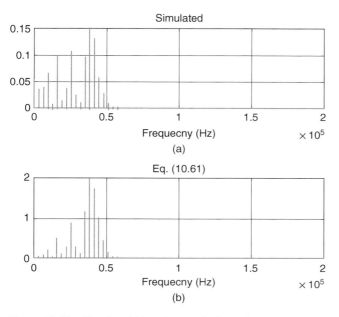

Figure 10.12 Simulated (a) and numerically evaluated (b), single-sided spectrum of $J_0(\tau_0)$. The numerical evaluation was performed for the first 50 harmonics of F_{sens} according to Eq. (10.61). Note that no scaling has been applied to the bottom waveform.

10.10   Expectation of the Absolute Value of a Gaussian R.V

Assuming $x \sim \mathcal{N}(0, \sigma^2)$ then,

$$
\begin{aligned}
\mathbb{E}[|x|] &= \int_{-\infty}^{+\infty} |x| \frac{1}{\sqrt{2\pi\sigma^2}} e^{-\frac{x^2}{2\sigma^2}} dx \\
&= \int_{-\infty}^{0} (-x) \frac{1}{\sqrt{2\pi\sigma^2}} e^{-\frac{x^2}{2\sigma^2}} dx + \int_{0}^{+\infty} x \frac{1}{\sqrt{2\pi\sigma^2}} e^{-\frac{x^2}{2\sigma^2}} dx \\
&= \frac{2}{\sqrt{\pi}} \int_{0}^{+\infty} \frac{x}{\sqrt{2\sigma^2}} e^{-\frac{x^2}{2\sigma^2}} dx \overset{\zeta = \frac{x}{\sqrt{2\sigma^2}}}{=} \frac{2}{\sqrt{\pi}} \int_{0}^{+\infty} \zeta e^{-\zeta^2} \sqrt{2\sigma^2} d\zeta \\
&= \frac{2\sqrt{2\sigma^2}}{\sqrt{\pi}} \left[\lim_{\xi \to +\infty} \left(-\frac{e^{-\zeta^2}}{2} \Big|_0^{\xi} \right) \right] \\
&= \frac{2\sqrt{2\sigma^2}}{\sqrt{\pi}} \left[\lim_{\xi \to +\infty} \left(-\frac{e^{-\xi^2}}{2} + \frac{1}{2} \right) \right] = \sqrt{\frac{2}{\pi}} \sigma
\end{aligned}
\tag{10.62}
$$

10.11   Probability of Outage Under Ambient Selection Diversity

Assume the setup illustrated in Figure 10.13, where the tag is simultaneously illuminated by multiple FM stations. d_i with $i \in \{1, 2, \ldots, L\}$ is the distance between the ith illuminating FM station and the tag. d_0 denotes the distance between the

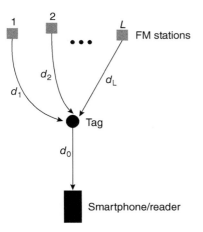

Figure 10.13   Tag is simultaneously illuminated by multiple stations, each offering different impinged RF power.

tag and the smartphone/reader. The power received at the smartphone from the ith station, $P_{\text{rec}}^{(i)}$ is given by:

$$P_{\text{rec}}^{(i)} = P_{\text{tx}}^{(i)} \frac{\tilde{k}_i}{d_i^{v_i}} \tilde{\gamma}_i \eta \frac{\tilde{k}_0}{d_0^{v_0}} \tilde{\gamma}_0 \tag{10.63}$$

where $\tilde{\gamma}_j$ is a Gamma distributed random variable with shape and scale parameters k_j, θ_j respectively. $P_{\text{tx}}^{(i)}$ is the transmission power of the ith FM station and η the scattering efficiency of the tag.

By setting $\gamma_i = P_{\text{tx}}^{(i)} \frac{\tilde{k}_i}{d_i^{v_i}} \tilde{\gamma}_i$ and $k(d_0) = \frac{\tilde{k}_0}{d_0^{v_0}}$, the received power at the smartphone, from a specific FM station i (i.e., the smartphone is tuned at the frequency that the backscattered signal, resulting from illumination by station i, is centered) is given by:

$$P_{\text{rec}}^{(i)} = \gamma_i \eta k(d_0) \tilde{\gamma}_0 \tag{10.64}$$

with $\gamma_i \sim \text{Gamma}(k_i, \theta_i)$ and $\mathbb{E}[\gamma_i] = k_i \theta_i$ the average power impinged at tag from station i. The probability density function (PDF) of the Gamma distribution is given by:

$$f_{\gamma_i}(\gamma_i) = \frac{1}{\Gamma(k_i)\theta^{k_i}} \gamma_i^{k_i-1} e^{-\frac{\gamma_i}{\theta_i}} \tag{10.65}$$

where $\Gamma(x)$ the Gamma function. Setting $\gamma_0 = \eta k(d_0) \tilde{\gamma}_0$, the received power can be defined as $P_{\text{rec}}^{(i)} = \gamma_i \gamma_0$ with $\gamma_i \perp \gamma_0$. The probability of the maximum among the received powers for each station i, to drop below a threshold Θ_{RF} is given by:

$$\Pr\left[\max_{i \in \{1,\dots,L\}} \{P_{\text{rec}}^{(i)}\} < \Theta_{\text{RF}} \right] = \Pr\left[P_{\text{rec}}^{(1)} < \Theta_{\text{RF}} \cap P_{\text{rec}}^{(2)} < \Theta_{\text{RF}} \cap \dots \cap P_{\text{rec}}^{(L)} < \Theta_{\text{RF}} \right]$$

$$= \Pr[\gamma_1 \gamma_0 < \Theta_{\text{RF}} \cap \dots \cap \gamma_L \gamma_0 < \Theta_{\text{RF}}]$$

$$\overset{\gamma_i \perp \gamma_j, \gamma_0}{=} \mathbb{E}_{\gamma_0}\left[\Pr\left(\gamma_1 < \frac{\Theta_{\text{RF}}}{\gamma_0} \cap \dots \cap \gamma_L < \frac{\Theta_{\text{RF}}}{\gamma_0} \mid \gamma_0 \right) \right]$$

$$\overset{\gamma_i \perp \gamma_j}{=} \mathbb{E}_{\gamma_0}\left[\prod_{i=1}^{L} \Pr\left(\gamma_i < \frac{\Theta_{\text{RF}}}{\gamma_0} \mid \gamma_0 \right) \right] \tag{10.66}$$

The cumulative distribution function (CDF) of γ_i is defined as:

$$F_{\gamma_i}(x) = \Pr(\gamma_i < x) = \int_0^x f_{\gamma_i}(\gamma_i) d\gamma_i = \frac{1}{\Gamma(k_i)\theta^{k_i}} \int_0^x \gamma_i^{k_i-1} e^{\frac{\gamma_i}{\theta_i}} d\gamma_i \tag{10.67}$$

Using Eqs. (10.67) and (10.65), Eq. (10.66) becomes:

$$\Pr\left[\max_{i \in \{1,\dots,L\}} \{P_{\text{rec}}^{(i)}\} < \Theta_{\text{RF}} \right] = \mathbb{E}_{\tilde{\gamma}_0}\left[\prod_{i=1}^{L} \Pr\left(\gamma_i < \frac{\Theta_{\text{RF}}}{\gamma_0} \mid \gamma_0 \right) \right]$$

$$= \int_0^{+\infty} f_{\gamma_0}(\gamma_0) \prod_{i=1}^{L} F_{\gamma_i}\left(\frac{\Theta_{\text{RF}}}{\gamma_0} \right) d\gamma_0 \tag{10.68}$$

The lower incomplete gamma function is defined as $\gamma(s,x) = \int_0^x t^{s-1}e^{-t}dt$. Using as second argument $\frac{x}{\theta}$:

$$\gamma\left(s, \frac{x}{\theta}\right) = \int_0^{\frac{x}{\theta}} t^{s-1}e^{-t}dt \overset{*}{=} \int_0^x \left(\frac{t}{\theta}\right)^{s-1} e^{-t/\theta}\frac{1}{\theta}dt = \frac{1}{\theta^s}\int_0^x t^{s-1}e^{\frac{-t}{\theta}}dt$$

(10.69)

where in point *, variable substitution was performed. That way the CDF of Eq. (10.67) becomes $F_{\gamma_i}(x) = \frac{1}{\Gamma(k_i)}\gamma\left(k_i, \frac{x}{\theta_i}\right)$ and Eq. (10.68):

$$\Pr\left[\max_{i\in\{1,\ldots,L\}}\{P_{\text{rec}}^{(i)}\} < \Theta_{\text{RF}}\right] = \frac{1}{\theta^{k_0}}\frac{1}{\prod\limits_{j=0}^{L}\Gamma(k_j)}\int_0^{+\infty}\gamma_0^{k_0-1}e^{\frac{\gamma_0}{\theta_0}}\prod_{i=1}^{L}\gamma\left(k_i, \frac{\Theta_{\text{RF}}}{\gamma_0\theta_i}\right)$$

(10.70)

which concludes the proof.

Acknowledgment

The research work was supported by the Hellenic Foundation for Research and Innovation (HFRI) and the General Secretariat for Research and Technology (GSRT), under the HFRI PhD Fellowship grant (GA. no. 2263).

References

1 Stockman, H. (1948). Communication by means of reflected power. *Proceedings of the IRE,* (October 1948), pp. 1196–1204.

2 Alevizos, P.N., Bletsas, A., and Karystinos, G.N. (2017). Noncoherent short packet detection and decoding for scatter radio sensor networking. *IEEE Transactions on Communications* 65 (5): 2128–2140.

3 Kampianakis, E., Kimionis, J., Tountas, K. et al. (2014). Wireless environmental sensor networking with analog scatter radio and timer principles. *IEEE Sensors Journal* 14 (10): 3365–3376.

4 Konstantopoulos, C., Koutroulis, E., Mitianoudis, N., and Bletsas, A. (2016). Converting a plant to a battery and wireless sensor with scatter radio and ultra-low cost. *IEEE Transactions on Instrumentation and Measurement* 65 (2): 388–398.

5 Vannucci, G., Bletsas, A., and Leigh, D. (2008). A software-defined radio system for backscatter sensor networks. *IEEE Transactions on Wireless Communications* 7 (6): 2170–2179.

6 Alevizos, P.N., Tountas, K., and Bletsas, A. (2017). Multistatic scatter radio sensor networks for extended coverage. *IEEE Transactions on Wireless Communications* 17 (7): 4522–4535. submitted.

7 Fasarakis-Hilliard, N., Alevizos, P.N., and Bletsas, A. (2015). Coherent detection and channel coding for bistatic scatter radio sensor networking. *IEEE Transactions on Communications* 63: 1798–1810.

8 Kimionis, J., Bletsas, A., and Sahalos, J.N. (2014). Increased range bistatic scatter radio. *IEEE Transactions on Communications* 62 (3): 1091–1104.

9 Daskalakis, S.N., Assimonis, S.D., Kampianakis, E., and Bletsas, A. (2016). Soil moisture scatter radio networking with low power. *IEEE Transactions on Microwave Theory and Techniques* 64 (7): 2338–2346.

10 Ensworth, J. F. and Reynolds, M. S. (2015). Every smart phone is a backscatter reader: modulated backscatter compatibility with bluetooth 4.0 low energy (BLE) devices. *Proceedings of IEEE RFID*, San Diego, CA (15–17 April 2015), pp. 78–85.

11 Iyer, V., Talla, V., Kellogg, B. et al. (2016). Inter-technology backscatter: towards internet connectivity for implanted devices. *Proceedings of ACM SIG-COMM*, Florianopolis, Brazil (22–26 August 2016), pp. 356–369.

12 Talla, V., Hessar, M., Kellogg, B. et al. (2017). Lora backscatter: enabling the vision of ubiquitous connectivity. *Proceedings of the ACM on Interactive, Mobile, Wearable and Ubiquitous Technologies* 1 (3): 105:1–105:24.

13 Vougioukas, G., Daskalakis, S. N., and Bletsas, A. (2016). Could battery-less scatter radio tags achieve 270-meter range? *Proceedings of IEEE Wireless Power Transfer Conference (WPTC)*, Aveiro, Portugal (5–6 May 2016), pp. 1–3.

14 Liu, V., Parks, A., Talla, V. et al. (2013). Ambient backscatter: wireless communication out of thin air. *Proceedings of ACM SIGCOMM*, Hong Kong, China (12–16 August 2013), pp. 39–50.

15 Parks, A. N., Liu, A., Gollakota, S., and Smith, J. R. (2014). Turbocharging ambient backscatter communication. *Proceedings of ACM SIGCOMM*, Chicago, Illinois, USA (17–22 August 2014), pp. 619–630.

16 Wang, A., Iyer, V., Talla, V. et al. (2017). FM backscatter: Enabling connected cities and smart fabrics. *USENIX Symposium on Networked Systems Design and Implementation*, Boston, MA, USA (27–29 March 2017).

17 Vougioukas, G. and Bletsas, A. (2017). 24 µW 26 m range batteryless backscatter sensors with FM remodulation and selection diversity. *Proceedings of IEEE RFID Technology and Applications (RFID-TA)*, Warsaw, Polland (20–22 September 2017).

18 Daskalakis, S.N., Kimionis, J., Collado, A. et al. (2017). Ambient backscatterers using FM broadcasting for low cost and low power wireless applications. *IEEE Transactions on Microwave Theory and Techniques* 99: 1–12.

19 Darsena, D., Gelli, G., and Verde, F. (2017). Modeling and performance analysis of wireless networks with ambient backscatter devices. *IEEE Transactions on Communications* 65 (4): 1797–1814.

20 Qian, J., Gao, F., Wang, G. et al. (2017). Semi-coherent detection and performance analysis for ambient backscatter system. *IEEE Transactions on Communications* 65 (12): 5266–5279.

21 Gang Yang, Ying-Chang Liang, Rui Zhang, and Yiyang Pei. *Modulation in the Air: Backscatter Communication Over Ambient OFDM Carrier*, https://arxiv.org/pdf/1704.02245.pdf. 2017.

22 N. Van Huynh, D. T. Hoang, X. Lu, D. Niyato, P. Wang, and D. In Kim (2017) Ambient backscatter communications: a contemporary survey, https://arxiv.org/pdf/1712.04804.pdf.

23 Proakis, J.G. and Salehi, M. (2001). *Communication Systems Engineering*, 2e. Upper Saddle River, NJ: Prentice-Hall.

24 Kimionis, J., Bletsas, A., and Sahalos, J. N. (2012). Design and implementation of RFID systems with software defined radio. *Proceedings of IEEE European Conference on Antennas and Propagation (EuCAP)*, Prague, Czech Republic (26–30 March 2012), pp. 3464–3468.

25 Sakrison, D.J. (1968). *Communication Theory : Transmission of Waveforms and Digital Information*. New York: Wiley.

26 Rasmussen, C.E. and Williams, C.K.I. (2006). *Gaussian Processes for Machine Learning*. The MIT Press.

27 Willsky, A. S., Wornell, G. W., and Shapiro, J. H. (2003). *Stochastic processes detection and estimation*. MIT 6.432 Course Note.

28 Abramowitz, M. and Stegun, I.A. (eds.) (1964). *Handbook of Mathematical Functions with Formulas, Graphs, and Mathematical Tables*. Washington, D.C.: U.S. Government Printing Office.

29 Taub, H. and Schilling, D.L. (1986). *Principles of Communication Systems*, 2e. McGraw-Hill Higher Education.

30 Carson, J.R. (1922). Notes on the theory of modulation. *Proceedings of the Institute of Radio Engineers* 10 (1): 57–64.

31 Daskalakis, S. N., Assimonis, S. D., Kampianakis, E., and Bletsas, A. (2014). Soil moisture wireless sensing with analog scatter radio, low power, ultra-low cost and extended communication ranges. *Proceedings of IEEE Sensors Conference (Sensors)*, Valencia, Spain (3–5 November 2014), pp. 122–125.

32 Kampianakis, E., Kimionis, J., Tountas, K. et al. (2013) Backscatter sensor network for extended ranges and low cost with frequency modulators: application on wireless humidity sensing. *Proceedings of IEEE Sensors Conference (Sensors)*, Baltimore, MD, USA (3–6 November 2013).

33 MathWorkInc. SNR function documentation. Last visited on 13/2/2018, https://www.mathworks.com/help/signal/ref/snr.html.

34 Eric, W. Weisstein. Trigonometric power formulas. From MathWorld – A Wolfram Web Resource. Last visited on 13/2/2018, http://mathworld.wolfram.com/TrigonometricPowerFormulas.html.

11

Design of an ULP-ULV RF-Powered CMOS Front-End for Low-Rate Autonomous Sensors

Hugo García-Vázquez[1], Alexandre Quenon[2], Grigory Popov[2], and Fortunato Carlos Dualibe[2]

[1]*Electronics Department, Instituto de Astrofísica de Canarias (IAC), Canary Islands, Spain*
[2]*Analogue and Mixed-Signal Design Group, Electronics and Microelectronics Unit, University of Mons (UMONS), Mons, Belgium*

11.1   Introduction

Internet of Things (IoTs) [1, 2] and wireless sensor networks (WSN) have had a great impact on many aspects of our society, including industrial control, building automation, monitoring of large geographical areas, home automation, defense systems, air pollution monitoring, fire detection, water-quality monitoring, and natural disaster prevention. The global crisis of the economy has also affected the WSN markets. However, new opportunities are appearing that still make this market very attractive. Each node of the network has a radio transceiver, a microcontroller, sensors, and an energy source, usually a battery. One of the key issues in a WSN (see Figure 11.1) is the autonomy, as many wireless nodes are powered by batteries. For this reason, the key idea in this work is to design an ultra-low power and ultra-low voltage radio-frequency (RF)-powered complementary metal-oxide semiconductor (CMOS) radio transceiver for low-rate autonomous sensors.

This work takes a holistic approach to this problem, facing different aspects of the design of an ultra-low power RF front-end powered by energy harvesting for WSNs [3–7]. As usual in a WSN, the terminal node remains inactive most of the time so that it is in sleep mode, with zero or near-zero power consumption.

Although electrochemical batteries offer a relatively high-energy density at low cost with no moving parts, replacing batteries on a regular basis is impractical due to the maintenance costs, the difficulty of physical access to the sensing node, and the impact on the environment due to battery disposal. Also, sustainable development is required for all policies of the European Union. For this reason, it is

Wireless Power Transmission for Sustainable Electronics: COST WiPE - IC1301,
First Edition. Edited by Nuno Borges Carvalho and Apostolos Georgiadis.
© 2020 John Wiley & Sons, Inc. Published 2020 by John Wiley & Sons, Inc.

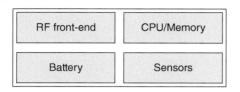

Figure 11.1 Block diagram of a WSN node.

desirable to replace the batteries with energy harvesters that will scavenge energy from the environment and convert it into available electrical power.

The most common sources of power scavenging are light energy, thermal energy, mechanical energy, and RF energy. Power supply can be directly obtained by rectifying the incoming RF signal (or even electromagnetic noise) to feed a supply capacitor [8–10]. This principle has been widely exploited in radio frequency identification (RFID) systems, where power levels of a few milliwatts can be obtained. By using power-harvesting technologies, the power consumption of the electronic systems can be minimized. Modern very large-scale integration (VLSI) technologies have contributed to such power minimization and also to the cost and size reduction of the WSN.

The final developed architecture has been designed in order to reduce the power consumption as much as possible. To do that, several low-voltage, low-power design techniques have been used. Figures 11.2 and 11.3 show the complete architecture of the system and the operation cycle flow diagram. The developed system includes the receiver and the transmitter (RF blocks), the harvester and power management circuits, as well as the control circuits including the control unit. Note that the sensor and the low-dropout (LDO) regulator are external blocks in this work.

As shown in Figure 11.2, when the 2.4 GHz on-off keying (OOK) signal arrives to the antenna, it is rectified and amplified by a voltage multiplier (charge pump) in order to obtain a specific DC level at the external capacitor. This voltage level must be high enough for supplying the complete system (0.5 V). It is detected by means of a voltage reference (VREF) and a self-biased hysteresis comparator (COMP1) which wakes the system up. Once the external capacitor voltage reaches the desired DC level, the control unit and the comparator (COMP2, used as an envelope detector) are powered. From the envelope detector, the stream containing the identification code (see Figure 11.3) is read by the control unit. If the identification matches the code assigned to the network node, the power management unit turns the sensor on. Then, when data from the sensor is ready the control unit reads its 8-bit output data, and after being serialized, it is sent through the transmitter. This consists of an OOK-modulated LC-VCO tuned at 2.4 GHz and a high-efficiency Class-E power amplifier, which includes a built-in driver.

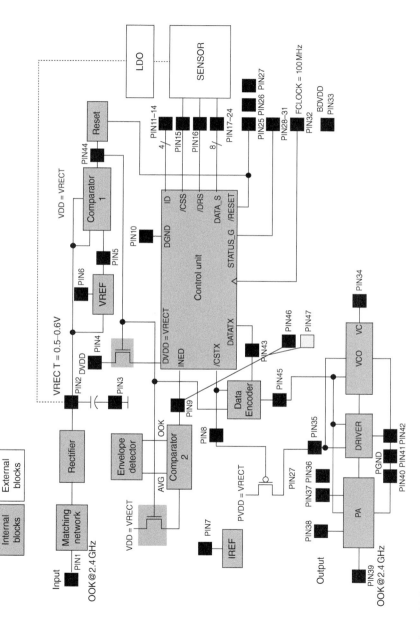

Figure 11.2 Block diagram of the developed system. The driver's enable transistor is only symbolic. It is actually enabled/disabled through internally distributed small-size zero-power transistors.

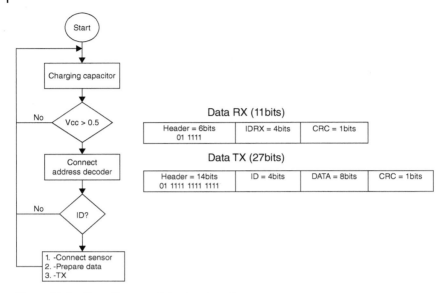

Figure 11.3 Flow diagram of the developed system.

11.2 Characterization of the Technology

One of the most important aspects for designing an integrated circuit (IC) is to have the technology characterized. This section presents the different steps followed in order to obtain characterization of the technology.

11.2.1 g_m/I_D Curves

The g_m/I_D method is a technique used for sizing the transistors of the circuits [11–15]. It is quite simple and different from other techniques that need many equations for properly sizing each transistor of the circuit. A CMOS 65-nm process was selected for designing the different blocks of this work. In order to use this methodology, the first thing required is to obtain the g_m/I_D curves for the technology used. These curves are obtained by using the equation:

$$\frac{g_m}{I_D} = \frac{1}{I_D}\frac{\delta I_D}{\delta V_{GS}} = \frac{\delta\left(\ln I_D\right)}{\delta V_{GS}} \tag{11.1}$$

Figures 11.4 and 11.5 show the g_m/I_D curves for the negative-channel metal-oxide semiconductor (NMOS) and positive-channel metal-oxide semiconductor (PMOS) transistors, the ones that are used in this work. The useful side of the g_m/I_D method is highlighted in these figures, where the g_m/I_D values are plotted on the y-axis and either the ID/(W/L) ratio or V_{GS} is plotted on the x-axis

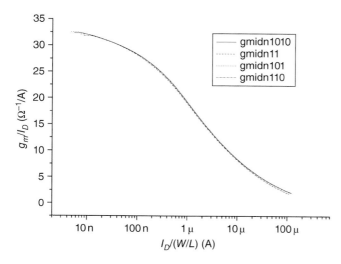

Figure 11.4 Curves (NMOS) g_m/I_D versus ID/(W/L) for $(W/L) = 10\,\mu m/10\,\mu m$, $(W/L) = 1\,\mu m/1\,\mu m$, $(W/L) = 10\,\mu m/1\,\mu m$, and $(W/L) = 1\,\mu m/10\,\mu m$.

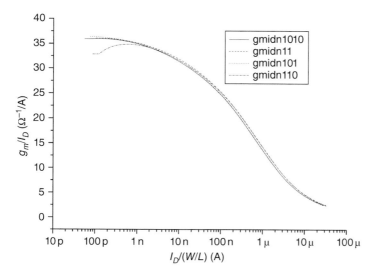

Figure 11.5 Curves (PMOS) g_m/I_D versus ID/(W/L) for $(W/L) = 10\,\mu m/10\,\mu m$, $(W/L) = 1\,\mu m/1\,\mu m$, $(W/L) = 10\,\mu m/1\,\mu m$, and $(W/L) = 1\,\mu m/10\,\mu m$.

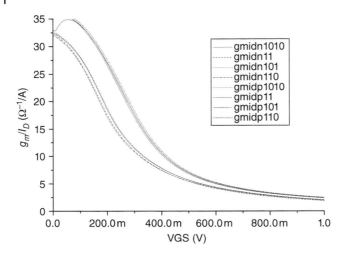

Figure 11.6 g_m/I_D versus V_{GS} for $(W/L) = 10\,\mu m/10\,\mu m$, $(W/L) = 1\,\mu m/1\,\mu m$, $(W/L) = 10\,\mu m/1\,\mu m$, and $(W/L) = 1\,\mu m/10\,\mu m$.

(see Figure 11.6). By using these curves, it is easy to design a circuit by choosing the working region, the size, and the current for each transistor. The g_m/I_D method is advantageous because, instead of using many equations, it only uses the characteristics illustrated earlier. A metal-oxide semiconductor field-effect transistor (MOSFET), apart from working in the cut-off, triode, or saturation region, depending on the value of the V_{GS}, it can also be operating in three other main regions:

- *Weak inversion*: when $V_{GS} < V_{TH}$, there are very few carriers in the channel beneath the gate. We enter this region for values of $g_m/I_D > 25$.
- *Strong inversion*: when $V_{GS} \gg V_{TH}$, there is a well-formed conduction channel. We enter this region for $g_m/I_D < 10$.
- *Moderate inversion*: it is a mid-region between the other two.

The g_m/I_D methodology consists of first deciding in which region to use the transistors implemented within the circuit; by using the value of g_m/I_D in the characteristic curve obtained for each transistor earlier, it can then obtain either the ID/(W/L) or the V_{GS} corresponding value. Therefore, there are three variables in total that can be chosen: transconductance, drain current, and size. The value of g_m/I_D that represents the operative region of the transistor is also chosen from other factors such as gain, power dissipation, frequency, stability criteria. In this work, the g_m/I_D method is employed since it is a very immediate, simple, and efficient way of designing.

11.2.2 C_{OX} and μC_{OX}

The technological parameters C_{OX} and μC_{OX} have been extracted from simulations. These parameters are necessary for the g_m/I_D methodology. Figure 11.7 shows the circuit that was used for obtaining the capacitance C_{OX}.

In the simulation, the metal-oxide semiconductor (MOS) transistor is used as a capacitor. By doing an AC simulation and drawing on Eq. (11.2) that expresses the total capacitance seen from the gate, a certain pole, and its relative frequency are found $(1/2\pi RC)$. From this value, knowing the resistor R and the dimensions W and L, C_{OX} can be obtained. The result is presented in Table 11.1. The values of the components used in Figure 11.7 are the following:

- $R = 10\,\text{k}\Omega$
- PMOS: $W = L = 50\,\mu\text{m}$, $V_{\text{IN}} = 1\,\text{V (AC)}$, $V_X = 0.5\,\text{V(DC)}$
- NMOS: $W = 250\,\mu\text{m}$, $L = 10\,\mu\text{m}$ $V_{\text{IN}} = 0.5\,\text{V(DC)} + 1\,\text{V(AC)}$

$$C = C_{\text{OX}}WL \tag{11.2}$$

Figures 11.8 and 11.9 show the values of the μC_{OX} for the NMOS and PMOS. They are directly obtained from the simulation of the betaeff for different W/L ratios.

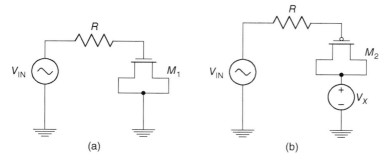

(a) (b)

Figure 11.7 Simulation setup for obtaining C_{OX}. (a) NMOS and (b) PMOS transistors.

Table 11.1 Technological parameters.

	PMOS	NMOS
C_{OX}	12.22 fF/μm²	
V_{TH} (mV)	100	163

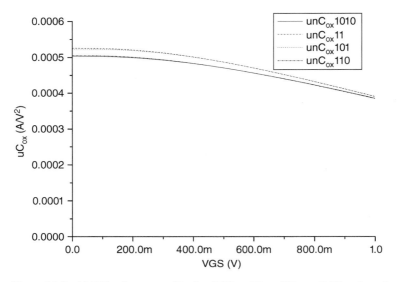

Figure 11.8 NMOS: μC_{ox} versus V_{GS} for $(W/L) = 10\,\mu m/10\,\mu m$, $(W/L) = 1\,\mu m/1\,\mu m$, $(W/L) = 10\,\mu m/1\,\mu m$, and $(W/L) = 1\,\mu m/10\,\mu m$.

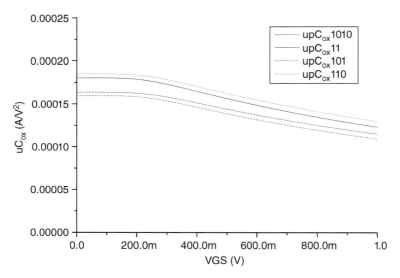

Figure 11.9 PMOS: μC_{ox} versus V_{GS} for $(W/L) = 10\,\mu m/10\,\mu m$, $(W/L) = 1\,\mu m/1\,\mu m$, $(W/L) = 10\,\mu m/1\,\mu m$, and $(W/L) = 1\,\mu m/10\,\mu m$.

11.2.3 Early Voltage (V_A)

The early voltage V_A is another important parameter that must be known. V_A is strongly dependent on the length of the transistor. Therefore, it is necessary to characterize both NMOS and PMOS transistors regarding the V_A dependence on the L of the channel.

Figure 11.10 shows the setups for NMOS ($V_{DS} = 0.5\,V$, VGS = 240 mV, W and L minimum size) and PMOS ($V_{SD} = 0.5\,V$, $V_{SG} = 340\,mV$, W and L minimum size). From this simulation, the typical characteristic ID versus V_{DS} curves is obtained. From these curves, the currents when the transistors enter in saturation are extracted. Then the final step for obtaining the V_A is to apply the following equation:

$$V_A = \frac{I_D}{\frac{\delta}{\delta V}\left(\frac{I_D}{V_{DS}}\right)} \tag{11.3}$$

From this simulation, a current IDSAT of 2.8 μA is obtained for the PMOS transistor with the minimum length $L = 140\,nm$. Applying the Eq. (11.3), an early voltage (V_A) of 1.96 V is obtained (based on the simulation). The same calculation is done with a factor of 10 for (W/L) respect to the minimum size. The drain current is found to be 844.16 nA, and the $V_A = 57.2\,V$. In order to have a better understanding of the evolution of the V_A, the same calculation but considering a factor of 5 the minimum size was done. Table 11.2 summarizes these values.

The procedure was repeated for the NMOS transistor. Figure 11.11 shows the evolution of the early voltage regarding the length for NMOS and PMOS transistors, where it can be shown that the relationship of the V_A with the L (in μm) is linear and it is given by the following two equations:

$$V_A\,NMOS = 29.44 \cdot 106L - 0.92 \tag{11.4}$$

$$V_A\,PMOS = 43.84 \cdot 106L - 4.17 \tag{11.5}$$

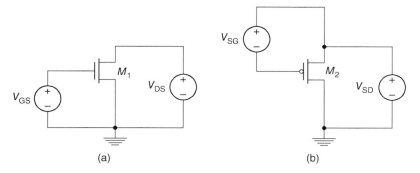

Figure 11.10 Simulation setup for obtaining V_A. (a) NMOS and (b) PMOS transistors.

Table 11.2 V_A for the PMOS transistor.

	$(W/L)_{MIN}$	$5(W/L)$	$10(W/L)$
I_{DSAT}	2.8 µA	1 µA	844.16 nA
V_A	1.96 V	28 V	57.2 V

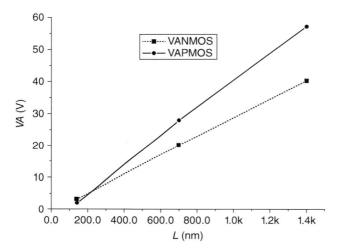

Figure 11.11 V_A versus length for NMOS and PMOS transistors.

11.3 Ultra-Low Power and Ultra-Low Voltage RF-Powered Transceiver for Autonomous Sensors

This section presents the ultra-low power and ultra-low voltage RF-powered transceiver for autonomous sensors. This is divided into three subsections: power management (PM) and receiver (RX), control unit (CU), and transmitter (TX).

11.3.1 Power Management (PM) and Receiver (RX)

The power management unit and the receiver are responsible for the internal voltage regulation and the optimization of the total power consumption of the chip and for obtaining the data from the OOK modulation. As shown in Figure 11.12, this unit comprises the following subcircuits: a rectifier with input matching network, an external capacitor, a V_{REF} circuit, a comparator to activate the switch on/off control circuits, and a comparator to extract the information from the OOK modulation. These circuits are explained in the following subsections.

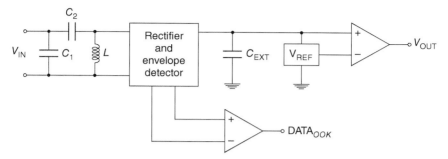

Figure 11.12 Simplified block diagram of the PM and the RX.

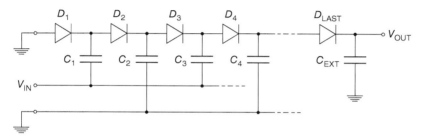

Figure 11.13 Dickson charge pump cell for RF signals.

11.3.1.1 Rectifier

The rectifier is based on the Dickson charge pump [16] connected as shown in Figure 11.13. It is designed to obtain a specific DC voltage from the 2.4 GHz AC signal sensed by the antenna. To ensure maximum power transfer, it is necessary to match the rectifier's input impedance to that of the antenna (typically 50 Ω). For this reason, a passive LC matching network was integrated in the input of the rectifier. Also, by using this input matching network topology, the level of the signal is increased. Most common energy-harvesting implementations (i.e. RFID tags) use the Dickson rectifier topology with the input matching network.

Figures 11.14 and 11.15 show the used topologies for the diodes [17] that were employed for implementing the rectifier. The chosen CMOS technology offers low-threshold voltage transistors that are obviously advantageous for ultra-low voltage designs. Nevertheless, the structure of the rectifier must be adapted for reducing the transistors' high-leakage currents. By using this topology, the rectifier improves its sensitivity, which is −26 dBm in order to obtain a V_{OUT} of 0.6 V.

Figure 11.16 shows the charge and the discharge of the external capacitor C_{EXT}. It was sized in agreement with the total consumption for ensuring an unregulated voltage supply for the chip greater than the minimum operational voltage (0.5 V) during the necessary time that the circuit remains awake.

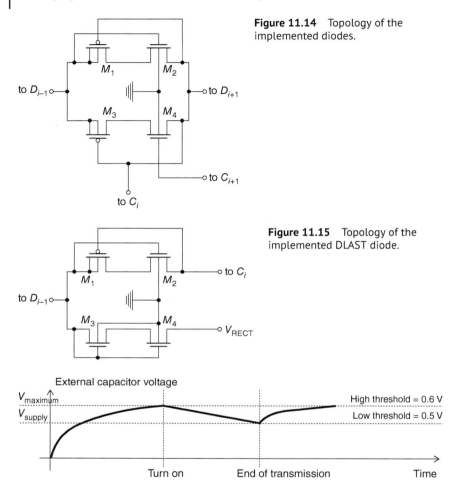

Figure 11.14 Topology of the implemented diodes.

Figure 11.15 Topology of the implemented DLAST diode.

Figure 11.16 Charge and discharge of the external capacitor C_{EXT}.

First, it is charged up to a voltage called the high threshold (0.6 V). Once the latter is reached, the control unit turns on and works while consuming the energy stored in the capacitor (comparing the IDs, reading data from the sensor and transmitting). The control unit goes to sleep either at the end of data transmission or when the capacitor voltage reaches the low threshold, which is the minimal operational voltage supply (0.5 V). During the control unit idle mode, the external capacitor is charged again toward the high threshold voltage.

Assuming isotropic antennas (GRX = GTX = 0 dB) and a 100 mW at 2.4 GHz emitting source, the Friis transmission equation [18, 19] gives rise to a maximum distance of 2 m to the receiver.

11.3.1.2 Voltage Reference (VREF) Circuit

The V_{REF} circuit was designed with a diode formed by the $p-n$ junction of a PMOS transistor (MOS compatible bipolar) and a resistor as shown in Figure 11.17. The voltage V_{REF} will increase at the same rate as the input voltage V_{RECT} from the external capacitor C_{EXT} until it reaches the necessary level for the diode (around 0.5 V). Then, this voltage level is maintained at V_{REF}, as long as the input voltage V_{RECT} continues to be higher than 0.5 V.

11.3.1.3 Comparator for Power Management (COMP1)

Figure 11.18 shows the schematic of the used comparator with hysteresis and output stage [20]. This comparator allows two different thresholds depending on if the

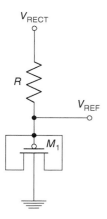

Figure 11.17 Schematic of the V_{REF} circuit.

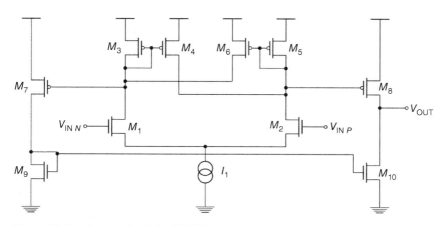

Figure 11.18 Schematic of the COMP1.

voltage signal in the input is increasing or decreasing. It is employed as a threshold detector, which turns on or puts the system on stand-by according to the capacitor voltage level, taking into account the voltage of V_{REF}. In order to avoid wrong behavior during the period of time that the DC level of V_{REF} is smaller than 0.5 V, it is necessary to add a certain offset in the negative input of the comparator. In this way, the output of the comparator will be activated only when the voltage of the capacitor C_{EXT} is higher than the low threshold.

11.3.1.4 Current Reference Circuit (IREF)

Figure 11.19 shows a simplified schematic of the current reference circuit that was designed for this chip. The current reference circuit (IREF) is a block that provides all the current references to the different blocks (i.e. comparators, VCO, driver, etc.). Also, this block can turn on/off some of them depending on the operation that the front-end is carrying out in this precise moment (comparing the ID, reading the sensor data, transmitting, etc.). This strategy allows power saving, which is critical in this application. In order to avoid the deviation of the resistance value of the integrated resistors due to their tolerance, an external resistor with a small tolerance connected to the voltage supply will be used to fix the current I_{EXT}.

11.3.1.5 Comparator for the Demodulation (COMP2)

This comparator (see Figure 11.20) is used to obtain the data from the OOK signal. It is a rail-to-rail comparator with an output stage for increasing the gain. The circuit was designed using the g_m/I_D methodology and trying to reduce the power consumption as much as possible.

11.3.2 Control Unit (CU)

The control unit is the digital circuit responsible for decoding the information from the receiver, verifying the identification, communicating with the sensors, and obtaining the measured value, controlling the different blocks (power on/off) and encoding the information to the transmitter. The description of the circuit was

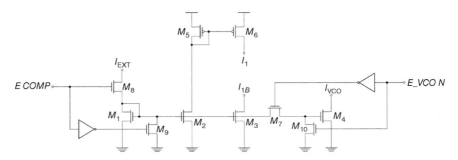

Figure 11.19 Schematic of the I_{REF} circuit.

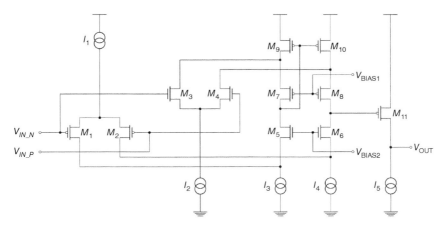

Figure 11.20 Schematic of the COMP2.

made with VHDL (VHSIC-HDL, Very High Speed Integrated Circuit Hardware Description Language), then passed to Verilog using Design Compiler and the technology. In order to obtain the layout of the complete circuit, the software INNOVUS was used. The clock frequency of this finite state machine (FSM) is 100 MHz.

The control unit receives a frame (RX: 11 bits) composed of the header (6 bits), the identification code (4 bits) and the cyclic redundancy check (CRC) (1 bit). This frame can be identified (/INED signal) in Figure 11.21. In this simulation, the frame corresponds to "01111101010." If the received ID matches the ID of the node, the FSM activates the sensor (/CSS signal), then when the data from the sensor is ready (/DRS signal), the FSM passes the sensor's data to the transmitter (/DATA PLL signal). For the transmission, the control unit encodes a transmission frame (TX: 27 bits) composed of the header (14 bits), the identification code (4 bits), data from the sensor (8 bits), and the CRC (1 bit). It is sent while there is enough energy in the external capacitor (C_{EXT}). The total time of the received and the transmitted frames are 220 and 540 ns, respectively, due to the time for 1 bit being 20 ns. Figure 11.22 shows the layout of the control unit with the distribution of the input and output ports. The dimensions of this digital block are 54.8 μm × 51.2 μm. The total RMS current root mean square current (IRMS) of the control unit is 50.81 μA for a voltage supply of 0.5 V.

11.3.3 Transmitter (TX)

11.3.3.1 Voltage-controlled oscillator (VCO)
The VCO depicted in Figure 11.23 is the first building block of the transmitter after the control unit. Its oscillation frequency is controlled by the voltage

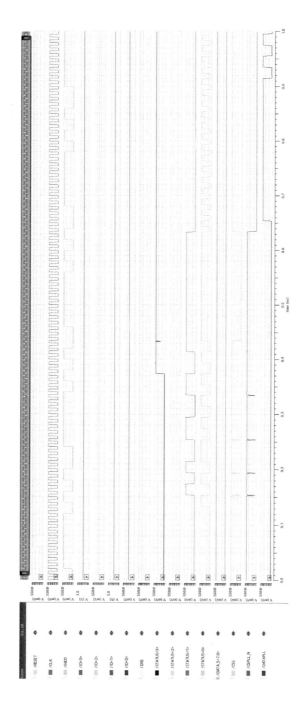

Figure 11.21 Control unit behavior.

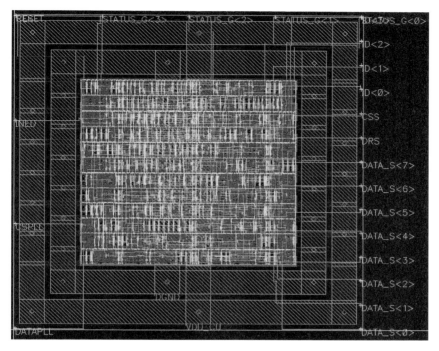

Figure 11.22 Layout of the control unit.

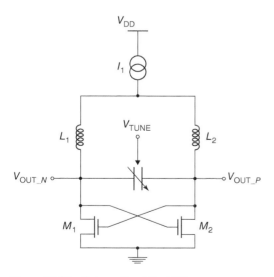

Figure 11.23 Schematic of the VCO.

V_{TUNE}, which selects the desired 2.4 GHz signal. This control voltage biases the varactor that makes up part of the VCO's LC tank. The dimensions of the VCO are 500 μm × 200 μm. The circuit occupies a substantial area due to the size of the inductors. The dimensions of each inductor ($L1 = L2 = 5$ nH) are 200 μm × 200 μm. The RMS current I1 of the VCO is 410 μA for a voltage supply of 0.5 V. This VCO provides an output peak-to-peak voltage of 230 mV centered at 130 mV with a frequency of 2.4 GHz.

11.3.3.2 Power amplifier (PA) with built-in driver

The driver is used to control the PA by charging and discharging the PA as a function of the output of the VCO. In this work, a chain of inverters incrementing the (W/L) ratio with respect to the previous one was used in order to charge the PA (see Figure 11.24). Also, power off/on control circuits were included. Due to the output signal of the VCO being differential and the level of the signal being only a few millivolts, a differential to single-ended circuit was added in the input of the driver. Figure 11.25 shows different signals of the driver and the PA. With a voltage supply of 0.7 V, the driver can reach a current I_{RMS} of 90 mA.

The power amplifier is the last building block of the front-end of the transmitter to be connected to the antenna. Taking into account the modulation, the frequency, and the necessary power, the class-E [18, 19] topology was chosen for the power amplifier in order to increase the efficiency. Figure 11.26 shows the schematic of the chosen power amplifier. The PA was designed for an output impedance of 50 Ω. The size of the main transistor (M1) is 8000 μm × 0.14 μm. The inductors $L1$ and $L2$ must be external components due to the high currents that they have to endure. The RMS current I_{RMS} of the PA is 210 mA for a voltage

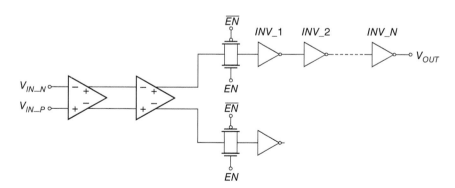

Figure 11.24 Simplified schematic of the driver.

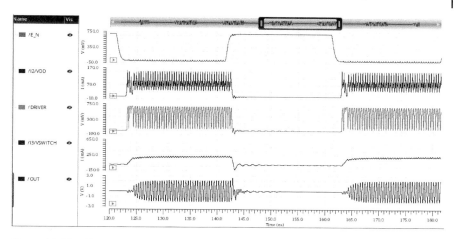

Figure 11.25   Simulation of the PA.

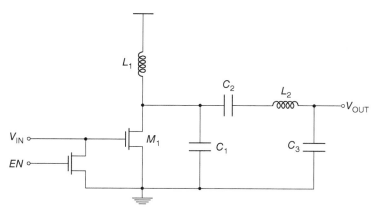

Figure 11.26   Schematic of the PA.

supply of 0.7 V. This PA provides an output peak-to-peak voltage of 4 V with a frequency of 2.4 GHz (/OUT signal in Figure 11.25).

11.4   Experimental Results

Figure 11.27 Layout of the chip. Figure 11.27 shows the layout of the complete chip. The dimensions of the 52-pad chip are 976 µm × 1018 µm, including pads.

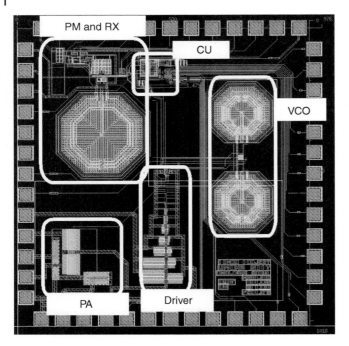

Figure 11.27 Layout of the chip.

These were full-custom designed aiming to minimize parasitic capacitances. The prototype was provided with several test pads in order to check relevant analogue and digital voltage nodes.

The analogue blocks including the rectifier, the input matching network, the voltage and current references, and the comparators are situated on the top-left of the layout. The digital blocks including the control unit are placed in the top-center. The radio frequency blocks are placed in two different places in function of the power consumption. The VCO is located on the center-right of the layout and the driver and the PA on the bottom-left.

Figure 11.28 shows a simulation with the behavior of the chip. The external capacitor (C_{EXT}) is charged ($/V_{RECT}$ signal) from time = 0 ms to time = 1.2 ms in order to achieve the high threshold, then the transmission starts ($/V_{OUT}$ signal) for a very small period of time compared with the time required to charge the capacitor. The transmission continues while the voltage decreases until it reaches the low threshold stopping the transmission and restarting the charge of the capacitor.

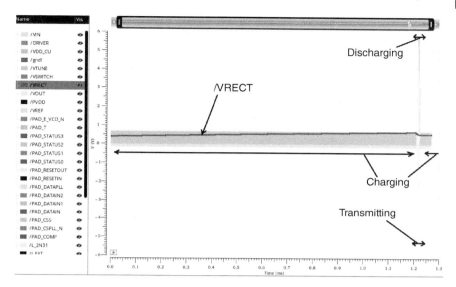

Figure 11.28 Charge and transmission of the chip.

11.5 Conclusion

In this work, the design and simulation results of an ultra-low power and ultra-low voltage RF-Powered CMOS front-end for autonomous sensors were presented. This circuit is able to harvest electromagnetic energy, validate the input stream identification code, and transmit the information of a sensor to a hub or master node in a WSN. In order to reduce the power consumption as much as possible, several techniques have been applied. The use of low-threshold voltage transistors available in the chosen CMOS 65-nm technology allows us to achieve a system capable of operating with an ultra-low voltage supply of only 0.5 V. This has significantly reduced the power consumption and increased the rectifier sensitivity. The simulation results of this system are promising and all performance figures will be checked with a prototype which is being manufactured at the time of writing.

Acknowledgments

This work was supported by the Walloon (Belgium) Region DGO6 BEWARE Fellowships program (1410164-POHAR, co-funded by the European Commission through the COFUND programme of the Marie Curie Actions) and the Fond National pour la Recherche Scientifique (F.R.S.-FNRS) of Belgium.

Thanks to William Van Hoeck, Daniel Binon and Nathalie Durieux for their valuable technical and administrative support. Thanks to Sarah Grief for her support with relation to the technical English review of this work. Thanks to Thierry Dutoit, the UMONS/NUMEDIART Institute, Thierry Delmot and nSILITION SPRL for their support.

This work is a tribute to our colleague Grigory Popov who suddenly passed away on 2 December 2016.

References

1 Mumtaz, S., Alsohaily, A., Pang, Z. et al. (2017). Massive internet of things for industrial applications: addressing wireless IoT connectivity challenges and ecosystem fragmentation. *IEEE Industrial Electronics Magazine* 11 (1): 28–33.

2 Wollschlaeger, M., Sauter, T., and Jasperneite, J. (2017). The future of industrial communication: automation networks in the era of the internet of things and industry 4.0. *IEEE Industrial Electronics Magazine* 11 (1): 17–27.

3 Vullers, R.J.M., Schaijk, R.v., Visser, H.J. et al. (2010). Energy harvesting for autonomous wireless sensor networks. *IEEE Solid-State Circuits Magazine* 2 (2): 29–38. https://doi.org/10.1109/MSSC.2010.936667.

4 Wang, A., Kwong, J., and Chandrakasan, A. (2012). Out of thin air: energy scavenging and the path to ultralow-voltage operation. *IEEE Solid-State Circuits Magazine* 4 (2): 38–42. https://doi.org/10.1109/MSSC.2012.2193073.

5 Papotto, G., Carrara, F., Finocchiaro, A., and Palmisano, G. (2014). A 90-nm CMOS 5 Mbps crystal-less RF-powered transceiver for wireless sensor network nodes. *IEEE Journal of Solid-State Circuits* 49 (2): 335–346. https://doi.org/10 .1109/JSSC.2013.2285371.

6 Reinisch, H. et al. (2011). An electro-magnetic energy harvesting system with 190 nW idle mode power consumption for a BAW based wireless sensor node. *IEEE Journal of Solid-State Circuits* 46 (7): 1728–1741. https://doi.org/10.1109/ JSSC.2011.2144390.

7 Masuch, J., Delgado-Restituto, M., Milosevic, D., and Baltus, P. (2013). Co-integration of an RF energy harvester into a 2.4 GHz transceiver. *IEEE Journal of Solid-State Circuits* 48 (7): 1565–1574. https://doi.org/10.1109/JSSC .2013.2253394.

8 Le, T., Mayaram, K., and Fiez, T. (2008). Efficient far-field radio frequency energy harvesting for passively powered sensor networks. *IEEE Journal of Solid-State Circuits* 43 (5): 1287–1302. https://doi.org/10.1109/JSSC.2008.920318.

9 Popov, G., Dualibe, F.C., Moeyaert, V. et al. (2016, 2016). A 65-nm CMOS battery-less temperature sensor node for RF-powered wireless sensor networks. *IEEE Wireless Power Transfer Conference (WPTC)*: 1–4. Aveiro. doi: https://doi .org/10.1109/WPT.2016.7498806.

10 Cost Action IC1301 Team (2017). Europe and the future for WPT: European contributions to wireless power transfer technology. *IEEE Microwave Magazine* 18 (4): 56–87. https://doi.org/10.1109/MMM.2017.2680078.

11 Silveira, F., Flandre, D., and Jespers, P.G.A. (1996). A gm/ID based methodology for the design of CMOS analog circuits and its application to the synthesis of a silicon-on-insulator micropower OTA. *IEEE Journal of Solid-State Circuits* 31 (9): 1314–1319. https://doi.org/10.1109/4.535416.

12 Jespers, P. (2010). *The gm/ID Methodology, a Sizing Tool for Low-Voltage Analog CMOS Circuits: The Semi-Empirical and Compact Model Approaches*, 1e. Springer.

13 Castagnola, J.L., Dualibe, F.C., and Garcia-Vazquez, H. (2016). Using scattering parameters and the gm/ID MOST ratio for characterisation and design of RF circuits. *IEEE 59th International Midwest Symposium on Circuits and Systems (MWSCAS), Abu Dhabi*: 1–4. https://doi.org/10.1109/MWSCAS.2016.7870068.

14 Fiorelli, R., Silveira, F., and Peralias, E. (2014). MOST moderate-weak-inversion region as the optimum design zone for CMOS 2.4-GHz CS-LNAs. *IEEE Transactions on Microwave Theory and Techniques* 62 (3): 556–566. https://doi.org/10 .1109/TMTT.2014.2303476.

15 Garcia-Vazquez, H., Dualibe, F.C., and Popov, G. (2017). A 0.5 V fully differential transimpedance amplifier in 65-nm CMOS technology. *IEEE 60th International Midwest Symposium on Circuits and Systems (MWSCAS), Boston, MA*: 763–766. https://doi.org/10.1109/MWSCAS.2017.8053035.

16 Mingliang, L. and Heights, R. (2006). *Demystifying Switched-Capacitor Circuits*. Elsevier. ISBN: 978-0-7506-7907-7.

17 Flandre, D., Bulteel, O., Gosset, G. et al. (2011). Disruptive ultra-lowleakage design techniques for ultra-low-power mixed-signal microsystems. *Faible Tension Faible Consommation (FTFC), Marrakech*: 1–4. https://doi.org/10.1109/ FTFC.2011.5948908.

18 Baker, R.J. (2007). CMOS circuit design, layout, and simulation. In: *The Design of CMOS Radio-Frequency Integrated Circuits*, Revised 2e. Wiley.

19 Lee, T.H. *The Design of CMOS Radio-Frequency Integrated Circuits*, 2e. Cambridge: Cambridge University Press.

20 Allen, P.E. and Holberg, D.R. (1987). *CMOS Analog Circuit Design*. Oxford: Oxford University Press.

12

Rectenna Optimization Guidelines for Ambient Electromagnetic Energy Harvesting

Erika Vandelle[1], Simon Hemour[2], Tan-Phu Vuong[1], Gustavo Ardila[1], and Ke Wu[3]

[1] *IMEP-LAHC, Institut Polytechnique de Grenoble (Grenoble INP), Université Grenoble Alpes, Grenoble, France*
[2] *IMS, University of Bordeaux, Bordeaux, France*
[3] *Poly-Grames, Ecole Polytechnique de Montréal, Montréal, Canada*

12.1   Introduction

Energy harvesting is a promising solution to develop self-powered electronics. Many different applications such as the Internet-of-Things, smart environments, military or agricultural monitoring are based on the deployment of sensor networks that require a multitude of small and scattered electronic devices. The low-power operation of those distributed devices enables the collection of wireless energy from their environment as a mean of supply in order to obtain sustainable, autonomous, and maintenance-free systems.

Among different kinds of energy, the electromagnetic (EM) energy can be found in the environment within a large spectrum from the sun light to the radio waves with sufficient amount of energy so that it can be harvested and converted into useful electrical energy. The robustness of radio frequency (RF) energy and its omnipresence mostly due to the human activities, embracing TV/radio broadcasting and wireless communications and sensing makes it a good candidate for supplying electrical power to low-power and low-duty cycle electronic devices. RF power is usually intentionally transferred, during the RF communications, by the base station to the receiver, in which an energy harvester is embedded. In this case, the source of power is well-known, and directive antennas can be used to ensure the reception of a high amount of power. However, when power-constrained sensor networks with slow activity rates, are at stake, the less costly and simpler solution is to recycle the unused RF power that is present in the environment over a certain time.

Wireless Power Transmission for Sustainable Electronics: COST WiPE - IC1301,
First Edition. Edited by Nuno Borges Carvalho and Apostolos Georgiadis.
© 2020 John Wiley & Sons, Inc. Published 2020 by John Wiley & Sons, Inc.

Major challenges remain to be overcome in order to design optimal ambient energy harvesters. The collection and the conversion of RF power into DC power, compatible with the battery operation, is performed by a so-called rectenna that is an antenna followed by an RF-to-DC rectifier. For some decades, rectennas have been designed and optimized mostly for point-to-point transmission involving sent-on-purpose high power microwave radiations [1–3], and consequently, high efficiencies were successfully reached. However, ambient RF energy requires the consideration of very low-density energy-harvest technologies [4–6] since the ambient power densities are limited especially for humans' safety reasons. Consequently, the rectifiers tend to operate inefficiently, and this imposes essential optimizations at the rectifier, the matching network and the antenna levels. Furthermore, the power density, the frequency spectrum, the polarization, and the direction of arrival of the available EM energy are not constant over time, or over space. This is due to multiple reasons such as multipath propagation, varying demand of the users, arbitrary position of the harvesters and/or of the energy source, humans' movements, etc. Those characteristics of the incident signals impose constraints on the design of rectenna for ambient EM energy harvesting. Optimization processes must be performed accordingly leading to different specifications of the antenna than those in communication systems.

This chapter reviews the optimization process to address the low power operation of ambient energy harvesting rectennas. Due to the essential role of the antenna to deliver power to the rectifier part, the probability of collecting energy in the ambient environment is discussed. Accordingly, a new metric is derived to assess the performance of a rectifying antenna for ambient microwave energy harvesting. This new metric is evaluated for different rectennas from the state-of-the-art.

12.2 Rectennas Under Low Input Powers

Radio wave communication has transformed RF energy into an undeniably ubiquitous energy in our environment and especially in urban areas. Although the omnipresence and the momentary superposition's of some of the RF radiations of different frequencies tend to enhance the available power densities, the amount of energy will always be limited for safety reasons. Table 12.1 reports some ambient RF power densities measurements [7–11], mostly performed in urban areas. These results must be examined carefully since the quantities of energy vary a lot with time and location as the number of users and their activities differ from one place/period-of-time to another. Moreover, the rules about the electromagnetic expositions strongly vary from one country to another as well as the frequency bands allocated to the different applications. Nevertheless, in general, between

Table 12.1 Ambient power densities measurements.

	Frequency band				
Power density (nW/cm²)	DTV 400–800 MHz	GSM900 900 MHz	GSM1800 1800 MHz	UMTS 3G 2100 MHz	Wi-Fi 2400 MHz
Average [7] London, UK	0.89	36	84	12	0.18
Average [8] Covilha, Portugal	0.34	2.79	1.2	1.25	—
Median [9], France	68	44	44	51	—
Average [10] Austria, Germany and Hungary, 25–100 m from base stations	—	10–100	—	—	—

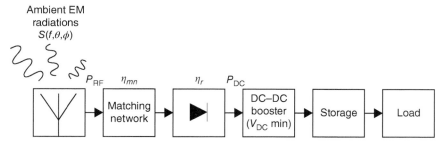

Figure 12.1 Block diagram of a rectenna for low-power energy harvesting.

10 and 100 nW/cm² can be expected at some frequencies of RF communications, most often the GSM900 (downlink in Europe: 935–960 MHz), the GSM1800 (downlink in Europe: 805.2–1879.8 MHz), and the Wi-Fi (2.4–2.5 GHz) (in indoor environments particularly) bands.

Figure 12.1 shows the composition of a rectenna designed for low-power energy harvesting. The ambient EM energy, after its capture by the antenna and its passage through the matching network, is delivered to the rectifier to be converted into electrical DC power. Since the DC voltage and power output is too low to directly power a load, a DC–DC converter is used to boost the DC voltage, and the output power is stored until it is sufficient to be supplied to the load. Hence, the minimum power that can be harvested $S_{harv_{min}}$ can be deduced from the antenna gain G, the signal wavelength λ_o in free space and the rectifier sensitivity $P_{RF_{sensitivity}}$, which is the minimum input RF power that results in the minimum DC voltage V_{DC} required by the DC–DC booster, with the following expression:

$$S_{harv_{min}} = P_{RF_{sensitivity}} \cdot \frac{4\pi}{\lambda_o^2 \cdot G} \tag{12.1}$$

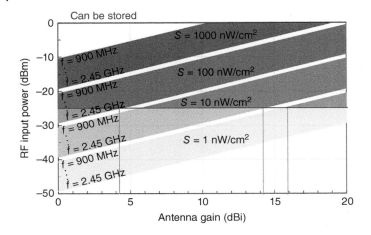

Figure 12.2 Harvestable power density (nW/cm^2) given as a function of the RF power (dBm) delivered to the rectifier and the antenna gain (dBi).

Figure 12.2 illustrates the power density bands that can be harvested as a function of the RF power that is delivered to the rectifier and of the antenna gain. The higher line in each band corresponds to the frequency band of 900 MHz, while the lower band represents the frequency of 2.45 GHz. As an example, the power density that can be stored is colored for a rectifier sensitivity of −25 dBm. We can see that the less sensitive the rectifier is, the lower the antenna gain must be to allow the storage of ambient RF power densities (see the darker region in Figure 12.2). Thus, the rectifier should be optimized to convert efficiently low ambient RF power levels into DC power, and the gain of the antenna must be high enough to allow for the storage of this power.

In this section, we discuss the optimization steps of the different components of the rectenna from the rectifier to the antenna under low power operation. The rectifier topologies and the minimum rectifier point operation of the state-of-the-art are given. The optimization of the matching network is considered by examining the Bode–Fano criterion and the matching efficiency. Finally, due to the limitation of the rectification efficiency, we mention the constraints imposed on the antenna's power captured by the low power operation and the characteristics of ambient waves.

12.2.1 Rectifier Optimization

The RF-to-DC conversion of ambient signals is performed by a rectifier composed of a nonlinear device followed by a smoothing capacitor, acting a dc pass filter, and a resistive load. The rectification process has been fully treated in [5], in which an overview of the rectification technologies is given, and the performance criteria of rectifiers for energy harvesting are derived.

The power conversion efficiency (PCE) is defined as the ratio between the output DC power of the rectifier and the RF input power originating from the antenna (before its passage through the matching network). The PCE of the rectifiers features a good performance evaluation and is the Figure-of-Merit (FoM) usually used to characterize the rectifier. Under low-power operation, it has been demonstrated that the rectification efficiency can be expressed as [12]

$$\eta_r = B_0 \cdot \sum_j P_{\omega_j} \cdot \eta_{mn_j} \qquad (12.2)$$

where B_0 is a constant depending on the diode parameters such as thermal voltage, series resistance, junction resistance, and ideality factor [5]. P_{ω_j} and η_{mn_j} represent the RF input power originating from the antenna and the matching efficiency at the frequency f_j. Because of the dependency of the PCE on the input power, it is inevitably limited under low input powers. The low barrier Schottky diode is, for now, the most common technology used for rectification process at ambient power levels due to its highest PCE, typically around 10% at $-30\,\text{dBm}$ [13–16]. The choice of the Schottky diode is naturally essential to obtain the most efficient power conversion. Several investigations [13, 14, 17] have concluded on the dominance of the performance of some Schottky diodes, such as the SMS7630 and the HSMS285X, in the low power rectification mostly due to their low threshold voltage.

A rectifier is usually composed of one or several diodes organized in different possible configurations as shown in Figure 12.3. The simplest and most efficient topology at low input power is the half-wave rectifier [5, 18, 19] that rectifies only the positive (or negative) cycle of the wave with a series diode [20–23] or a shunt diode [24, 25] as shown in Figure 12.3a,b, respectively. Since the signal goes through only one diode, the losses due to the passage of the wave in the diode are minimized. The voltage doubler [26–29] or the modified Greinacher [15, 30, 31] (Figure 12.3c,d) configurations have also been largely developed to reach higher DC voltage levels thanks to the full-wave rectification (both cycles are rectified) while limiting the losses due to the passage through the diodes. Because of the

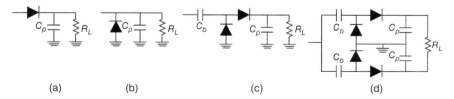

(a)	(b)	(c)	(d)

Figure 12.3 Most common rectifier topologies: (a) single series diode, (b) single shunt diode, (c) voltage doubler, and (d) modified Greinacher.

nonlinearity of the diode, the input impedance of the rectifier varies with the frequency, the input power, and the resistive load. Therefore, many different optimizations can be performed for a given rectifier topology. In general, the rectifiers for ambient energy harvesting are designed for low input power levels in the range of −30 to −20 dBm. Although, the best efficiency is desired, the rectifier has to be able to deliver enough DC voltage so that the DC–DC booster can operate. The DC–DC booster generally performs a maximum power point tracking (MPPT) so that the load value taken by the converter and seen from the rectifier output corresponds to the optimal load that gives the best efficiency. Under low input power, the rectifier acts approximately as a linear device. Thus, the optimal load corresponds to the zero-bias junction resistance of the diode and is proportional to the number of diodes [13]. Then, the DC voltage is equivalent to half of the open circuit voltage.

Figure 12.4 reviews the rectification efficiency of different designs reported in the literature, under optimal load, at the minimum input power (sensitivity), which is the input power that gives the minimum open-circuit voltage necessary to cold-start an ultra-low-power DC–DC converter. An open-circuit voltage of 330 mV is considered here, which is the typical voltage required t0 cold-start the BQ25504 DC–DC converter [34]. Theoretically, the DC voltage across the optimal load is equal to half of the open-circuit voltage, so the performance of the reported rectifiers is given for an output DC voltage of 165 mV. This is the

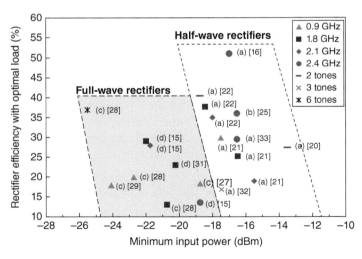

Figure 12.4 State-of-the-art of the efficiency (%) of some rectifiers (with matching network) at the minimum input power (dBm) that leads to an output DC voltage of 165 mV with the optimal load (theoretically corresponding to a 330-mV open-circuit output voltage).

minimum point of operation above which energy from the ambient can actually be stored. Most of the time, the rectifiers are matched for very low input powers around −30 or −20 dBm, and therefore, become less matched at higher input power levels. However, the rectifier should be optimized at its sensitivity since no energy will be stored under that threshold. The letter given for each point refers to the rectifier topology described in Figure 12.3. The sensitivities of half-wave rectifiers are lower than that of full-wave rectifiers, mostly due to their lower resistive load, which results, on the other hand, in higher rectification efficiencies. Consequently, when optimized at their sensitivity for a given ambient power density, the half-wave rectifiers can reach higher efficiencies than full-wave rectifiers at the condition that they are associated to antennas with higher gains, which is not practically appropriate for ambient energy harvesting, as it will be discussed in Section 12.3.3.

The full-wave rectifiers are often designed in a multibranch configuration to give a multiband operation and increase the DC voltage/power when several frequency bands are simultaneously harvested [15, 27, 28, 32, 35, 36]. As suggested by (12.2), the multiband operation of a single rectifier results in higher PCE when simultaneous frequency bands are rectified at the same time, which is performed with multitone rectifiers [20–23, 37, 38]. Note that the multiband operation of multitone rectifiers is usually limited to a small number of frequency bands because of the complexity of the impedance matching.

12.2.2 Low Power Matching Network Optimization

12.2.2.1 The Bode-Fano Criterion

The matching network maximizes the power transfer between the antenna and the rectifier and must therefore be carefully optimized [39–41] in order to obtain optimal rectenna's efficiency and output power. The rectifier, composed of one or several diodes, is equivalent to a complex load made up of a shunt resistor R_e and a shunt capacitor C_e. For a complex impedance, modeled by a parallel RC load impedance, the reflection coefficient over the bandwidth of a lossless matching network is limited by the Bode-Fano criterion:

$$|\Gamma_{mn}| \geq e^{\frac{-1}{2BC_e R_e}} \tag{12.3}$$

Equation (12.3) states that the bandwidth B can increase only at the expense of a higher reflection coefficient Γ_{mn}. Hence, a null reflection can only be obtained at discrete frequencies. Furthermore, higher Q circuits (high R or C) are intrinsically harder to match than lower Q circuits since a small bandwidth or a high reflection coefficient is obtained. The product of C_e and R_e depends on the input power P_{RF} and the resistive load R_L. The capacitance C_e varies proportionally with the number of diodes, but it is independent on the output load resistance R_L and on the input power P_{RF}. The resistance R_e increases with the resistive load, but it is inversely proportional to the number of diodes, the RF input power, and the center frequency. Thus, the maximum frequency bandwidth that allows for a reasonable reflection coefficient increases with the number of diodes, the input power, and the frequency but decreases with the resistive load. As the number of diodes and the available RF input power are limited in energy harvesting, this bandwidth is usually narrow, as shown in Figure 12.5 where the minimum reflection coefficient is given as a function of the frequency bandwidth for an ideal single series diode rectifier with the Schottky diode SMS7630-061L and an output capacitor $C_p = 10\,\text{nF}$.

Therefore, the design of broadband rectifiers is to be avoided when low powers are involved. As a result, the aggregate spectrum of rectennas is usually increased by implementing multiband rectifiers with multibranch where one or two frequency bands are allocated to one rectifier [28, 29].

12.2.2.2 Matching Network Efficiency
The minimum reflection coefficient at the input of a matching network is not systematically achieved since the matching network, itself, introduces insertion losses. Those losses depend on the quality factor of the elements that compose a matching network but also on the quality factor of its topology.

Two elements are sufficient to match a source to a load: one in series and one in parallel (*L* topology). If we consider a resistive source R_{so} and a load R_L, the *L* matching network forms an *RLC* circuit with the load. The reactance of the load can be absorbed with a reactive element combined to the *L* matching network

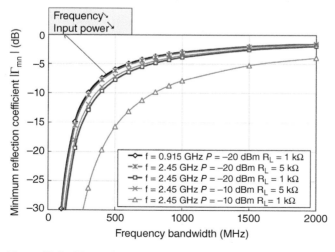

Figure 12.5 Theoretical reflection coefficient of an ideal single serial diode rectifier composed of the Schottky diode SMS7630-061L with an output capacitor $C_p = 10\,\text{nF}$.

elements. Hence, considering a purely resistive load, the L matching network with the resistive load acts like a resonant circuit that achieves the source resistance at the frequency of resonance. The impedance matching is realized thanks to a voltage gain (and current drop) equal to

$$k = \sqrt{m} \tag{12.4}$$

with $m = R_{hi}/R_{lo}$. R_{hi} is the higher resistance to match to R_{lo} the lower resistance. The voltage gain is achieved at the output of the matching network if $R_{so} > R_L$ and at the input if $R_{so} < R_L$. Since low Q matching networks introduce lower insertion losses, several L matching networks can be cascaded in order to reduce the quality factor. Multiple combined L matching networks match the source resistance to intermediate resistances and these intermediate resistances to the load resistance. Hence, by freely choosing the intermediate resistances, we can reduce the quality factor of a single-L matching network, and the total quality factor is the sum of the quality factors of the L matching networks and is given by [42]

$$Q = \sqrt{(m)^{1/N} - 1} \tag{12.5}$$

From the quality factor of the matching network, the matching efficiency (that represents the insertion losses) of a L matching network with N stages can be derived as [42]

$$\eta_{mn} = \frac{1}{1 + N\frac{Q}{Q_c}} \tag{12.6}$$

where Q_c is the net quality factors of the components of the matching networks, function of the inverse of the parasitic resistance (attenuation constant in the case of transmission lines). Figure 12.6 shows the matching efficiency computed for net quality factors of 10, 30, and 50 and different numbers of L matching networks. The best configuration is when the antenna impedance matches perfectly the rectifier impedance, but usually, the antenna impedance is chosen to be 50 Ω so that its parameters can be easily measured and compared to other work. As an example, the power match factors for a 50 Ω-antenna associated to the low barrier Schottky diodes SMS7630-061L in the single series and in the voltage doubler configurations with optimal load and an output capacitor $C_p = 10\,\text{nF}$ (high enough to act as a short circuit for RF signals up to 2.45 GHz) appear in Figure 12.6 for the frequency 2.45 GHz and input powers lower than $-20\,\text{dBm}$. When the diode is only followed by the capacitor and a resistive load, the power match factor tends to increase when the RF input power decreases. Still, a relatively low power match (lower than 10) is usually obtained. Note that, in reality, because the matching network utilized to match the antenna's resistance to the rectifier's input resistance, varies with the power match factor, so does the net quality factor Q_c of the matching network (especially with transmission lines). Thus, rigorously, the net quality factor should be re-evaluated for each power match factor, as it has been done in [39].

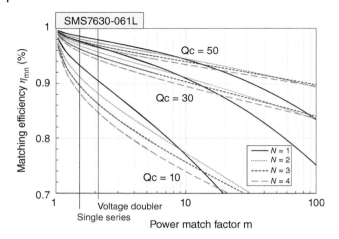

Figure 12.6 Theoretical matching efficiency as a function of the power match factor *m* for different net quality factors Qc and different numbers of cascaded *L* matching networks. As an example, the power match factors for a 50 Ω-antenna associated to an ideal single series rectifier and to an ideal voltage doubler rectifier ($C_b = 10\,nF$), with the Schottky diode SMS7630-061L, optimal load, and a capacitor $C_p = 10\,nF$, are given for the frequency 2.45 GHz and an input power of −20 dBm.

12.2.3 Low-Power Antenna Optimization

The antenna has an essential role in the process of energy harvesting since it is the transducer that converts the ambient available electromagnetic signals into electrical signals, and the optimization of this interface can lead to higher power levels at the rectifier input. As mentioned previously, broadband [15, 27, 28, 32, 37] or multiband antennas [20, 21, 35, 38] should be privileged to be able to harvest a large spectrum and increase the rectification efficiency and DC output power. The polarization of the antenna [43, 44] is also a critical point and will be discussed in Section 12.3.2. One of the main constraint on the antenna for ambient RF energy harvesting is that the rectenna has to be insensitive to the position of the energy sources, thus it is essential to use omnidirectional [3, 20, 31, 45, 46] or multidirectional [33, 47–51] antennas in order to overcome the spatial diversity of the ambient energy. Finally, the antenna part usually has to be designed so that a large amount of power is captured, and sufficient DC power can be supplied to the load [24, 30, 33, 37, 45, 52–59].

For fixed frequency of operation and polarization, the DC output power and the efficiency of the rectenna can be both increased with the enhancement of the power captured by the antenna part. As explained in this section, this can be performed with various options like increasing the antenna directivity, the number of antennas or optimizing the antenna efficiency.

12.2.3.1 Enhancement of the Output DC Power

The RF power entering the rectifier is preliminary captured by the antenna and is defined in the spherical coordinate as

$$P_{RF_0}(\theta, \phi) = S(\theta, \phi) \cdot A_e(\theta, \phi) = S(\theta, \phi) \cdot \frac{\lambda_0^2 \cdot D_a(\theta, \phi) \cdot \eta_{rad}}{4\pi} \tag{12.7}$$

in which $S(\theta, \phi)$ is the ambient available power density of the RF signal arriving on the harvester with the angle of incidence (θ, ϕ), A_e is the antenna effective aperture, D_a is the antenna directivity, and η_{rad} is the radiation efficiency of the antenna. The directivity in the direction of maximum radiation intensity (θ_o, ϕ_o) is defined by

$$D_a(\theta_o, \phi_o) = \frac{4\pi}{\Omega} \tag{12.8}$$

where Ω is the solid angle described by the radiation pattern of the antenna at frequency f_o.

Since the radiation pattern represents the field distribution in space normalized with respect to the radiation of an isotropic antenna, it is understandable, as suggested by Eqs. (12.7) and (12.8), that the more concentrated is the radiation of an antenna, the more power it will receive but in a very limited space region. For a fixed power flux, Eqs. (12.2) and (12.7) suggest that the performance of a rectenna improves when the effective aperture of the antenna increases. However, in reality, because a highly directive antenna intercepts RF waves only coming from a reduced portion of directions, as shown by (12.8), it is not appropriate to harvest ambient RF energy. In order to increase the power delivered to one or several rectifiers without reducing the angular beam-width of the harvester, two methods exist as shown in Figure 12.7.

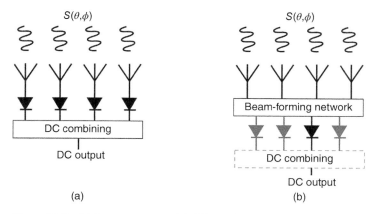

Figure 12.7 (a) Rectenna array with DC combination. (b) Antenna array with a beam-forming network (RF combination) associated to rectifying elements.

12.2.3.2 Rectenna Array

The first method consists of a rectenna array (see Figure 12.7a) where a number of antennas (with the same radiation pattern) are associated to the same number of rectifiers, and the DC outputs of each of them are combined [37, 52, 53, 60, 61]. This solution of multiple rectennas allows for the interception of more RF energy but the solid angle, that is the spatial coverage of the harvesting system, is not reduced. Assuming an incident power density $S(\theta, \phi)$ and $M \times N$ rectennas, with similar effective aperture $A_e(\theta, \phi)$, capturing RF power as in (12.7), then the total power collected by the rectenna array is

$$P_{\text{RF}_1}(\theta, \phi) = M \cdot N \cdot P_{\text{RF}_0}(\theta, \phi) \tag{12.9}$$

Each rectifier receives the same amount of power $P_{\text{RF}_0}(\theta, \phi)$ so the total rectification efficiency of the rectenna array, given by (12.10), is the product of the rectification efficiency of a single rectifier $\eta_r(P_{\text{RF}_0})$ with the efficiency of the DC combination of the $M \times N$ rectifiers' outputs $\eta_{\text{DC1}_{MN}}$

$$\eta_1(\theta, \phi) = \eta_r(P_{\text{RF}_0}(\theta, \phi)) \cdot \eta_{\text{DC1}_{MN}} \tag{12.10}$$

Assuming a low RF power, the rectification efficiency can be considered proportional to the input power, as in (12.2), and the total DC power collected can be written as

$$P_{\text{DC}_1}(\theta, \phi) = P_{\text{RF}_1}(\theta, \phi) \cdot \eta_1(\theta, \phi) = M \cdot N \cdot B'_0 \cdot P^2_{\text{RF}_0}(\theta, \phi) \cdot \eta_{\text{DC1}_{MN}} \tag{12.11}$$

where B'_0 depends on the diode's parameters and the matching efficiency between the antennas and the rectifiers [5].

12.2.3.3 Antenna Array with BFN

The other strategy consists of designing a multiple-port antenna array using passive beam-forming techniques [33, 54], as shown in Figure 12.7b. Considering an array composed of $M \times N$ identical antennas distributed along the x- and y-axis, a passive beam-forming network composed of several ports can associate for each of its input ports i ($i = 1, \ldots, M \times N$) specific phase differences (β^i_x, β^i_y) between its output ports connected to the antenna elements, spaced apart of distances d^i_x and d^i_y, along the x- and y-axis, respectively. Hence, each port i corresponds to a radiation pattern composed of a high gain beam steered in a particular direction. The directivity D^i_B of each beam B_i is the product of the directivity $D_a(\theta, \phi)$ of the antenna element with the array factor $AF_i(\theta, \phi)$ and is given by

$$D_B^i(\theta, \phi) = D_a(\theta, \phi) \cdot AF_i(\theta, \phi)$$

$$= D_a(\theta, \phi) \cdot \left\{ \frac{1}{M} \cdot \frac{\sin\left(\frac{M}{2}\psi_x^i\right)}{\sin\left(\frac{\psi_x^i}{2}\right)} \right\} \cdot \left\{ \frac{1}{N} \cdot \frac{\sin\left(\frac{N}{2}\psi_y^i\right)}{\sin\left(\frac{\psi_y^i}{2}\right)} \right\} \quad (12.12)$$

where $\psi_x^i = kd_x^i \sin\theta \cos\phi + \beta_x^i$ and $\psi_y^i = kd_y^i \sin\theta \cos\phi + \beta_y^i$

The maximum radiation intensity of the beam corresponding to the port i occurs at a scan angle (θ_s^i, ϕ_s^i) given by

$$\theta_s^i = \sin^{-1}\left(\sqrt{\left(\frac{\beta_x^i}{d_x^i}\right)^2 + \left(\frac{\beta_y^i}{d_y^i}\right)^2}\right) \quad (12.13a)$$

$$\phi_s^i = \tan^{-1}\left(\frac{\beta_x^i d_x^i}{\beta_y^i d_y^i}\right) \quad (12.13b)$$

Thus, the directivity of the beam B_i is maximum at (θ_s^i, ϕ_s^i) and is equivalent to:

$$D_B^i(\theta_s^i, \phi_s^i) = D_a(\theta_o, \phi_o) \cdot M \cdot N \cdot \eta_{BFN} \cdot \eta_{SC}^i \quad (12.14)$$

where η_{BFN} represents the insertion losses in the beam-forming network, and η_{SC}^i mirrors the scanning losses, in the direction (θ_s^i, ϕ_s^i), equivalent to

$$\eta_{SC}^i = \frac{|D_a(\theta_s^i, \phi_s^i)|}{|D_a(\theta_o, \phi_o)|} \quad (12.15)$$

with $D_a(\theta_o, \phi_o)$ and $D_a(\theta_s^i, \phi_s^i)$, the directivity of the antenna element in the direction of maximum intensity and in the direction of scan, respectively.

The maximum directivity of the antenna array is the product of the directivity of the antenna element with the number of antennas composing the array and, as stated in (12.8), this implies a division of the beamwidth. That is why, by permitting the presence of multiple beams, the multiport beam-forming network results in an aggregate solid angle equivalent to that of the single element as shown in Figure 12.8. The outputs of the rectifiers can be combined with a DC combining circuit, or, in the case of low-cost sensors, it can also be envisaged to attach one sensor to each of the rectifier's outputs.

The total power entering the rectifier attached to the BFN's port i is given by

$$P_{RF_2}^i(\theta, \phi) = P_{RF_0}(\theta, \phi) \cdot AF_i(\theta, \phi) \cdot \eta_{BFN} \quad (12.16)$$

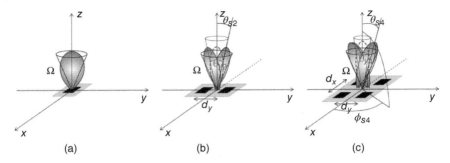

Figure 12.8 Directional radiation pattern with solid angle Ω of a (a) single antenna element, (b) two antennas aligned along the y-axis with a two port beam-forming network, and (c) 2×2 antennas aligned along the x- and y-axis with a four-port beam-forming network.

Because the RF power is combined before the rectification process, each rectifier receiving energy tends to operate more efficiently than with the first method (as long as the product of the array factor with the BFN's efficiency is higher than 1). The rectification efficiency of each rectifier is given by

$$\eta_2^i(\theta, \phi) = \eta_r(P_{RF_2}^i(\theta, \phi)) \tag{12.17}$$

Thus, similarly as in (12.11), the total output DC power of the rectenna can be expressed as

$$P_{DC2}(\theta, \phi) = \eta_{DC2_{MN}} \sum_{i=1}^{M \times N} P_{DC_2}^i(\theta, \phi)$$

$$= \eta_{BFN}^2 \cdot B_0' \cdot P_{RF_0}^2(\theta, \phi) \cdot \eta_{DC2_{MN}} \sum_{i=1}^{M \times N} (AF_i(\theta, \phi)^2) \tag{12.18}$$

where $\eta_{DC2_{MN}}$ is the combination efficiency of the DC outputs of the $M \times N$ rectifiers, when a DC combining circuit is attached to the rectifiers' outputs. From the comparison of (12.11) with (12.18), when an incident wave arrives from the direction (θ, ϕ), it can be deduced than the RF combination performed by a beam-forming network results in a higher DC power than the DC combination of multiple rectennas, only if the following condition is respected:

$$\eta_{BFN}^2 \cdot \eta_{DC2_{MN}} \sum_{i=1}^{M \times N} (AF_i(\theta, \phi)^2) > M \cdot N \cdot \eta_{DC1_{MN}} \tag{12.19}$$

If we compare both methods in their direction of maximum radiation, then, for each beam B_i, (12.19) reduces to

$$\eta_{BFN}^2 \cdot \eta_{SC}^{i}{}^2 \cdot \eta_{DC2_{MN}} \cdot M \cdot N > \eta_{DC1_{MN}} \tag{12.20}$$

12.2.3.4 Optimization of the Antenna Efficiency

The radiation efficiency is a key parameter in the design of the antenna in order to optimize the power captured and the size. The radiation efficiency represents the power lost in the conductor and in the dielectric instead of being radiated. The conductor and dielectric losses that cohabit in the antenna can be represented by the resistance R_{cd}, and the radiation efficiency can be expressed as [62]

$$\eta_{\text{rad}} = \left[\frac{R_r}{R_{cd} + R_r} \right] \tag{12.21}$$

where R_r is the radiation resistance. Special care should be attributed to the design of high efficient antennas in order to optimize the received power, which is essential in RF energy harvesting.

As with the radiation efficiency, the aperture efficiency should be as high as possible so that the size of the antenna remains the smallest possible. From (12.7), we can see that the aperture efficiency is directly proportional to the radiation efficiency since it is the ratio between the physical size A_p of the antenna and the effective aperture A_e as shown by (12.22).

$$\eta_a = \frac{A_p}{A_e} \tag{12.22}$$

In order to obtain high aperture efficiency, besides increasing the radiation efficiency, the directivity of an antenna can be increased by intensifying the electric field responsible for the radiation without enlarging the size of the antenna. This is important because, by minimizing the antenna size, the number of antennas can be increased resulting in higher collected DC power.

12.3 The Chance of Collecting Ambient Electromagnetic Energy with a Specific Antenna

We understand from Section 12.2 that rectification efficiency is limited at low-power operation, which results in low output DC voltage and power. It has been demonstrated in the literature that the rectification efficiency can be increased with multitone operation, optimal matching, efficient antennas and with RF power combination at the antenna level. Besides, DC output voltage can be enhanced with multiple diode rectifiers, and sufficient DC power can generally be exploited from the environment by enhancing the size of the antenna part.

Nevertheless, because of the variability of the frequency, polarization, and direction of arrival of the incident RF waves on the harvester, two rectennas with equivalent performance can have different chance of intercepting ambient energy over time. Indeed, because of various obstacles in the propagation channels and the varying demand of RF related devices users, the flux of ambient waves varies with time and space. This leads to difficulties to optimize their collection and to evaluate the performance of rectennas in real conditions. The PCE of rectifiers is not sufficient to assess the performance rectennas. Thus, in order to add another dimension to that FoM, this section discusses the concept of chance that a rectenna actually intercepts a signal depending on its frequency, its polarization, and its angle of incidence.

12.3.1 Frequency Spectrum

As mentioned in Section 2.1, the frequency spectrum of the ambient energy present in our environment is composed of the frequency bands exploited for the mobile communications, the TV broadcasting, and the WiFi transmissions. Although the narrowness of those frequency bands is compatible with the design of narrowband rectifiers, the diversity of the available frequency spectrum in the ambient can justify the use of multiband rectifiers. Broadly speaking, a rectenna should be able to collect energy from several frequency bands so that over a certain time, it has more chance of harvesting energy. Moreover, it has been shown that the instantaneous superposition of several frequency bands in the rectenna's environment induces an increase of the rectification efficiency as suggested by Eq. (12.2), even if this condition cannot be assumed at any time. The chance of collecting energy can be first characterized by the frequency mismatch factor over a given time between the rectenna and the ambient radiations that can be expressed as

$$\Gamma_f = \sum_{j=1}^{A} \gamma_j = \sum_{j=1}^{A} F_j \cdot \delta(f_j) \tag{12.23}$$

with

$$\sum_{j=1}^{A} F_j = 1 \tag{12.24}$$

F_j is the frequency coefficient that represents the proportion of energy received from a specific frequency band f_j, while A represents the total number of the different frequency bands of the incoming waves, over time. $\delta(f_j)$ informs on the frequencies that the rectenna can harvest or not, by taking the value 1 or 0. The evaluation of the frequency mismatch factor of a rectenna requires some information on the rectenna's location (indoor, outdoor, continent) to estimate the frequency bands that have to be considered and with which proportion.

12.3.2 Polarization

The polarization of an electromagnetic wave is the direction of its electric field. When this term is attributed to an antenna, it refers to the polarization of the radiation that the antenna is designed to radiate or receive. The polarization of the ambient RF signals that rectenna intercepts can vary in time and with the location or orientation of the rectenna, so it is supposed to be unknown. The ambient energy is captured by the rectenna only when the polarization of the incident waves matches the polarization of the rectenna. Thus, the chance of collecting energy can be evaluated by defining a polarization mismatch Γ_p over a certain time during which multiple scenarios occur.

Let us assume that P_k is the proportion coefficient of incident RF waves that arrive with a specific polarization, and N represents the total number of the different polarizations, over time that do not occur simultaneously, and we have

$$\sum_{k=1}^{N} P_k = 1 \tag{12.25}$$

Then, the polarization mismatch between the incident waves and the harvester that contains M ports, each of them associated to a specific polarization, is given by

$$\Gamma_p = \sum_{k=1}^{N} \left[P_k \cdot \sum_{l=1}^{M} \frac{1 + |\boldsymbol{\rho}_{ik}|^2 \cdot |\boldsymbol{\rho}_{rl}|^2 + 2|\boldsymbol{\rho}_{ik}| \cdot |\boldsymbol{\rho}_{rl}| \cdot \cos(2\theta)}{(1 + |\boldsymbol{\rho}_{ik}|^2) \cdot (1 + |\boldsymbol{\rho}_{rl}|^2)} \right] \tag{12.26}$$

where θ is the angular misalignment between the major axes of the different polarizations, and $|\boldsymbol{\rho}_{ik}|$ and $|\boldsymbol{\rho}_{rl}|$ are the circular-polarization (CP) ratios of the incident signal and the receiving rectenna, respectively.

The circular polarization ratio is defined as the ratio between the electric fields along the right- and left-handed orthogonal unit vectors, and its magnitude can be expressed as:

$$|\rho| = \frac{a_r - 1}{a_r + 1} \quad \text{for left-handed (LH) polarization} \tag{12.27a}$$

$$|\rho| = \frac{a_r + 1}{a_r - 1} \quad \text{for right-handed (RH) polarization} \tag{12.27b}$$

where a_r is the axial ratio. Note that the circular polarization ratio of a linearly polarized signal is equal to 1. Table 12.2 gives the polarization mismatch factor, for and $M = 1$ and $N = 1$, between an incident radiation and a receiving rectenna under different scenarios in which the polarization of the incident signal can be vertical (V), horizontal (H), oblique, dual polarized (DP) ($V + H$ or oblique), right- or left-hand circularly or elliptically polarized (RHC, LHC, RHE, LHE). The polarization of the rectenna is supposed to be classical such that vertical, horizontal, DP, right- or left-handed circular polarized.

The evaluation of the polarization mismatch factor requires the estimation of the polarization of the incident radiations. In a general way, in conventional communications, radiations are more often transmitted with vertical polarization but,

Table 12.2 Polarization mismatch.

		V	H	DP (V + H)	RHC	LHC
				Polarization of the rectenna $M = 1$		
Polarization of the incident signal $N = 1$	V	1	0	1/2	1/2	1/2
	H	0	1	1/2	1/2	1/2
	Oblique	$cos^2(\theta)$	$cos^2(\theta)$	$cos^2(\theta) + cos^2(\pi/2 - \theta)$	$cos^2(\theta) + cos^2(\pi/2 - \theta)$	$cos^2(\theta) + cos^2(\pi/2 - \theta)$
	DP (V + H)	1/2	1/2	1	1	1
	RHC	1/2	1/2	1/2	1	0
	LHC	1/2	1/2	1/2	0	1
	RHE	B_{ik}^*	B_{ik}^*	B_{ik}^*	$\dfrac{|\rho_{ik}|^2}{1 + |\rho_{ik}|^2}$	$\dfrac{1}{1 + |\rho_{ik}|^2}$
	LHE	B_{ik}^*	B_{ik}^*	B_{ik}^*	$\dfrac{|\rho_{ik}|^2}{1 + |\rho_{ik}|^2}$	$\dfrac{1}{1 + |\rho_{ik}|^2}$

$$B_{ik}^* = \frac{1 + |\rho_{ik}|^2 + 2 \cdot |\rho_{ik}| \cdot \cos(2\theta)}{2(1 + |\rho_{ik}|^2)}$$

in a Rayleigh multipath propagation environment, it can be assumed that they will be found in two orthogonal polarizations with equal proportions [53]. Over time, this results in incident waves with oblique polarizations, with a misalignment θ with the major axe of the receiving antenna polarization. Thus, considering that the misalignment can take all value from 0 to π over a certain time, we can consider the average value of the mismatch factor given by (Table 12.3)

$$\overline{\Gamma}_p = \frac{1}{\pi} \int_0^\pi \left(\begin{array}{l} \dfrac{1}{2} \cdot \displaystyle\sum_{l=1}^M \dfrac{1 + |\boldsymbol{\rho}_{rl}|^2 + 2|\boldsymbol{\rho}_{rl}| \cdot \cos(2\theta)}{2 \cdot (1 + |\boldsymbol{\rho}_{rl}|^2)} \\ + \dfrac{1}{2} \cdot \displaystyle\sum_{l=1}^M \dfrac{1 + |\boldsymbol{\rho}_{rl}|^2 + 2|\boldsymbol{\rho}_{rl}| \cdot \cos\left(2\left(\theta - \frac{\pi}{2}\right)\right)}{2 \cdot (1 + |\boldsymbol{\rho}_{rl}|^2)} \end{array} \right) d\theta \qquad (12.28)$$

Table 12.2 gives the average value of the polarization mismatch over time for different polarizations of the receiving antenna. Table 12.2 suggests that DP or

Table 12.3 Polarization mismatch over time in a Rayleigh propagation environment.

Rectenna polarization $M = 1$	V	H	DP (V + H)	RHC	LHC
Average value of the polarization mismatch $\overline{\Gamma}_p$	1/2	1/2	1	1	1

circularly polarized antennas are more appropriate to harvest ambient waves in a multipath propagation environment from the polarization point of view.

However, it can be noted that because of orthogonal polarization of DP and CP antennas, the gain is divided by 2 compared to a linearly polarized antenna resulting in an equivalence of the product gain times polarization mismatch. Furthermore, as mentioned in Section 12.2, the increase in gain enhances the rectification efficiency, justifying the use of linearly polarized antenna for ambient RF energy harvesting when a single-port antenna is designed.

12.3.3 Spatial Coverage

The spatial coverage of the rectenna is essential for the evaluation of the chance of intercepting the ambient waves, with same frequency and same polarization than the rectenna, in a 3D space. In the far field region, the power collected by an antenna in a given direction defined by the angles θ and ϕ in spherical coordinates can be expressed by its radiation intensity [62], which also describes the radiation pattern of the antenna:

$$U(\theta, \phi) = B_o F(\theta, \phi) \qquad (12.29)$$

where B_o is a constant and $F(\theta, \phi)$ is related to the far-zone electric field of the antenna. Thus, the total power collected by the receiving antenna can be expressed as:

$$P_{rec} = B_o \iint F(\theta, \phi) \sin \theta \, d\theta \, d\phi = B_o \cdot F(\theta, \phi) \mid \max \cdot \Omega_a \qquad (12.30)$$

while $B_o \cdot F(\theta, \phi) \mid$ max gives the maximum radiation intensity, Ω_a informs on the region of space in which the power is captured assuming the radiation intensity is constant for any angles within Ω_a [62]. Furthermore, assuming that an incident signal have equal probability to arrive from any direction within a certain solid angle Ω_i, which can be defined by the position of the energy source or the orientation of the receiving antenna, the probability that an incident signal is captured by the antenna can be approximated by:

$$p((\theta_i, \phi_i) \in \Omega_a) = \frac{\Omega_a}{\Omega_i} \quad \text{if } \Omega_a \leq \Omega_i \text{ and } \Omega_a \in \Omega_i \qquad (12.31a)$$

$$p((\theta_i, \phi_i) \in \Omega_a) = 1 \quad \text{if } \Omega_a > \Omega_i \text{ and } \Omega_i \in \Omega_a \qquad (12.31b)$$

The solid angles of antennas can be approximated from the half power beam widths of the radiation pattern (HPBWs) Θ_1 and Θ_2 in two planes orthogonal to each other. When the HPBWs in two orthogonal planes of a directional radiation pattern are equivalent and equal to Θ, the surface is spherical, and the solid angle is easily determined as $\Omega_a = 2\pi(1 - \cos \Theta)$. When the HPBWs are different to each other, the solid angle is usually approximated by Eqs. (12.32) and (12.33) [62] for directional and omnidirectional patterns, respectively.

$$\Omega_{a\,\text{unidir}} \approx \Theta_1\Theta_2 \tag{12.32}$$

$$\Omega_{a\,\text{omnidir}} \approx \frac{4\pi}{\left(-172.4 + 191\sqrt{0.818 + \frac{180}{\Theta_1 \cdot \pi}}\right)} \tag{12.33}$$

12.3.4 Harvesting Capability

In order to define a FoM for rectenna operating in ambient energy harvesting scenarios, the chance for a rectifying antenna to intercept ambient RF radiations can be associated to its ability of rectifying them into DC power. In this way, the performance of rectennas to harvest ambient RF signals with low power levels, unknown polarization, and varying angle of incidence over a frequency band or a broadband spectrum can be further appreciated. The harvesting capability in %.steradian (%.sr) at a given input power can be defined combining all this parameters as follows:

$$C_{\text{har}} = \pi \sum_{j}^{A} \left[F_j \cdot \gamma(f_j) \cdot \Gamma_p^j \cdot \eta_r^j \cdot \frac{\Omega_a^j}{\Omega_i^j} \right] [\%.\text{sr}] \quad \text{if } \Omega_a^j \leq \Omega_i^j \text{ and } \Omega_a^j \in \Omega_i^j \tag{12.34a}$$

$$C_{\text{har}} = \pi \sum_{j}^{A} \left[F_j \cdot \gamma(f_j) \cdot \Gamma_p^j \cdot \eta_r^j \right] [\%.\text{sr}] \quad \text{if } \Omega_a^j > \Omega_i^j \text{ and } \Omega_i^j \in \Omega_a^j \tag{12.34b}$$

where A is the total number of frequency bands present in the environment. η_r^j is the rectification efficiency at the frequency f_j in the direction of maximum gain. Note that the maximum harvesting capability is 100π %.sr. In a particular case, where only one frequency band is considered, the harvesting capability of a rectenna in a Rayleigh multipath propagation environment, in which microwaves can arrive with any possible angle of incidence, the harvesting capability becomes

$$C_{\text{har}_o} = \overline{\Gamma_p} \cdot \eta_r \cdot \Omega_a / 4 \, [\pi\%.\text{sr}] \tag{12.35}$$

Figure 12.9 represents the harvesting capability C_{har_o} of state-of-the-art rectennas (single tone) for different frequencies at the minimum power density leading to an open-circuit voltage equivalent to 330 mV, which is the minimum DC voltage required to cold-start an ultra-low-power DC–DC booster [34]. This open-circuit voltage corresponds to a 165-mV DC voltage at the output of the rectenna across the optimal load. Figure 12.9 gives, at the frequency of operation of the rectenna, the efficiency of the rectenna, its chance to collect RF power over time (assuming that RF waves can come from everywhere with arbitrary polarization), and

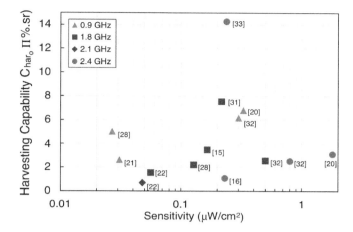

Figure 12.9 Harvesting capability (π %.sr) (12.35) of state-of-the-art rectennas at the minimum power density (μW/cm^2) resulting in an output DC voltage of 165 mV across the optimal load.

the minimum power density required at the rectenna level so that ambient energy can be stored.

12.4 Conclusion

Ambient energy harvesting involves very low-power levels, and consequently, trade-offs and optimization processes must be made on the different components of the RF energy harvester. The rectifier topology and rectifying element must be chosen accordingly to the output DC voltage necessary to cold-start a DC–DC booster, allowing for the storage of the harvested energy, while maximizing the efficiency. Due to only one rectifying element, the half-wave rectifiers account for lower losses than full-wave rectifiers at low power levels and reach higher rectification efficiencies but lower output voltages. Consequently, the lower output voltage obtained with the half-wave rectifiers, because of their low resistive load, results in a lower sensitivity. In general, the low sensitivity of the rectifiers requires an optimization of the power entering the rectifier by minimizing the losses and increasing the number of antennas. The optimization of the power that goes through the matching network shows that the broadband operation of the rectifier, coming from the impedance matching over a wide frequency band, results in a low reflection coefficient at the input of the rectifier under low-power operation. Thus, multiband rectifier structures should be favored at the expense of a possible increase in the number of sensors attached to the rectenna. Moreover, the matching efficiency is maximum when the antenna's impedance is the complex conjugate of the rectifier's impedance. Otherwise, due to low power match factors when a 50 Ω antenna is directly matched to the rectifier, the single stage L topology matching network gives the highest matching efficiency. The overall rectenna's sensitivity can be improved by enhancing the RF power transmitted to the rectifier with the networking of multiple rectennas and the combination of their DC outputs, or with the RF power combination before rectification performed by an antenna array associated to a beam forming network. These two methods allow for the enhancement of the RF power captured without reducing the chance of collecting incoming signals from any direction. Depending on the losses of the different DC and RF power combination circuits, one of the two methods will deliver higher DC power. The chance of collecting energy, which take into consideration the frequency, the polarization, and the direction of arrival of the incident ambient signals, can be considered to evaluate the performance of the rectennas designed for ambient RF energy harvesting. When all these arbitrary parameters are considered, the DC power collected by a rectenna over time can be estimated. This consideration reveals that multiband and dual-polarized rectennas should be favored to face the frequency and polarization diversities of the ambient signals, while the use of directive rectennas should be avoided to face the spatial diversity.

References

1 Brown, W.C. (1984). The history of power transmission by radio waves. *IEEE Transactions on Microwave Theory and Techniques* 32 (9): 1230–1242.

2 Shinohara, N. and Matsumoto, H. (1998). Experimental study of large rectenna array for microwave energy transmission. *IEEE Transactions on Microwave Theory and Techniques* 46 (3): 261–268.

3 Suh, Y.-H. and Chang, K. (2002). A novel dual frequency rectenna for high efficiency wireless power transmission at 2.45 and 5.8 GHz. *2002 IEEE MTT-S International Microwave Symposium Digest (Cat. No.02CH37278)* 2: 1297–1300.

4 Visser, H.J. and Vullers, R.J.M. (2013). RF energy harvesting and transport for wireless sensor network applications: principles and requirements. *Proceedings of the IEEE* 101 (6): 1410–1423.

5 Hemour, S. and Wu, K. (2014). Radio-frequency rectifier for electromagnetic energy harvesting: development path and future outlook. *Proceedings of the IEEE* 102 (11): 1667–1691.

6 Popović, Z., Falkenstein, E.A., Costinett, D., and Zane, R. (2013). Low-power far-field wireless powering for wireless sensors. *Proceedings of the IEEE* 101 (6): 1397–1409.

7 Pinuela, M., Mitcheson, P.D., and Lucyszyn, S. (2013). Ambient RF energy harvesting in urban and semi-urban environments. *IEEE Transactions on Microwave Theory and Techniques* 61 (7): 2715–2726.

8 Barroca, N., Saraiva, H.M., Gouveia, P.T. et al. (2013). Antennas and circuits for ambient RF energy harvesting in wireless body area networks. *2013 IEEE 24th Annual International Symposium on Personal, Indoor, and Mobile Radio Communications (PIMRC)*, London, United Kingdom (8–11 September 2013), pp. 532–537.

9 ANFR (2015). Exposition du public aux ondes, Etude de l'exposition du public aux ondes radioélectriques, 21 June 2019, pp. 1–18 [Online]. https://www.anfr.fr/controle-des-frequences/exposition-du-public-aux-ondes/ (accessed 16 September 2019).

10 Visser, H.J., Reniers, A.C.F., and Theeuwes, J.A.C. (2008). Ambient RF energy scavenging: GSM and WLAN power density measurements. *2008 38th European Microwave Conference*, Amsterdam, The Netherlands (28–30 October 2008), pp. 721–724.

11 Tavares, J., Barreca, N., and Saraiva, H.M. et al. (2013). Spectrum opportunities for electromagnetic energy harvesting from 350 mHz to 3 gHz. *2013 7th International Symposium on Medical Information and Communication Technology (ISMICT)*, Tokyo, Japan (6–8 March 2013), pp. 126–130.

12 Lorenz, C.H.P., Hemour, S., Liu, W. et al. (2015). Hybrid power harvesting for increased power conversion efficiency. *IEEE Microwave and Wireless Components Letters* 25 (10): 687–689.

13 Hemour, S., Zhao, Y., Lorenz, C.H.P. et al. (2014). Towards low-power high-efficiency RF and microwave energy harvesting. *IEEE Transactions on Microwave Theory and Techniques* 62 (4): 965–976.

14 Lorenz, C.H.P., Hemour, S., and Wu, K. (2016). Physical mechanism and theoretical foundation of ambient RF power harvesting using zero-bias diodes. *IEEE Transactions on Microwave Theory and Techniques* 64 (7): 2146–2158.

15 Song, C., Huang, Y., Zhou, J. et al. (2015). A high-efficiency broadband rectenna for ambient wireless energy harvesting. *IEEE Transactions on Antennas and Propagation* 63 (8): 3486–3495.

16 Sun, H., Guo, Y.-x., He, M., and Zhong, Z. (2012). Design of a high-efficiency 2.45-GHz rectenna for low-input-power energy harvesting. *IEEE Antennas and Wireless Propagation Letters* 11: 929–932.

17 M. Pinuela, P. D. Mitcheson, and S. Lucyszyn 2010. Analysis of scalable rectenna configurations for harvesting high frequency ambient radiation, *The 10th International Workshop on Micro and Nanotechnology for Power Generation and Energy Conversion Applications (PowerMEMS 2010)*, Leuven, Belgium.

18 Marian, V., Allard, B., Vollaire, C., and Verdier, J. (2012). Strategy for microwave energy harvesting from ambient field or a feeding source. *IEEE Transactions on Power Electronics* 27 (11): 4481–4491.

19 Chen, Y. and Chiu, C. (2017). Maximum achievable power conversion efficiency obtained through an optimized rectenna structure for RF energy harvesting. *IEEE Transactions on Antennas and Propagation* 65 (5): 2305–2317.

20 Niotaki, K., Kim, S., Jeong, S. et al. (2013). A compact dual-band rectenna using slot-loaded dual band folded dipole antenna. *IEEE Antennas and Wireless Propagation Letters* 12: 1634–1637.

21 Shen, S., Chiu, C., and Murch, R.D. (2017). A dual-port triple-band L-probe microstrip patch rectenna for ambient RF energy harvesting. *IEEE Antennas and Wireless Propagation Letters* 16: 3071–3074.

22 Sun, H., Guo, Y.-x., He, M., and Zhong, Z. (2013). A dual-band rectenna using broadband Yagi antenna array for ambient RF power harvesting. *IEEE Antennas and Wireless Propagation Letters* 12: 918–921.

23 Belo, D., Georgiadis, A., and Carvalho, N.B. (2016). Increasing wireless powered systems efficiency by combining WPT and electromagnetic energy harvesting. *2016 IEEE Wireless Power Transfer Conference (WPTC)*, Aveiro, Portugal (5–6 May 2016), pp. 1–3.

24 Sun, H. and Geyi, W. (2017). A new rectenna using beamwidth-enhanced antenna array for RF power harvesting applications. *IEEE Antennas and Wireless Propagation Letters* 16: 1451–1454.

25 Lee, T.J., Patil, P., Hu, C.Y. et al. (2015). Design of efficient rectifier for low-power wireless energy harvesting at 2.45 GHz. *2015 IEEE Radio and Wireless Symposium (RWS)*, San Diego, CA (25–28 January 2015), pp. 47–49.

26 Georgiadis, A., Andia, G.V., and Collado, A. (2010). Rectenna design and optimization using reciprocity theory and harmonic balance analysis for electromagnetic (EM) energy harvesting. *IEEE Antennas and Wireless Propagation Letters* 9: 444–446.

27 Kuhn, V., Lahuec, C., Seguin, F., and Person, C. (2015). A multi-band stacked RF energy harvester with RF-to-DC efficiency up to 84%. *IEEE Transactions on Microwave Theory and Techniques* 63 (5): 1768–1778.

28 Song, C., Huang, Y., Carter, P. et al. (2016). A novel six-band dual CP rectenna using improved impedance matching technique for ambient RF energy harvesting. *IEEE Transactions on Antennas and Propagation* 64 (7): 3160–3171.

29 Assimonis, S.D., Daskalakis, S., and Bletsas, A. (2016). Sensitive and efficient RF harvesting supply for batteryless backscatter sensor networks. *IEEE Transactions on Microwave Theory and Techniques* 64 (4): 1327–1338.

30 Olgun, U., Chen, C.C., and Volakis, J.L. (2011). Investigation of rectenna array configurations for enhanced RF power harvesting. *IEEE Antennas and Wireless Propagation Letters* 10: 262–265.

31 Zeng, M., Andrenko, A.S., Liu, X. et al. (2017). A compact fractal loop rectenna for RF energy harvesting. *IEEE Antennas and Wireless Propagation Letters* 16: 2424–2427.

32 Palazzi, V., Hester, J., Bito, J. et al. (2018). A novel ultra-lightweight multiband rectenna on paper for RF energy harvesting in the next generation LTE bands. *IEEE Transactions on Microwave Theory and Techniques* 66 (1): 366–379.

33 Vandelle, E., Bui, D.H.N., Vuong, T. et al. (2019). Harvesting ambient RF energy efficiently with optimal angular coverage. *IEEE Transactions on Antennas and Propagation* 67 (3): 1862–1873.

34 BQ25504 Ultra Low-Power Boost Converter with Battery Management for Energy Harvester Applications – Nano-Power Management – TI.com [Online]. http://www.ti.com/product/BQ25504 (accessed: 16 September 2019).

35 Keyrouz, S., Visser, H.J., and Tijhuis, A.G. (2013). Multi-band simultaneous radio frequency energy harvesting. *2013 7th European Conference on Antennas and Propagation (EuCAP)*, Gothenburg, Sweden (8–12 April 2013), pp. 3058–3061.

36 Parks, A.N. and Smith, J.R. (2014). Sifting through the airwaves: efficient and scalable multiband RF harvesting. *2014 IEEE International Conference on RFID (IEEE RFID)*, Orlando, FL (8–10 April 2014), pp. 74–81.

37 Hagerty, J.A., Lopez, N.D., Popovic, B., and Popovic, Z. (2000). Broadband rectenna arrays for randomly polarized incident waves. *2000 30th European Microwave Conference*, Paris, France (2–5 October 2000), pp. 1–4.

38 Suh, Y.-H. and Chang, K. (2002). A high-efficiency dual-frequency rectenna for 2.45- and 5.8-GHz wireless power transmission. *IEEE Transactions on Microwave Theory and Techniques* 50 (7): 1784–1789.

39 Lorenz, C.H.P., Hemour, S., and Wu, K. (2015). Modeling and influence of matching network insertion losses on ambient microwave power harvester. *2015 IEEE MTT-S International Conference on Numerical Electromagnetic and Multiphysics Modeling and Optimization (NEMO)*, Ottawa, Ontario, Canada (11–14 August 2015), pp. 1–3.

40 Kimionis, J., Collado, A., Tentzeris, M.M., and Georgiadis, A. (2017). Octave and decade printed UWB rectifiers based on nonuniform transmission lines for energy harvesting. *IEEE Transactions on Microwave Theory and Techniques* 65 (11): 4326–4334.

41 Niotaki, K., Georgiadis, A., Collado, A., and Vardakas, J.S. (2014). Dual-band resistance compression networks for improved rectifier performance. *IEEE Transactions on Microwave Theory and Techniques* 62 (12): 3512–3521.

42 Niknejad, A.M. (2007). *Electromagnetics for High-Speed Analog and Digital Communication Circuits*, 1e. Cambridge, NY: Cambridge University Press.

43 Lu, P., Yang, X.S., Li, J.L., and Wang, B.Z. (2015). A polarization-reconfigurable rectenna for microwave power transmission. *2015 International Workshop* *on Antenna Technology (iWAT)*, Seoul, South Korea (4–6 March 2015), pp. 120–122.

44 Sun, H. and Geyi, W. (2016). A new rectenna with all-polarization-receiving capability for wireless power transmission. *IEEE Antennas and Wireless Propagation Letters* 15: 814–817.

45 Zhang, B., Kovitz, J.M., and Rahmat-Samii, Y. (2016). A hemispherical monopole rectenna array for multi-directional, multi-polarization, and multi-band ambient RF energy harvesting. *2016 IEEE International Symposium on Antennas and Propagation (APSURSI)*, Fajardo, PR (26 June–1 July 2016), pp. 603–604.

46 Fantuzzi, M., Prete, M.D., Masotti, D., and Costanzo, A. (2017). Quasi-isotropic RF energy harvester for autonomous long distance IoT operations. *2017 IEEE MTT-S International Microwave Symposium (IMS)*, Honolulu, HI (4–9 June 2017), pp. 1345–1348.

47 Kimionis, J., Georgiadis, A., Isakov, M. et al. 3D/inkjet-printed origami antennas for multi-direction RF harvesting. *2015 IEEE MTT-S International Microwave Symposium*, Phoenix, AZ (17–22 May 2015), pp. 1–4.

48 Chen, Y. and You, J. (2018). A scalable and multidirectional rectenna system for RF energy harvesting. *IEEE Transactions on Components, Packaging and Manufacturing Technology* 8 (12): 2060–2072.

49 Bui, D.H.N., Vuong, T., Verdier, J. et al. (2019). Design and measurement of 3D flexible antenna diversity for ambient RF energy scavenging in indoor scenarios. *IEEE Access* 7: 17033–17044.

50 Fezai, F., Menudier, C., Thevenot, M. et al. (2016). Multidirectional receiving system for RF to dc conversion signal: application to home automation devices. *IEEE Antennas and Propagation Magazine* 58 (3): 22–30.

51 Bjorkqvist, O., Dahlberg, O., Silver, G. et al. (2018). Wireless sensor network utilizing radio-frequency energy harvesting for smart building applications [education corner]. *IEEE Antennas and Propagation Magazine* 60 (5): 124–136.

52 Ren, Y.-J. and Chang, K. (2006). 5.8-GHz circularly polarized dual-diode rectenna and rectenna array for microwave power transmission. *IEEE Transactions on Microwave Theory and Techniques* 54 (4): 1495–1502.

53 Popović, Z., Korhummel, S., and Dunbar, S. (2014). Scalable RF energy harvesting. *IEEE Transactions on Microwave Theory and Techniques* 62 (4): 1046–1056.

54 Lee, D., Lee, S., Hwang, I. et al. (2017). Hybrid power combining rectenna array for wide incident angle coverage in RF energy transfer. *IEEE Transactions on Microwave Theory and Techniques* 65 (9): 3409–3418.

55 Hu, Y.Y., Xu, H., Sun, H., and Sun, S. (2017). A high-gain rectenna based on grid-array antenna for RF power harvesting applications. *2017 10th Global Symposium on Millimeter-Waves*, Hong Kong (24–26 May 2017), pp. 161–162.

56 Basta, N.P., Falkenstein, E.A., and Popovic, Z. (2015). Bow-tie rectenna arrays. *2015 IEEE Wireless Power Transfer Conference (WPTC)*, Boulder, CO (13–15 May 2015), pp. 1–4.

57 Volakis, J.L., Olgun, U., and Chen, C.-C. (2012). Design of an efficient ambient WiFi energy harvesting system. *IET Microwaves, Antennas & Propagation* 6 (11): 1200–1206.

58 Erkmen, F., Almoneef, T.S., and Ramahi, O.M. (2018). Scalable electromagnetic energy harvesting using frequency-selective surfaces. *IEEE Transactions on Microwave Theory and Techniques* 66 (5): 2433–2441.

59 Takhedmit, H., Cirio, L., Costa, F., and Picon, O. (2014). Transparent rectenna and rectenna array for RF energy harvesting at 2.45 GHz. *The 8th European Conference on Antennas and Propagation (EuCAP 2014)*, The Hague, Netherlands (6–11 April 2014), pp. 2970–2972.

60 Shinohara, N. and Matsumoto, H. (1998). Dependence of dc output of a rectenna array on the method of interconnection of its array elements. *Electrical Engineering in Japan* 125 (1): 9–17.

61 Gutmann, R.J. and Borrego, J.M. (1979). Power combining in an array of microwave power rectifiers. *1979 IEEE MTT-S International Microwave Symposium Digest*, Orlando, FL (30 April–2 May 1979), pp. 453–455.

62 Balanis, C.A. (2005). *Antenna Theory: Analysis and Design*, 3e. Hoboken, NJ: Wiley-Blackwell.

Index

Wireless Power Transmission for Sustainable Electronics: COST WiPE - IC1301,
First Edition. Edited by Nuno Borges Carvalho and Apostolos Georgiadis.
© 2020 John Wiley & Sons, Inc. Published 2020 by John Wiley & Sons, Inc.